RECENT ADVANCES IN ROBOTIC SYSTEMS

Edited by **Guanghui Wang**

Recent Advances in Robotic Systems
http://dx.doi.org/10.5772/61677
Edited by Guanghui Wang

Contributors

Nils Gageik, Christian Reul, Sergio Montenegro, Thor I Fossen, Martin Šipoš, Jan Rohac, Tor Arne Johansen, Jakob Mahler Hansen, Mizuho Shibata, Daniel Rodrigo Ramirez Rebollo, Pedro Ponce, Arturo Molina, Antoni Grau, Edmundo Guerra, Yolanda Bolea, Rodrigo Munguia, Christopher Kitts, Rafhael Milanezi De Andrade, Antônio Bento Filho, Arnaldo Gomes Leal Junior, Yuqing He, Junfeng Xiong, Feng Gu, Decai Li, Jianda Han, Mohamad El Hariri, Patrick Henaff, Hayssam Serhan, Hadi Behzadnia, Ahmed Zahedi, Hassan Ghanbari, Hamed Tabatabaei, Gerald Seet, Viatcheslav Iastrebov, Wee-Ching Pang, Quang Huy Dinh

Published by InTech

Janeza Trdine 9, 51000 Rijeka, Croatia

Publishing Process Manager Danijela Duric

Technical Editor SPi Global

Cover InTech Design team

First published September, 2016

Printed in Croatia

Additional hard copies can be obtained from orders@intechopen.com

Recent Advances in Robotic Systems, Edited by Guanghui Wang

p. cm.

Print ISBN 978-953-51-2570-9
Online ISBN 978-953-51-2571-6

For EU product safety concerns:
IN TECH d.o.o., Prolaz Marije Krucifikse Kozulić 3, 51000 Rijeka, Croatia,
info@intechopen.com or visit our website at intechopen.com.

Meet the editor

Dr. Guanghui Wang is currently an assistant professor at the University of Kansas, USA. He is also with the National Laboratory of Pattern Recognition, Chinese Academy of Sciences, China, as an adjunct professor. His current research interests include computer vision, image analysis, machine learning, and robotics. He authored one book "Guide to Three Dimensional Structure and Motion Factorization" at Springer-Verlag, and has published over 80 papers in peer-reviewed journals and conferences. He served as associate editor or on the editorial board of four journals, as an area chair or TPC member of 20+ conferences, and as a reviewer of three research organizations, 20+ journals, and 20+ conferences.

Contents

Preface

Robotics is a rapidly growing, interdisciplinary research area. This book presents some state-of-the art research ideas and findings in a wide range of robotics. The book contains 12 chapters written by experts and researchers in respective fields. The topics cover from design, modeling, control, learning, to applications of UAVs, USVs, mobile robots, humanoid robots, and underwater robots. The book is suitable for graduate students, researchers, and industrial practitioners in the field of robotics. The main contents of this book are as follows.

The first chapter describes the design and implementation of a fully autonomous quadrocopter, which is capable of self-reliant search, count, and localization of a predefined object; and chapter 2 designs a flooding disaster oriented unmanned surface vehicle to handle unknown dynamics and trajectory tracking control in complicated surroundings.

Chapter 3 presents a compliance effect of dynamic humanoid robot walking developed using muscles in the knees of a 3D simulated biped robot; and a kinematic analysis is introduced in chapter 4 for the triangle-star robot with telescopic arm and kinematics chains.

Chapter 5 describes an implementation methodology to improve the feature initialization process in monocular SLAM; a state estimator is studied in chapter 6 for robust GNSS-aided navigation; and chapter 7 presents a multi-robot formulation for optimal control of mobile robots.

Chapter 8 describes a technique of using laser outline graphics to augmented reality applications in a shared humans and robots environment; a real-time dynamic and optimized bandwidth management algorithm is proposed in chapter 9 for teleoperated collaborative robots; and chapter 10 analyzes the series elastic actuator for dynamic robot applications in a non-structured environment.

Chapter 11 describes a small-sized underwater robot, with detailed specification and control mechanism, to improve the spatial resolution of undersea measurements; and the last chapter proposes a cognition algorithm for reinforcement learning and applies it to a humanoid hand grasping tasks.

Acknowledgements

The editor would like to thank all authors who have invested tremendous time in writing these high quality chapters. I really appreciate all reviewers for their time and professional comments and suggestions, especially Dr. Yao Sui, Mr. Yuanwei Wu, and Mr. Sushil Pratap Bharati. The book could not have been completed without the excellent support of Ms. Danijela Duric and Ms. Viktorija Zgela at InTech. I also wish to show my gratitude to everyone who helped throughout the preparation of this book.

Dr. Guanghui Wang
Electrical Engineering and Computer Science
University of Kansas

Chapter 1

Autonomous Quadrocopter for Search, Count and Localization of Objects

Nils Gageik, Christian Reul and Sergio Montenegro

Additional information is available at the end of the chapter

http://dx.doi.org/10.5772/63568

Abstract

This chapter describes and evaluates the design and implementation of a new fully autonomous quadrocopter, which is capable of self-reliant search, count and localization of a predefined object on the ground inside a room.

A camera attached to the quadrocopter and directed at the ground is used to find the searched objects and to determine its positions during the autonomous flight in real time. Hence, objects that fulfil the scanning parameters can be found in different positions. Based on its own known position and the position of the object in the picture of the camera, the position of the detected objects can be determined. Thus, repeated detections of objects can be excluded. Consequently, objects can be counted and localized autonomously.

The position of the object is transferred to the ground station and compared with the true position to evaluate the system. Two different search situations and two different strategies, breadth first search (BFS) and depth first search (DFS), are investigated and their results are compared. The evaluation shows the potential, constraints and drawbacks of this approach just as the effects of the search strategy, and the most important parameters and indicators such as field of view (FOV), masking area (MA) and minimal object distance. Moreover, the accuracy, performance and completeness of the search are discussed. The entire system is composed of low-cost components and constructed from scratch. Its integration in the innovative real-time operating system RODOS (Real-time Onboard Dependable Operating System) developed by the German Aerospace Centre is described in detail. RODOS has been developed for embedded systems such as satellites and comparable aerospace systems.

Keywords: autonomous UAV, quadrocopter, quadrotor, search and rescue, count, object localization

1. Introduction

Equipping unmanned aerial vehicles (UAVs) such as quadrocopters with more and more autonomous abilities is an interesting field of research. Furthermore, it is a requirement for challenging autonomous search and rescue missions, which are still a field of interest [1–14]. Especially, fully autonomous systems are challenging since they cannot rely on external systems like Global Positioning System (GPS) or optical tracking for accurate positioning. State of the art is the usage of a laser scanner for obstacle detection, collision avoidance and via a simultaneous localization and mapping (SLAM) algorithm for positioning [15, 16]. But laser scanners are heavy, expensive and fail in some situations like a smoking environment. Other approaches are vision based, but the high computational burden often requires an external computer for computation [17–19].

Therefore, a solution for a fully autonomous system is presented using a new hardware design combining optical and PMD cameras with infrared and ultrasonic distance holders for a reliable system capable of search and rescue missions. This chapter focuses on the concept, implementation and evaluation of the search, count and localization of red balls (example search targets) with the mentioned autonomous system based on an innovative new hardware design.

In a preliminary calibration scan, the parameters of the object are defined: a red ball is used as an example object. The scan determines the colour and radius of the ball. The implementation and principles of the object recognition and search will be described in detail. After determining the scanning parameters, the autonomous search can be executed. This is done autonomously by the quadrocopter, which uses inertial, infrared, ultrasonic, pressure and optical flow sensors to determine and control its orientation and position in six DOF (degree of freedom).

This research is part of the AQopterI8 Project of the Chair Aerospace Information from the University of Würzburg [20].

2. Terms and background

To clarify different terms, parameters and algorithms, which will be used later, those are defined in this chapter. The main idea of the presented search approach is that the quadrocopter uses a camera directed at the ground and by flying through the search area; it scans all possible locations on the floor for a target (red ball). If a target is detected, it is added to the list of detected targets unless a target has already been detected at this position. Thus, the whole area can be searched for targets and the amount of targets as well as their positions can be determined.

The most significant parameters for the performance of the search, the virtual field of view (VFOV), the masking area (MA) and the search strategy are investigated, and therefore need to be defined.

http://dx.doi.org/10.5772/63568

In this case, the field of view (FOV) is defined as the area on the floor which the camera covers (**Figure 1**). Using computer vision object detection, a target can be found on this single picture of the floor. The field of view is specified by the camera (hardware), whereas the virtual field of view is the area which the search strategy uses in order to cover the whole search area at least once. For VFOV, a smaller value than the true FOV may be used to leave no room for inaccuracy. Detection might fail if the quadrocopter does not fly exactly as expected by the search strategy or if the target is located between two pathways and cannot be seen completely. A smaller VFOV leads to a higher coverage and a longer search pathway (compare **Figures 2** and **3**).

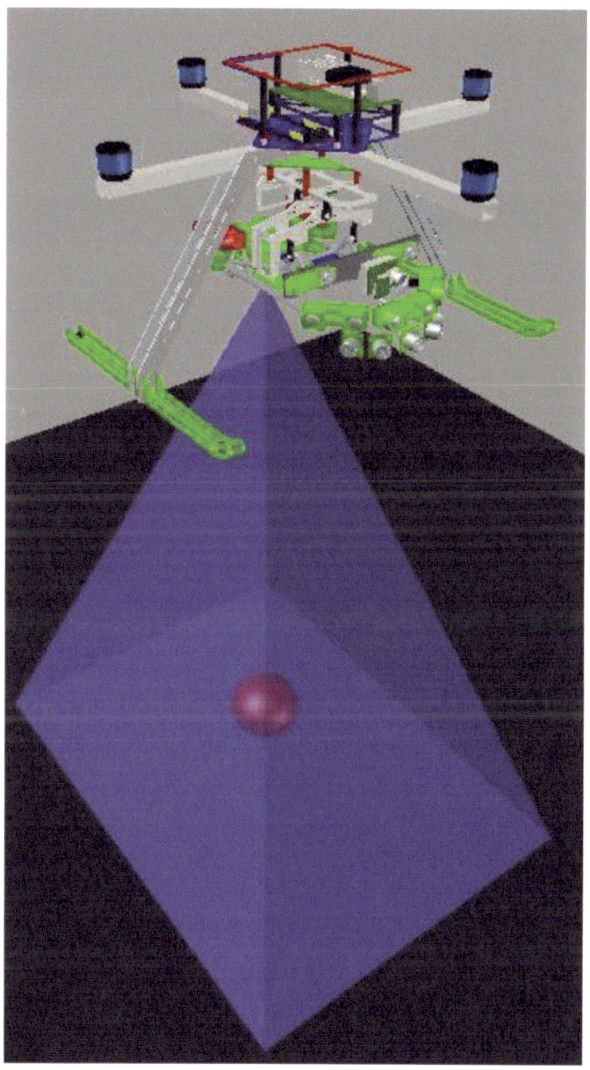

Figure 1. Field of view (FOV).

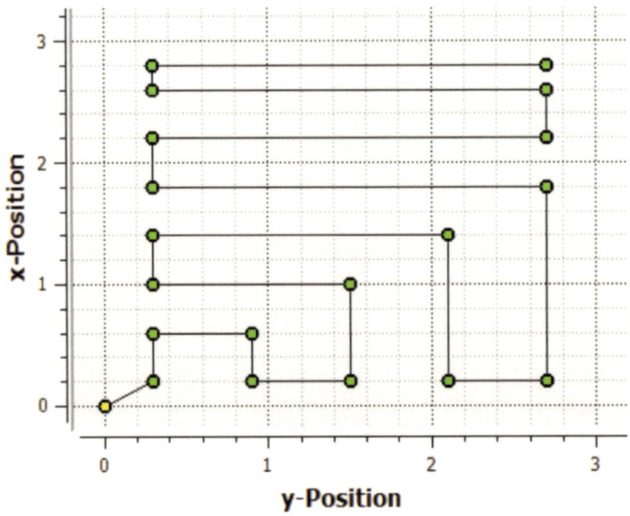

Figure 2. BFS waypoint list (VFOV 40 × 60).

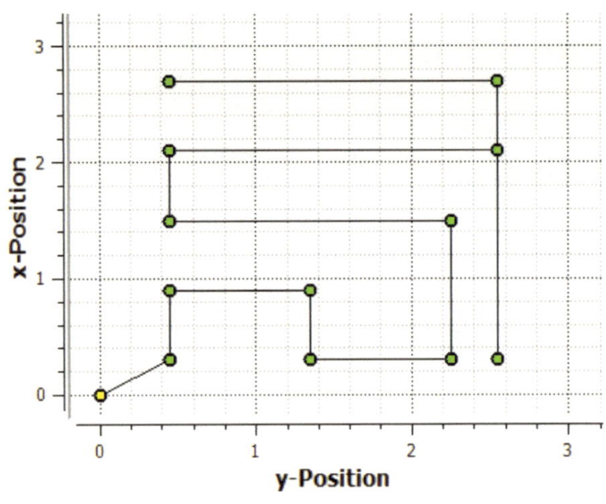

Figure 3. BFS waypoint list (VFOV 60 × 90).

Two different search strategies, which are referred to as BFS (breadth-first-search) and DFS (depth-first-search) later, are investigated. They correspond to the original algorithms, which are used to search nodes in a graph. For reasons of simplification, all search algorithms start at the bottom left corner of the search area.

The idea of the BFS strategy is shown in **Figure 2**. This strategy follows the general rule, which says that closer positions are reached before farther ones. In general, the used algorithm follows the iterative rule Up-Right–Down-Right–Up-Left.

In contrast to the BFS, the DFS does not search nearer but farther positions first. At first, the algorithm covers the sides of the search area and proceeds with smaller iterations until the complete search area is covered (compare **Figures 4** and **5**).

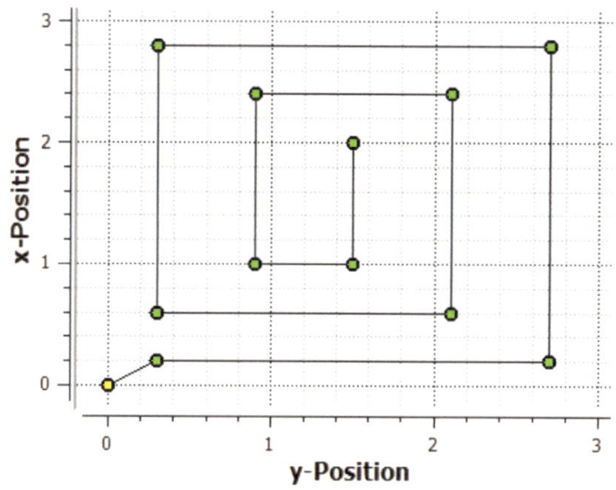

Figure 4. DFS waypoint list (VFOV 40 × 60).

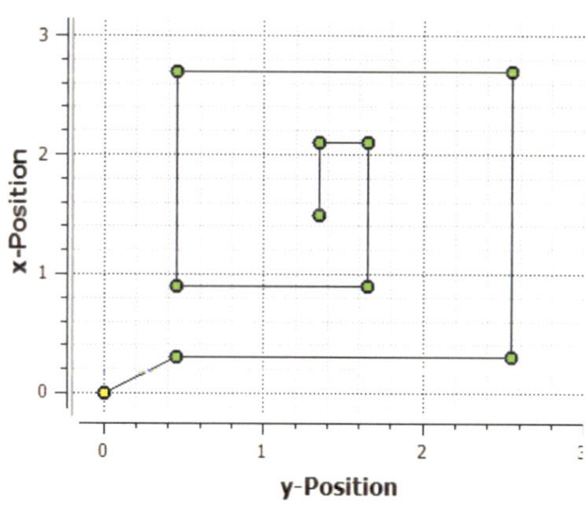

Figure 5. DFS waypoint list (VFOV 60 × 90).

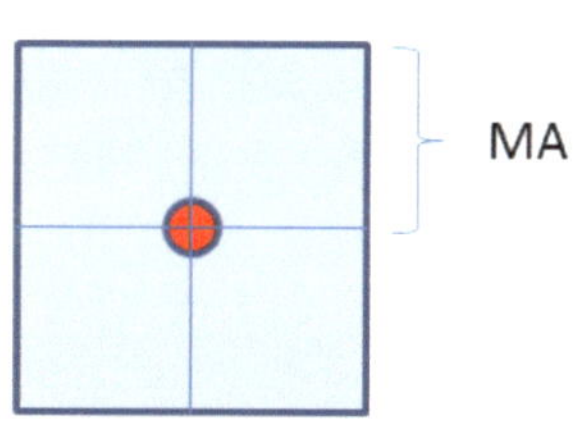

Figure 6. Masking area around an accepted target (red dot).

The masking area (**Figure 6**) determines a square which is set around a found object to avoid multiple detections of the same target. It is determined by a distance named MA. During the search a target might be seen several times from different positions. Because of errors and noise the target is never detected exactly twice at the same position, and therefore would be considered as a new object multiple times. The masking area is subtracted and added to the *X*-coordinate and the *Y*-coordinate of every accepted target and it is proved if the newly detected target is located within one of these coordinates. If so, the newly detected target is discarded, otherwise it is accepted. Instead of a circle a square masking area can be chosen for reasons of simplification and the FOV being also a square.

3. Concept

The concept of the search is separated into two parts: the object or target search and the flight search

3.1. Object search

The task of the object search is to determine the amount and positions of the targets by fusing the results of the object detection with the current position of the quadrocopter (**Figure 7**). It manages the list of found objects and adds new ones if necessary.

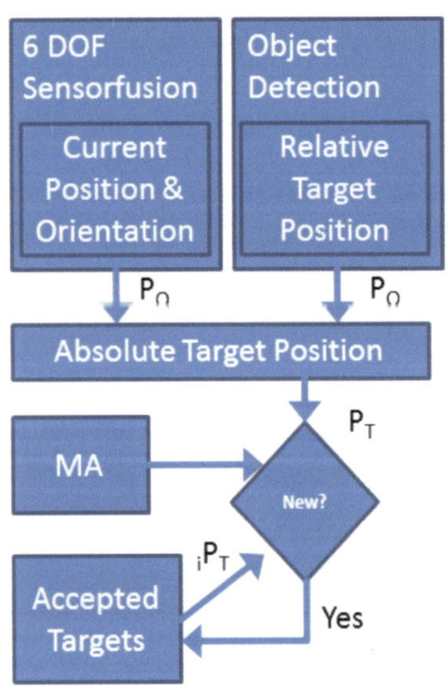

Figure 7. Object search concept.

Whenever the object detection has a hit, the absolute position of this new target is computed by Formula (1),

$$\boldsymbol{P}_T = \boldsymbol{P}_O + \boldsymbol{P}_Q + \boldsymbol{C}_O \tag{1}$$

where $\boldsymbol{C}_O$ is the offset between the camera and the centre of the quadrocopter or its position sensor, $\boldsymbol{P}_Q$ is the current position of the quadrocopter and $\boldsymbol{P}_O$ is the relative position of the found object determined by Formula (2)–(5):

$$\mathrm{P}_O = \frac{\mathrm{h}}{C_h} \cdot (\boldsymbol{M} - \boldsymbol{Z}) \cdot \boldsymbol{C}_w \tag{2}$$

$$\boldsymbol{M} = \begin{bmatrix} \frac{M_x}{R_x} & \frac{M_y}{R_y} \end{bmatrix} \tag{3}$$

$$\boldsymbol{Z} = \begin{bmatrix} 0.5 & 0.5 \end{bmatrix} \tag{4}$$

$$\boldsymbol{C}_w = \begin{bmatrix} C_x & 0 \\ 0 & C_y \end{bmatrix} \tag{5}$$

M_x and M_y in Formula (2) are the coordinates of the object's centre point determined by the object detection, C_x and C_y are the calibration width in X and Y, respectively, at a height of C_h, h is the current height and $\boldsymbol{Z}$ is a constant. C_x and C_y are determined by the true FOV of the camera. R_x and R_y are the resolution of the camera in the X - and Y-directions, respectively.

Next, the position P_T is compared with all positions ${}_iP_T$ with i indicating the index of the already accepted position. If the new position P_T occurs within the masking area of any target ${}_iP_T$, it is discarded, otherwise it is accepted.

3.2. Flight search

The task of the flight search is the waypoint generation. It ensures that the quadrocopter with a determined VFOV covers the whole search area at least once. The VFOV is a static parameter, which represents the FOV. A bigger FOV leads to less waypoints and a shorter flight search, while with a smaller FOV more waypoints and resulting pathways are created. In general, waypoints are not generated next to each other iteratively in small steps because of the bad flight performance of this approach [21], but with maximal distance according to the search strategy.

Then, the flight search is executed statically. That means the waypoint list is generated once at the beginning and it is not changed during the flight. The waypoint list is determined by the search strategy, the search area and the VFOV. **Figures 2–5** illustrate the effect of these parameters on the waypoint list.

4. Object recognition

The algorithm for object recognition is based on significant information about an object's shape and colour and determines if it is a target or not [22]. The implemented search algorithm (object recognition) expects targets, which are round and monochrome, such as red balls. Extending the system to enable a detection of more than one target colour at the time can easily be achieved. To provide the possibility of searching for other shapes like rectangles or human bodies the object recognition needs to be replaced or changed fundamentally. For the experiments described in this chapter, red balls with a diameter of approximately 7 cm are considered as search targets.

In the following sections, the necessary image processing fundamentals will be briefly discussed. Afterwards, the circle detection algorithm used to identify the balls will be introduced. Finally, the recognition procedure consisting of an initial scan to determine the search parameters and a subsequent search will be explained.

4.1. Image processing fundamentals

For implementation and guidance, the open source computer vision library OpenCV can be recommended [23]. It contains a variety of basic image processing core algorithms as well as advanced procedures for applications such as object recognition, feature extraction and machine learning.

4.1.1. Image representation

A standard way to represent a picture while using a PC is the RGB model. Each pixel is described by three intensity values: *r*ed, *g*reen and *b*lue. Here we assume a resolution of 8 bit. Therefore, the range for each value is 0–255 (=$2^8 - 1$).

For a bench of calculations, the RGB representation of a pixel is impractical and a single value per pixel is preferred. Using the so-called greyscale image the value grey of a pixel x can be determined by its original RGB values (Formula (6)):

$$grey(x) = 0.299 \cdot x.R + 0.587 \cdot x.G + 0.114 \cdot x.B \tag{6}$$

The simplest representation is the binary image in which only two values exist: 0 and 1. If a pixel meets certain requirements, for example if it has a specific RGB or a greyscale value, a value of 1 is assigned, otherwise it has a value of 0.

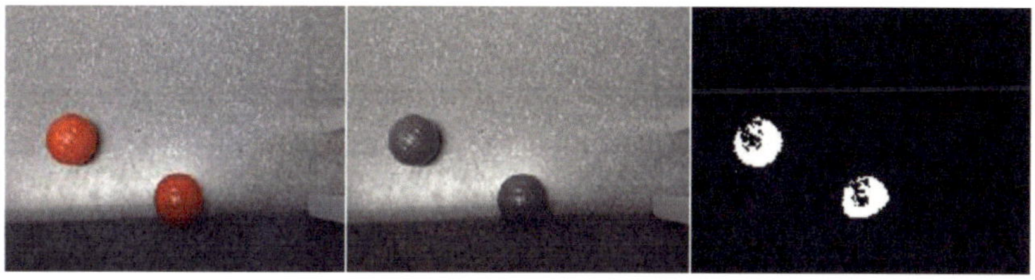

Figure 8. Image of two balls using its RGB (left), greyscale (middle) and binary (right) representation.

An example image of two balls and the corresponding greyscale image can be seen in **Figure 8**. Furthermore, a binary image was created by assigned value 1 to each pixel with a red value higher than 100 and with green and blue values lower than 50.

4.1.2. Filters

In contrast to the operations already introduced, filters use a variety of pixels and not just a single one to determine the new value of a pixel. The idea behind the filters is to perform 2D convolution: a so-called filter matrix is slid over the original image and simple multiplications of the filter elements with the underlying values of the image pixels are performed. The calculated outcomes are summed up and the result is stored as the new value of the pixel, which is located under the so-called hot spot, the centre of the filter matrix. The entire process is shown in **Figure 9**.

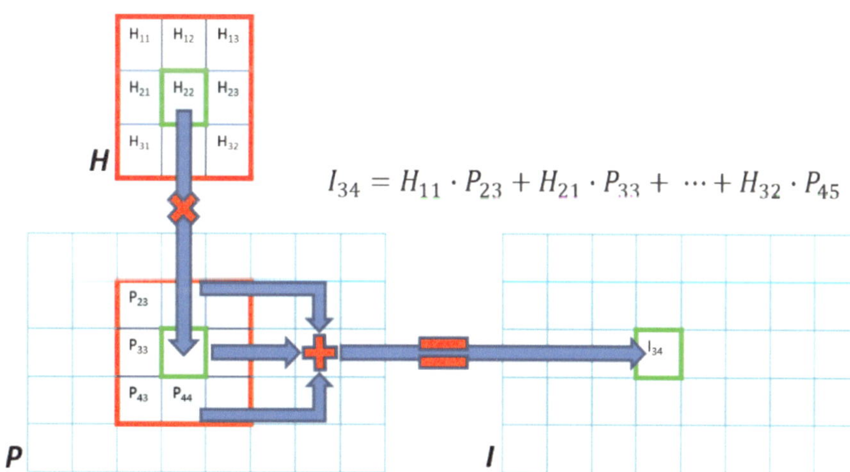

Figure 9. Mode of operation of a linear filter [24].

4.1.3. Edge detection

One of the most frequent applications of filters is edge detection [24]. Edges can be defined as regions in which big intensity changes occur in a certain direction. To detect those changes one or several filters have to be applied to the greyscale image. In most applications, the so-called Canny edge detector is used [25]: after deploying a Gaussian filter in order to remove noise, the intensity gradients are computed by applying the Sobel operator, which consists of two separate filter matrices. The first filter computes the gradient in the x-direction:

$$H_x^S = \begin{bmatrix} -1 & 0 & 1 \\ -2 & 0 & 2 \\ -1 & 0 & 1 \end{bmatrix} \tag{7}$$

The second one highlights the change of intensity in the y-direction:

$$H_Y^S = \begin{bmatrix} -1 & -2 & -1 \\ 0 & 0 & 0 \\ 1 & 2 & 1 \end{bmatrix} \tag{8}$$

The local edge strength E can then be calculated by combining the resulting images D_x and D_y for each pixel (u, v):

$$E(u,v) = \sqrt{\left(D_x(u,v)\right)^2 + \left(D_y(u,v)\right)^2} \tag{9}$$

Furthermore, the local edge orientation angle $\Phi(u, v)$ can be determined as

$$\Phi\left(u,v\right) = \tan^{-1}\left(\frac{D_y(u,v)}{D_x(u,v)}\right) \tag{10}$$

In general, the described procedure leads to blurry edges. Therefore, an edge thinning technique called non-maximum suppression is applied. The computed first derivatives are combined into four directional derivatives and the resulting local maxima are considered as edge candidates.

Finally, a hysteresis threshold operation is applied to the pixels. Two thresholds have to be defined: an upper one and a lower one. If the local edge strength of a pixel is higher than the upper one, the pixel immediately is accepted as an edge pixel. Pixels whose gradient is below the lower threshold are rejected. If the local edge strength is between the lower and the upper threshold, only pixels adjacent to pixels with gradients above the upper threshold are accepted. This process promotes the detection of connected contours. In this work, values of 20 and 60 were used for the lower and upper thresholds, respectively.

4.2. Hough circle transformation

A circle is defined by its centre $C(x_C, y_C)$ and its radius r. All points $P(x_P, y_P)$ on the outline of the circle satisfy the circle equation:

$$(x_C - x_P)^2 + (y_C - y_P)^2 = r^2 \tag{11}$$

Identifying circles from an edge image by using this equation and the simple approach of checking for every centre candidate, how many edges lie on a circle around it, is very inefficient and highly inadvisable. In the following sections two much faster and more robust approaches are presented.

4.2.1. Basic method

The basic idea behind the Hough transformation can be seen in **Figure 10**. Given a target circle with radius r. If circles with the same radius r are drawn around an edge point of the target circle, they will intersect. The main accumulation of intersection will occur in the centre of the target circle.

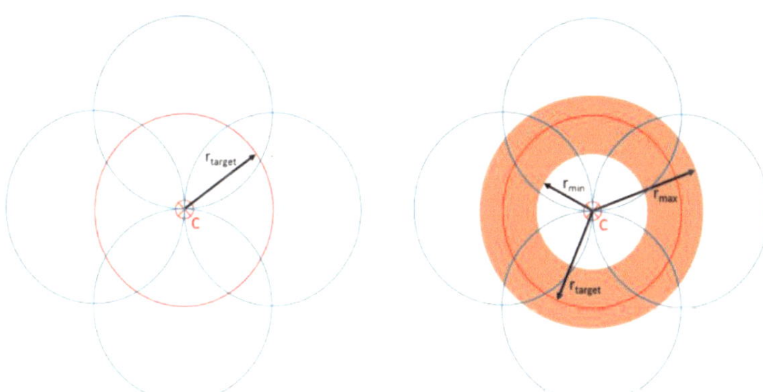

Figure 10. Intersecting circles (blue) drawn around the edge pixels of the target circle(s) (red) using one radius (left) and a range of radii (right) [26].

To identify a circle with known radius r in an edge image, a so-called accumulator array is used. Typically, it has the same dimension as the edge image or is scaled down by a low integer number. If the target radius is exactly known, a two-dimensional array is sufficient. After initializing every cell with zero the voting process starts. Each edge pixel is treated as a possible circle outline and all corresponding centre candidates in the accumulator array get a vote. This means that their value is incremented by 1. After voting all detected circles can be identified by checking, in which cells in the accumulator array earned enough votes. Consequently, a threshold is needed. As the value of each cell roughly corresponds to the number of circle outline pixels in the edge image, a useful threshold can be derived from the maximal number of votes, i.e., the circle's circumference [25].

However, in real-world application the target radius often is not exactly known and the algorithm has to search for a range of radii. This is implemented by extending the accumulator array to three dimensions: two for the already described two-dimensional arrays and one for each radius. During the voting process each edge pixel votes for all possible circle centres by incrementing the corresponding values in every radius plane.

It can easily be seen that the standard approach is not suitable to most real-world applications despite being very robust. First of all, it is slow because for every edge pixel approximately $2\pi r$ centre candidates have to be calculated per radius r. Another problem is that the accumulator arrays can be very memory intensive, especially if the input edge image resolution is high and the target radius is not exactly known. Hence, several improvements were introduced and will be discussed in the following section.

4.2.2. Gradient method

As shown in Formula (10), the local edge orientation angle can be easily determined. By exploiting this, the circle detection algorithm can be executed with much higher efficiency. The key observation is that all edges are perpendicular to the line that connects the edge pixel and the centre of the circle. Therefore, it is not necessary to calculate up to $2\pi r$ centre candidates for each edge pixel and vote for them in the accumulator array. Because of the edge orientation angle the amount of candidates can be narrowed down to only a few pixels. This is shown in **Figure 11**.

The vertical line in the centre of **Figure 11** shows the respective local edge orientation. If the radius is exactly known, in theory only two possible centres correspond to the given edge direction: C_1 and C_2. They are located on a line perpendicular to the edge direction and their distance to the considered edge pixel is r. If the algorithm is searching for a range of radii, the sets of possible centres $C_1[\,]$ and $C_2[\,]$ are located on aforementioned line and the distances of the centres to the initial edge pixel vary from r_{min} to r_{max}.

Hence, the accumulator array can be reduced to two dimensions even when searching for several radii [27]. During the voting process the location of each edge pixel that casts a vote is stored. After the vote the centre candidates are selected. To be taken into consideration the accumulator value of a valid centre candidate has to be above the given threshold and higher than the values of all its immediate neighbours. The approved centre candidates are sorted in descending order according to their accumulator values.

Now the best fitting radius has to be determined. For this, the previously stored edge pixels are considered. The distances between each of these pixels and the centre candidates are calculated. Using these distances, the best supported radius can be determined. Finally, it has to be checked, if the resulting centre is not too close to any previously accepted centre and if it is supported by a sufficient number of edge pixels.

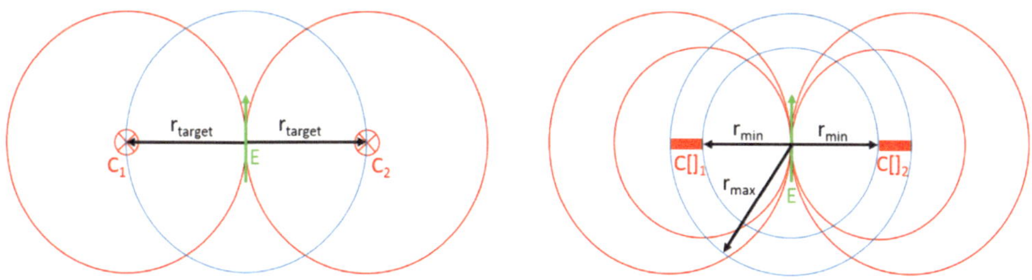

Figure 11. Possible locations of centres given a specific edge direction (green) using one radius (left) and a range of radii (right) [26].

4.2.3. *Run-time comparison*

In Ding *et al.* [22], the run times of the described basic method and the gradient method are compared by detecting red balls in an image using a resolution of 192 × 144 pixel. The achieved results were averaged over 10 measurements and are displayed in **Table 1**.

	1 ball	5 balls
Basic method	35.2 ms	63.4 ms
Gradient method	12.0 ms	12.5 ms

Table 1. Results of the run-time comparison.

When only one ball is detected, the gradient method is already about three times faster than the basic method. The difference becomes even more significant when the number of balls and therefore the amount of edges is increased.

4.3. Recognition procedure

The recognition procedure is split into two phases: the initial search to determine the target parameters pre-flight and the actual search which is performed mid-air by the quadrocopter.

4.3.1. *Initial scan*

Prior to the search, some parameters have to be predefined. Thus, a picture of the search target is taken on a plane background and from a height close to the flying height of the quadrocopter. The radius can be directly determined by detecting the ball and storing the radius, in which the maximum number of votes in the accumulator array was achieved.

Furthermore, the dominating colour within the detected circle is calculated by combining several of the most frequent red, green and blue values. It is notable that the average values are not used because they can be heavily influenced by bright spots on the search target which emerge because of unfavourable lighting conditions. The final target colour does not consist of exactly one set of RGB values but of ranges for each colour channel which are derived from the original values.

Furthermore, the algorithm searches for more than one radius. Therefore, the initially detected radius ±2 can be chosen as a target range to compensate for light variations of height during the flight. To allow bigger chances in flying altitude, the radius range would have to be adjusted according to the currently measured distance of the quadrocopter to the floor.

4.3.2. Search

The actual search is performed using a resolution of 192 × 144 pixels. This allows quick processing while still preserving all the information necessary for a successful detection.

After taking a picture it is converted into a greyscale image and the Hough detection is performed. The number and quality of detected circles heavily depend on the threshold used during the Hough circle detection. A good value for the required number of votes is 30% of the circumference of the smallest radius. With significantly higher values, target objects tend to get missed far too often because the constant movement of the quadrocopter tends to prevent the camera from taking sharp pictures.

All detected circles, i.e., all target candidates, are then analysed for their colour. For each pixel inside a candidate's enclosing circle, it is checked if its RGB values lie within a certain range. The range is determined during the initial scan. For a candidate to get confirmed as a target, a certain percentage of its pixels has to be target pixels. Setting this threshold to about 40% was a good value here.

5. Hardware design

Figure 12 depicts the hardware design of the quadrocopter (**Figure 13**). The brain of the system is composed of two processing units, an AVR 32 bit MCU (microcontroller unit) UC3A and the LP-180 providing an AMD-x86 processor and 2 × 1.6 GHz system clock [28].

The CPUs can be seen in the centre of the picture. The MCU interfaces all sensors except those connected via USB and performs the control part with real-time computing, while the task of the LP-180 contains all functions with a high computational burden such as object recognition and mapping.

The quadrocopter uses a couple of sensors for obstacle detection and is capable of distance-controlled collision avoidance [29]. For object recognition, the C270 camera from Logitech is used [30]. All processing is done on-board the quadrocopter, so it is capable of a fully autonomous flight.

http://dx.doi.org/10.5772/63568

Figure 12. Hardware design.

Figure 13. AQopterI8 picture.

6. Software implementation

6.1. Overview and background (RODOS)

The underlying software problem with multiple, here three, real-time processing units interacting with each other is a typical application of the real-time operation system RODOS (Real-time Onboard Dependable Operating System). An important aspect in the selection of RODOS is its integrated real-time middleware. Developing the control and payload software on the top of a middleware provides the maximum of modularity. Different functions can be developed independently and simultaneously and it is very simple to interchange modules later without worrying about side effects because all modules are encapsulated as building blocks (BB) and they interact only through well-defined interfaces.

RODOS was originally developed for space applications at DLR (German Space Agency) and is now distributed as open source for many applications such as Robotics [31, 32]. RODOS was designed for application domains demanding high dependability (e.g., space) and targets the irreducible complexity in all implemented functions. It follows the quote 'Perfection is achieved, not when there is nothing more to add, but when there is nothing left to take away.' from Antoine de Saint-Exupery.

The quadrocopter firmware—ranging from attitude control to route planning, and payload software, e.g., identification of objects—is implemented as a (software) network of communicating building blocks. A useful feature of the RODOS middleware is the location transparency of building blocks. BBs can interact and communicate in the same way independently of the location of communication partners. This includes the same computer, a different computer in the same vehicle, on a different vehicle or between vehicles and ground station (operator interface).

RODOS was designed as the brain of the Avionic system and introduced for the first time (2001) the NetworkCentric concept. A NetworkCentric core avionics machine consists of several harmonized components, which work together to implement dependable computing in a simple way. In a NetworkCentric system we have a software network of BBs and a hardware Network interconnecting vehicles (radio communication), computers inside of vehicles (buses and point-to-point links), intelligent devices (attached to buses) and simple devices attached to front end computers. To communicate with external units, including devices and other computing units, each node provides a gateway to the network and around the networks several devices may be attached to the system. The communication is asynchronous using the publisher–subscriber protocol. No fixed communication paths are established and the system can be reconfigured easily at run time. For instance, several replicas of the same software can run in different nodes and publish the result using the same topic without knowing each other. A voter may subscribe to that topic and vote on the correct result. Application can migrate from node to node or even to other vehicles without having to reconfigure the communication system. The core of the middleware distributes messages only locally, but using the integrated gateways to the NetworkCentric network, messages can reach any node and application in the network. The communication in the whole system includes software applications, computing nodes and even IO devices. Publishers make messages public under a given topic. Subscribers (zero, one or more) to a given topic get all messages which are published under this topic. As mentioned before, for this communication there is no difference in which node (computing unit or device) a publisher and subscribers is running, and beyond this they may be any combination of software tasks and hardware devices! To establish a transfer path, both the publisher and the subscriber must share the same topic. A topic is a pair consisting of a data type and an integer representing a topic identifier. Both the software middleware and the hardware network switch (called middleware switch) interpret the same publisher/subscriber protocol.

6.2. Software design

A simplified RODOS-based software design of the AQopterI8 quadrocopter can be seen in **Figure 14**. The different BBs exchange services by middleware topics. These BBs are located on three different CPUs, the on-board MCU UC3A, the on-board x86 PC LP-180 and the ground segment (GS) with an off-board CPU. The GS provides the user with a GUI and is used for commanding (**Figure 15**).

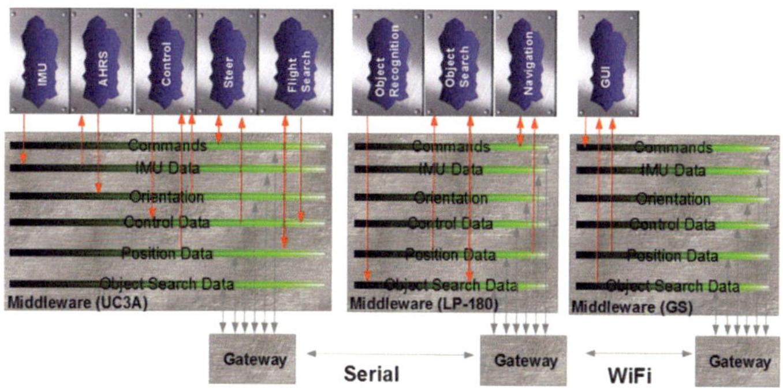

Figure 14. RODOS-based software design (simplified).

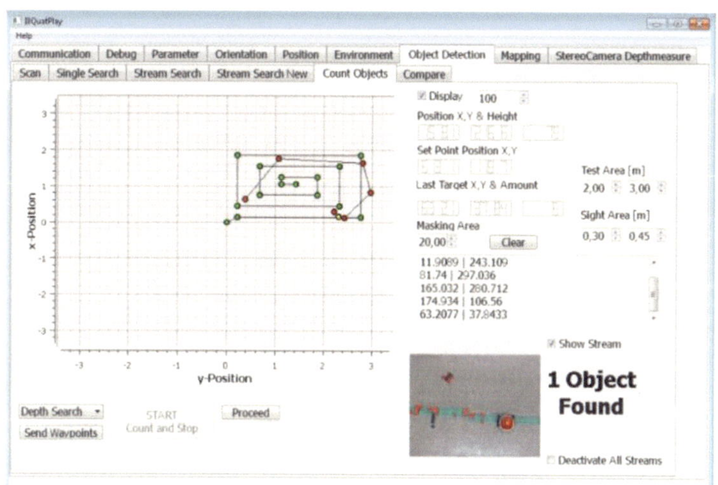

Figure 15. I8Quatplay (Qt-based Commanding Software GUI).

The IMU BB updates the IMU readings every 10 ms with already calibrated and conditioned sensor values. The attitude heading reference system (AHRS) computes from these data the 3D orientation of the quadrocopter using a complementary quaternion filter. The control BB performs the six degree of freedom (DOF) position control based on the position and orientation given by other BBs. The Steer BB drives the motors and executes the commands of the operator and navigation.

The position is further required by the Object Search BB and sent via the gateway using the serial communication link as well as to the GS via WiFi. Thanks to the Gateway and Middleware of RODOS, these data can be used in the same way on another device as on the same device. The kernel of RODOS provides support for thread execution, time management, synchronization and transparent access to external devices such as sensors by the HAL (hardware abstraction layer) interface.

7. Evaluation

7.1. Overview evaluation

To investigate the performance, accuracy and limitations of the proposed system and to compare both search strategies (DFS, BFS) as well as to discuss the optimal parameters for the masking area and VFOV, the results of 63 experiments from two setups are presented.

7.2. First setup

The first setup contained 42 experiments. The search area consisted of a 3 m × 2 m square with two randomly placed balls at the positions (50, 50) and (240, 140) according to **Figure 16**.

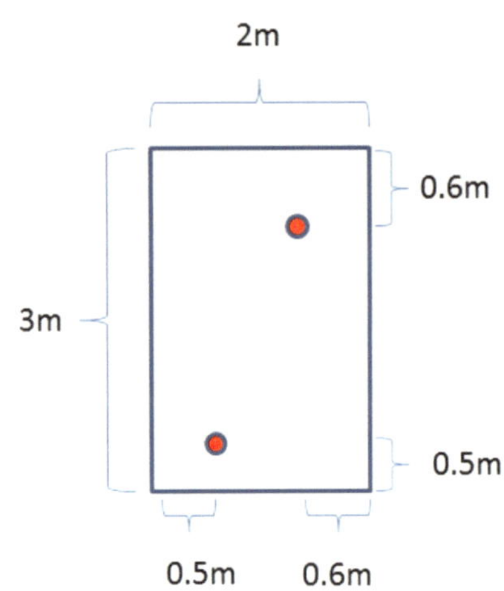

Figure 16. First setup.

In this setup the experiment was repeated for both search strategies, BFS and DFS, for four different masking areas with MA = 0.1 m, MA = 0.15 m, MA = 0.2 m and MA = 0.3 m and with three different VFOV: 0.3 m × 0.45 m, 0.40 m × 0.60 m and 0.60 m × 0.90 m. Then the computed position of the target was compared with the manually measured one, supposed to be the true position. For every single parameter setting, the average error dx in the X- and dy in the Y-direction was computed, first over both targets and then over the entire run together. Also the number of double detections D (fail positive) and misses M (fail negative) was counted (**Table 2**). In the second run, the experiments for MA = 0.3 m were skipped because MA = 0.2 m showed no problem in this setup.

Detection failures		DFS		BFS	
		M	D	M	D
VFOV	30–45	0	0	0	3
	40–60	0	1	1	0
	60–90	4	0	2	1
MA	10	2	1	1	3
	15	0	0	1	1
	20	0	0	1	0
	30	2	0	0	0

Table 2. Detection errors first setup: M missing and D double detections.

From these data no clear difference in accuracy between DFS and BFS or between the different parameter settings could be identified, but it could be concluded that the average error in one axis is less than 15 cm. This setup of randomly placed balls is predominantly affected by coincidence. It might be that one setting leading to one flight path fits well to the placement of the balls.

By taking a look at the detection failures (**Table 2**), clear conclusions can be made. The real FOV is about 65 cm × 45 cm and it can clearly be seen that a VFOV of 40 cm × 60 cm or higher leads to misses. The bigger the VFOV is, the more misses occur, as expected. A proper VFOV of 30 cm × 45 cm leads to no misses for both search strategies. The data show that a lower MA can lead to double detections. This is the case because a target might be seen several times. As the position error in one direction is about 15 cm, MA should be at least in the same range. Conclusively, it can be seen that the DFS performed better and also that there is still a dominating systematical error.

7.3. Second setup

Based on the outcome of the first setup, in the second setup more balls were placed to reduce the effect of coincidence. In addition, the search area was changed to a 2 m × 3 m square (**Figures 17** and **18**), which aimed to equalize the results between the two search strategies and to improve the results of the BFS. This time positions were selected, which should cause troubles for all settings (**Table 3**).

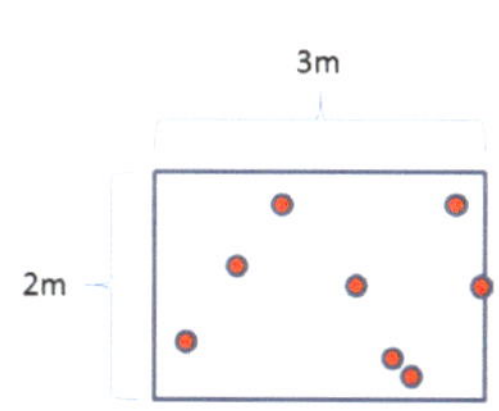

Figure 17. Second setup.

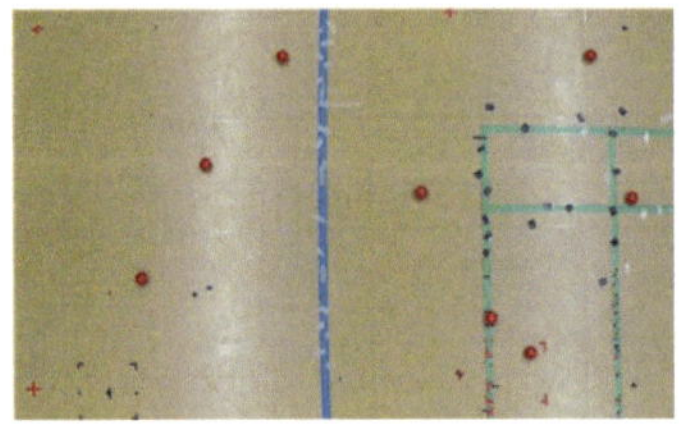

Figure 18. Picture of second setup from above.

Detection failures		DFS		BFS	
MA	VFOV	M	D	M	D
15	25–35	Skipped		2	4
	30–45	0	2	2	0
	40–60	0	0	1	0
	60–90	2	0	4	0
20	25–35	Skipped		3	2
	30–45	0	0	1	0
	40–60	0	0	2	0
	60–90	2	0	4	0
30	25–35	Skipped		2	0
	30–45	0	0	1	0
	40–60	1	0	1	0
	60–90	3	0	4	0

Table 3. Detection errors second setup: M missing and D double detections.

Figure 19 depicts the results shown in the QT Control-Software for a run with the settings MA = 20 cm and 30 × 45 cm for VFOV. It demonstrates that for these settings all targets were detected properly.

The second setup more clearly showed the effect of each parameter or setting and underlined the already expected results. More targets reduced the effect of coincidence, and therefore one run was seen to be enough.

The average position error for the DFS was 16 cm and for the BFS it was about 20 cm. According to these data the DFS can already be concluded as more accurate, but a clearer distinction between both search strategies can be made by taking the detection failures into account. For the DFS there are 10 detection errors in nine experiments compared to 20 detections errors of the BFS in the same setup. Considering these bad results, a value of 25 cm × 35 cm for VFOV was tested with the BFS, but this led to even worse results. There is no setting for the BFS without detection error, but there are four settings with no detection error for the DFS.

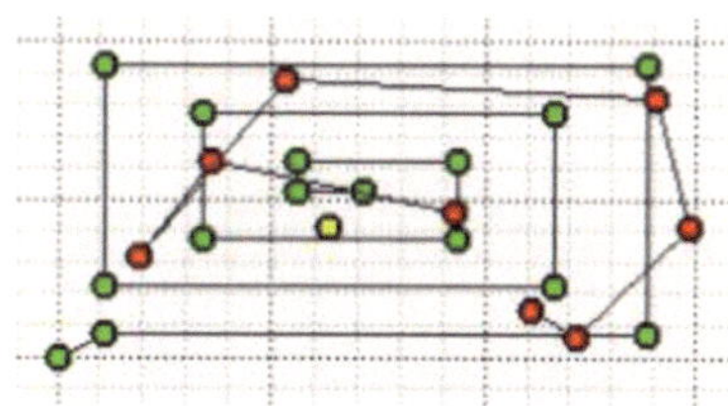

Figure 19. GUI picture of search result (20–30–45): red: found targets; green: waypoints; yellow: position.

7.4. Summary evaluation

To sum up, it can be said that all settings, the search strategy, the masking area and the VFOV have a significant effect on the performance of the search. Although still other, partly random parameters and circumstances have an important influence on the result, optimal values of these parameters are required. This is underlined by **Figures 20–22**, which show that the DFS with MA = 20 cm and a VFOV of 30 cm × 45 cm or 40 cm x 60 cm detected eight balls exactly and nothing else mistakenly. This means there exist settings, which solved this challenging setup. It shall be mentioned that an MA of 30 cm led to a miss in one of the two cases because

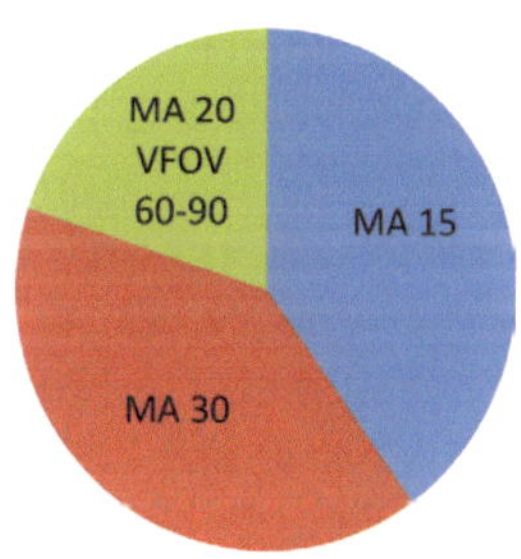

Figure 20. DFS detection failures (general distribution).

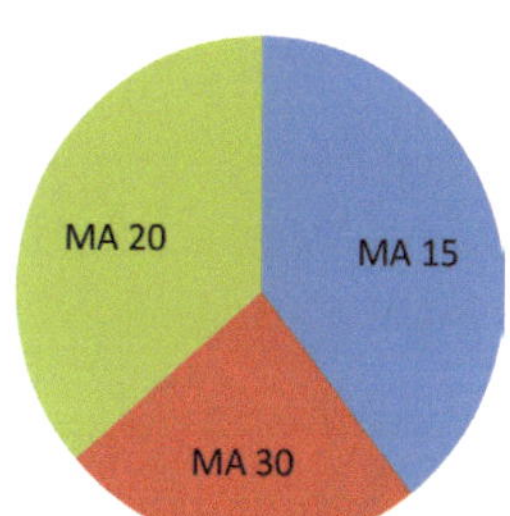

Figure 21. BFS detection failures (distribution after MA).

then one of the closest balls, which are 20 cm away, was not accepted. The BFS showed no good results here and only a VFOV of 30 cm × 45 cm and of 40 cm × 60 cm led to acceptable results. Altogether these results also made the BFS look worse than it was. Some balls were positioned in such way that the BFS failed them by a few centimetres.

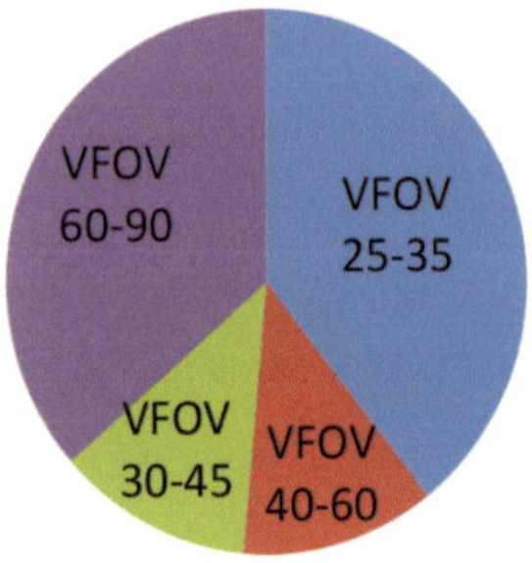

Figure 22. BFS detection failures (distribution after VFOV).

8. Conclusion and discussion

The evaluation demonstrated that the system is capable of autonomously detecting, counting and localization of objects with an accuracy of about 15–20 cm. It was proven that an optimal value for MA (20 cm) has to be a bit higher than the accuracy of the position system and that objects with the distance of 20 cm (MA) in each axis can still be distinguished. Also the coherence of the parameters MA and VFOV on the performance of the search and the detection errors was demonstrated. A smaller VFOV with a smaller MA leads to more double detections, while a too high MA leads to misses of nearby objects. As a general rule, too high VFOV leads to misses because some areas are not searched properly. In this context the acceptance tolerance, which was set to 25 cm in setup 2, is a parameter, which comes into effect. A waypoint is already marked as reached if the current position of the quadrocopter is within this tolerance. This can result in an incomplete cover of the search area and it explains why the BFS misses some targets at the side of the search area.

The best parameter for VFOV was 30 cm × 45 cm. This setting together with the best value for MA showed no detection error even in a challenging room with eight objects.

Furthermore, the evaluation proved that the DFS performed better than the BFS. The reason for that is the fact that smaller waypoint steps are less accurate than less big ones because of the set point jumps and the jump effect as well as the control and sensor system. These result simplified mean that also for a flying robot such as a quadrotor using the underlying on-board sensors less turns and commands are better. This could already be demonstrated in previous experiments [21]. The reason for that can be found in the dynamic of the quadrotor as an aerial vehicle with very little friction (air) and the on-board optical sensors, which are especially affected by the behaviour of the system. Rotations, which mainly occur after set point changes, are a source of error for the position determination.

http://dx.doi.org/10.5772/63568

Although the system was proven capable of performing autonomous and challenging search, count and localization missions, the evaluation of the system did not show a very high accuracy according to the determined positions and the fact that optical sensors were used, which generally can reach higher accuracies. There are multiple sources for accuracy errors, which start from the manually measured and placed target positions in a region of several centimetres. The next major source of error is the starting error, which means the wrong position measured by the optical flow during lift off and the wrong initial position and orientation or placement error of the quadrocopter on the starting position. An initial orientation error for yaw of only 1° leads to a position error of 5 cm after 3 m. It is most likely that the initial yaw orientation error was sometimes in the range of a few degrees. These are good explanations for the high systematical error, which can be seen in the data. A proof of this fact is given by a closer look at some raw data. They demonstrate that the accuracy for the closer object is much better than for the farer object, even if the closer object is detected later in some cases. The best explanation for this is an initial yaw orientation error or missing alignment.

In general, it can be concluded that for this setup proof of high accuracy is challenging and the accuracy of the system might be better than the data show, but at the same time this is not the presented work.

Other sources of error are wrong calibration values for the relative position of the detected object Po (Formula (2))) and simplifications of Formula (2), an incorrectly measured height, a wrong scaling factor for the optical flow and bad lighting and surface conditions, which lead to position errors measured by the optical flow sensor.

The current orientation of the quadrocopter is not considered in the computation of the position P_T. This was intended because the effect of an orientation error should be excluded from the evaluation. In some cases this led to double detection errors.

9. Perspective

Although the system performed quite well in general, there is potential for optimization. The effect of the already mentioned acceptance tolerance and an improved procedure for the waypoint navigation would allow higher values for VFOV. Furthermore, the system can be improved by using two phases. In the first phase, the object search just tries to find something with a low resolution reducing the computational burden and increasing the possible sample time. The focus of the first phase is to overlook nothing. If it has a hit, the quadrocopter suspends the waypoint search and flies to the position of the hit. Then, the second phase is executed using a high resolution and accuracy and only in this phase the accepted position is determined. Computational burden is unimportant in the second phase because the quadrocopter is on position hold.

A different approach with a moving camera and flexible height could also be investigated. In this case, the quadrocopter would possibly not need to search the whole area or at least the waypoint list could be much smaller. In our setup, the quadrocopter could simply fly 4 m up and could see the complete search area. But that will not be possible in every situation as usually rooms are not that high. However, it needs to be compared that which accuracies and detection performance could be achieved then. Taking obstacles and unknown limitations into account as well as the fact that objects might not be detected properly from a distance and at an angle, this approach is much more sophisticated, but also offers more potential and might save flight time, and therefore could reduce the energy consumption.

Another interesting improvement would be to use the obstacle detection sensors to improve the position computation, and therefore the accuracy of the localizations. A challenging part here is a reasonable distribution of the limited resources of the LP-180 to the different demanding tasks.

Acknowledgements

The author would like to thank Universitätsbund Würzburg, IHK Mainfranken, Simone Bayer, Barbara Tabisz and Sascha Dechend for their support and help on this work.

Author details

Nils Gageik[2*], Christian Reul[1] and Sergio Montenegro[2]

*Address all correspondence to: nils.gageik@uni-wuerzburg.de

1 Artificial Intelligence and Applied Computer Science, University of Würzburg, Würzburg, Germany

2 Aerospace Information Technology, University of Würzburg, Würzburg, Germany

References

[1] Nonami K., Autonomous Flying Robots, Springer. 2010, ISBN-10: 4431538550

[2] Aibotix GmbH, www.aibotix.de

[3] Microdrones GmbH, www.microdrones.com

[4] ArduCopter, http://code.google.com/p/arducopter

[5] HiSystems GmbH, www.mikrokopter.de

[6] Mellinger D. et al, Trajectory generation and control for precise aggressive maneuvers with quadrotors, The International Journal of Robotics Research, 31(5), 2012.

[7] Gageik N., Mueller T., Montenegro S., Obstacle detection and collision avoidance using ultrasonic distance sensors for an autonomous quadrocopter, UAVveek 2012.

[8] Autonomous Flying Robots, University Tübingen, Cognitive Systems, www.ra.cs.uni-tuebingen.de

[9] Gronzka S., Mapping, state estimation, and navigation for quadrotors and human-worn sensor systems, PhD Thesis, Uni Freiburg, 2011.

[10] Lange S., Sünderhauf N. Neubert P., Drews S., Protzel P., Autonomous Corridor Flight of a UAV Using a Low-Cost and Light-Weight RGB-D Camera, Advances in Autonomous Mini Robots, 2012, ISBN: 978-3-642-27481-7

[11] Ding W. et al, Adding Optical Flow into GPS/INS Integration for UAV navigation, International Global Navigation Satellite Systems Society, IGNSS 2009.

[12] Wang J. et al, Integration of GPS/INS/VISION Sensor to Navigate Unmanned Aerial Vehicle, The International Archives of Photogrammetry, Remote Sensing and Spatial Information Sciences, Vol. XXXVII Part B1, Beijing 2008.

[13] Blösch M.et al., Vision Based MAV Navigation in Unknown and Unstructured Environments, Robotics and Automation (ICRA), 2010 IEEE International Conference, 21–28.

[14] Corke P., An Inertial and Visual Sensing System for a Small Autonomous Helicopter, Journal of Robotic Systems 21(2), 43–51, 2004.

[15] Dryanovski I., Valenti R., Xiao J., Pose Estimation and Control of Micro-Air Vehicles, August 2012.

[16] Grzonka S. et al, A Fully Autonomous Indoor Quadrotor, IEEE Transactions on Robotics 28(1), 2012.

[17] Kendoul F. et al, Optical-flow based vision system for autonomous 3D localization and control of small aerial vehicles, Robotics and Autonomous Systems 2009, Elsevier.

[18] Herisse B. et al, Hovering flight and vertical landing control of a VTOL Unmanned Aerial Vehicle using Optical Flow, 2008 IEEE International Conference on Intelligent Robots and Systems.

[19] Zing S. et al, MAV Navigation through Indoor Corridors Using Optical Flow, 2010 IEEE International Conference on Robotics and Automation.

[20] Gageik, Reul, Montenegro, Autonomous Quadrocopter for Search, Count and Localization of Objects, UAV World 2013.

[21] Gageik N., Strohmeier M., Montenegro S., Waypoint Flight Parameter of an Autonomous UAV, IJAIA, 4(3), May 2013, pmdtechnologies GmbH, www.pmdtec.com

[22] Reul, C., Implementierung und Evaluierung einer Objekterkennung für einen Quadrocopter, BA Thesis, April 2013.

[23] Itseez. OpenCV [Internet]. Available from: http://www.opencv.org [Accessed: 30.01.2016]

[24] Burger, W. and Burge, M. Principles of Digital Image Processing—Fundamental Techniques. 1st ed. London: Springer; 2009. 260 pp.

[25] Rhody H. Lecture 10: Hough Circle Transform [Internet], 2005.

[26] Smereka M.. Duleba, I. Circular object detection using a modified hough transform. International Journal for Applied Mathematics and Computer Science. 2008; 18(1):85–91.

[27] Bradski G., Kaehler A. Learning OpenCV. 1st ed. Sebastopol, CA: O'Reilly Media; 2008. 555 pp.

[28] Commell, www.commell.com.tw.

[29] Gageik N., Benz P., Montenegro S. Obstacle detection and collision avoidance for a UAV with complementary low-cost sensors. IEEE Access 3: 2015; 599–609.

[30] Logitech, http://www.logitech.com.

[31] Rodos Wiki, wikipedia.org/wiki/Rodos_(operating_system).

[32] Rodos Framework, DLR Webpage, www.dlr.de/irs/desktopdefault.aspx/tabid-5976/9736_read-19576/.

Chapter 2

Design, Implementation and Modeling of Flooding Disaster-Oriented USV

Junfeng Xiong, Feng Gu, Decai Li, Yuqing He and Jianda Han

Additional information is available at the end of the chapter

http://dx.doi.org/10.5772/64305

Abstract

Although there exist some unmanned surface platforms, and parts of them have been applied in flooding disaster relief, the autonomy of these platforms is still so weak that most of them can only work under the control of operators. The primary reason is the difficulty of obtaining a dynamical model that is sufficient rich for model-based control and sufficient simple for model parameters identification. This makes them difficult to be used to achieve some high-performance autonomous control, such as robust control with respect to disturbances and unknown dynamics and trajectory tracking control in complicated and dynamical surroundings. In this chapter, a flooding disaster-oriented unmanned surface vehicle (USV) designed and implemented by Shenyang Institute of Automation, Chinese Academy of Sciences (SIA, CAS) is introduced first, including the hardware and software structures. Then, we propose a quasi-linear parameter varying (qLPV) model to approach the dynamics of the USV system. We first apply this to solve a structured modeling problem and then introduce model error to solve an unstructured modeling problem. Subsequently, the qLPV model identification results are analyzed and the superiority compared to two linear models is demonstrated. At last, extensive application experiments, including rescuing rope throwing using an automatic pneumatic and water sampling in a 2.5 m radius circle, are described in detail to show the performance of course keeping control and GPS point tracking control based on the proposed model.

Keywords: unmanned surface vehicles, quasi-linear parameter varying system, active modeling, unscented Kalman filter

1. Introduction

Floods are among the most major climate-related disasters and have resulted in substantial losses including enormous property damage and human casualties [1]. Numbers of casualties and losses could be larger in the future in response to global warming. The biggest challenge lying in rescuing operations is the low efficiency and the high risk of the rescuers, which is also a problem focused on by this work. Unmanned system is one of the solutions that replace people in the rescuing operations.

Unmanned surface vehicles (USVs), also called autonomous surface vehicles (ASVs), are often used to name the vehicles, which can run on the surface autonomously. Surface robot-assisted flood disaster rescue and inspection is a new research direction in the field of robotics. Here are some obvious advantages: (1) the smaller size allows the USVs to access to narrow and small space to get detailed information; (2) remote operation can avoid casualties of the rescuers caused by the unexpected potential dangers. After Hurricane Wilma in 2005, USVs have been used for emergency response by detecting damage to seawalls and piers, locating submerged debris, and determining safe lanes for sea navigation [2]. After the Fukushima nuclear accident in 2011, the United States and Japan have used robots to assess the damage jointly [3].

Also, in 2007, a new Trimaran unmanned surface vehicle (TUSV) as a test-bed to verify the robust motion control strategies has been designed in Shenyang Institute of Automation, Chinese Academy of Sciences (SIA, CAS) [4]. After that, in 2012, a water-jet propulsion USV equipped with different kind of sensors and ground control system has been designed and implemented in SIA to improve the performance of USV [5]. In 2015, to increase the reliability and real time of USV, the software architecture has been designed based on a real-time operating system QNX 6.5.0. Also, the selection of an appropriate platform and associated hardware as well as useful and sufficient sensors, and integrating these two entities has been taken into consideration. Modular design is adopted in the hardware and software structures to improve system scalability. The hardware structure comprises six sub-systems, including the on-board control computer sub-system, power sub-system, communication sub-system, sensor and perception sub-system, ground station sub-system, and execution sub-system. The software structure comprises six modules, the communicator module, GPS-IMU module, protocol module, tracker module, controller module, execute module, and engine module.

In general, the surface environment of flooding disasters, including fixed obstacles, floating obstacles, narrow canals, and the wind/wave/current disturbances, makes the target difficult to be inspected by an USV, because it presents a great limitation in trajectory tracking in complicated surroundings. The primary reason is the difficulty of obtaining accurate and applicable dynamical models. The hydrodynamic mechanism is very complex, and the dynamical model parameters change with Froude number $F_r = U / \sqrt{Lg}$, where U is the operating speed of USV, L is the overall length of USV (the submerged length of USV), and g is the acceleration of gravity [6]. When the Froude number is <0.5, the main fluid forces exerted on USV are the hydrostatic pressure by replacing water with respect to hydrodynamic pressure, called *displacement area*; when the Froude number is >0.5 but <1, the main fluid forces exerted on USV are hydrostatic and hydrodynamic pressure, called *semi-displacement area*; when the Froude number is >1, the main forces exerted on USV are hydrodynamic pressure, called *planning area*. Since the model structure and parameters will change greatly from *displacement area* to *planning area*, there is no unified dynamics model of a surface robot.

To approximate the hydrodynamics, we propose an active quasi-linear parameter varying (qLPV) model to approach the dynamics of the USV system. The LPV model concerns linear models whose state-space representations depend on state independent parameters [7]. The qLPV model is obtained by making the varying parameter of the LPV system a function of the state [8]. To accommodate the unstructured model error, the model error is introduced into the qLPV structured model as a complementation, and with the active modeling technique, the model error online estimation is used to improve the modeling accuracy. There are many available algorithms for active model online estimation such as the extended Kalman filter (EKF) [9] and epsilon-support vector regression (ε-SVR) [10]. In this chapter, the Unscented Kalman Filter (UKF) is utilized to obtain the unstructured model error.

Finally, to show the performance of the USV systems and the modeling methods, extensive experiments have been done including rescuing rope throwing by using an automatic pneumatic, rescuing of people, air–surface robots' cooperation, environment data collection, model parameters identification, communication distance testing, and water sampling in a 2.5 m radius circle.

2. System design of flooding disaster-oriented USV

Although the parts of USVs have been applied in reality, the autonomy of these platforms is still so weak that most of them can only work under the control of operators. This makes them difficult to be used to verify some high-performance autonomous control algorithm. Thus, a new USV system equipped with different kinds of sensors and ground control system is designed and introduced in this chapter.

Modular design is adopted in hardware and software structures, and the corresponding modules will be described in detail. Every hardware module is a separate subsystem which is connected together by waterproof aviation plugs. An automatic pumping system is equipped in the USV to drain water when the water level in the hull exceeds the warning level. Similar to the hardware system, every software module is a separate processor that is connected together by shared memories. Since all processors share a single view of data, the communication between processors can be fast as memory access. When the shared memories have been created, all the processors needed to do is to map the shared memory and initialize the read/write lock of thread in shared memory struct.

2.1. Hardware design

The USV system designed and implemented in SIA, CAS is shown in **Figure 1**. Its basic parameters are provided in **Table 1**.

Figure 1. The USV platform.

Length	Width	Height	Max velocity	Payload
2800 mm	700 mm	370 mm	36 km/h	70 kg

Table 1. Performance parameters of the USV.

The material of USV is fiber-reinforced polymer (FRP) which is a composite that is suitable for structures in corrosive environment and long-span lightweight structures due to its high-strength, light-weight, and anti-corrosive qualities [11].

To improve system scalability, the hardware structure adopts modular design. The hardware structure comprises six sub-systems, including the on-board control computer sub-system, power sub-system, communication sub-system, sensor and perception sub-system, ground station sub-system, and execution sub-system.

2.1.1. On-board control computer sub-system

The on-board control computer sub-system (**Figure 2**) contains an Advantech computer UNO-2170 with 2X LAN, 4X COM, 1X 32G Compact Flash (CF), and 2X PCM-3780. The Advantech computer UNO-2170 supports QNX Neutrino Real-time Operating System (RTOS) and can be used to record the experimental data in 0.01 s. The PCM-3780 is a general purpose multiple channel counter/timer card for the PC/104 bus. It provides two 16-bit counter channels which can be used to produce the required Pulse Width Modulation (PWM) wave to control the ignition/flame switch servo, steering rudder servo, engine throttle servo, and the selector switch servo. Using the Advantech computer UNO-2170, the GPS-IMU data, and the command data from ground station sub-system can be received though two COM serial ports.

Figure 2. On-board control computer sub-system.

2.1.2. Power sub-system

The whole on-board control computer sub-system, as well as all sensors, is powered by a 12 V battery jar. Besides, considering the possibility of overvoltage at the time of switching on power, Advantech PCM-3910 DC-DC power supply module is utilized to smooth the output voltage of the battery. Moreover, the engine can generate electricity to recharge the two batteries using a battery isolator to avoid the voltage dropping when the engine is starting. The working time of the USV system is larger than 2 h.

2.1.3. Communication sub-system

The communication sub-system mainly contains a Futaba receiver, two FGR2 900 MHz industrial radios, a wireless router, and an image transmission equipment. The Futaba receiver is used to receive the signal from the Futaba remote controller in emergency case. The industrial radio in ground station sub-system is used to transform command data from ground station sub-system to on-board control sub-system, while the other industrial radio in on-board control computer sub-system is used to transform feedback data from on-board control computer sub-system to ground station sub-system. The maximum communication distance from ground station sub-system to on-board control sub-system is 20 km. The wireless router is used to connect debugging computer with UNO-2170 computer since it is not convenient to use QNX SDP on a QNX Neutrino RTOS system for self-hosted development. The image transmission equipment is used to transfer the video of the IP camera.

2.1.4. Sensor and perception sub-system

The sensor sub-system contains a GPS-IMU system (see **Table 2**), an IP camera, a sonar, and a LIDAR. The GPS-IMU is used to locate the USV and obtain some inertial states such as attitude, velocity, and acceleration. The IP camera can monitor the environment of USV both in daytime and at night since it integrates infrared and visible light sensing device. Video from the IP camera is compressed based on the standard of H.264 and is transformed into the ground station sub-system through an image transmission equipment. The sonar and LIDAR sensors are also equipped in the USV system to detect the obstacles under and above the surface, respectively.

Specification	Value
Heading accuracy	0.2° (1 σ, base line ≥ 2 m)
Attitude accuracy	0.5° (1 σ)
Position accuracy	2 cm + 1 ppm (CEP)
Speed accuracy	0.1 m/s
Data updating rate	1 Hz/5 Hz/10 Hz/100 Hz
Gyro range	±100°/s (optional ± 300°/s)
Gyro zero offset	±100°/s

Table 2. Specification of GPS-INS (XW-GI5630).

2.1.5. Ground station sub-system

Ground station sub-system (**Figure 3**) is an important human computer interaction platform for information processing.

Figure 3. Ground station sub-system.

It is the main controller unit of the USV system except for the on-board control computer sub-system and plays an important role in assisting the operator to monitor the USV's state real-time. When the USV is in some emergency situations, the operator can take appropriate disposition to ensure the safety of the USV through the state displayed on the ground station sub-system.

The ground station sub-system includes a high-speed computer, two screens (one is data screen, the other is video screen), communication device and two joysticks (control the rudder angle and engine throttle of the USV system, respectively). It allows remote controller operates the USV such as ignition/flame, speed, and course keeping.

2.1.6. Execution sub-system

The execution sub-system (**Figure 4**) contains four servos: a rudder servo (used to control the rudder angle of USV), a throttle servo (used to control the throttle size), an ignition/flame servo (used to start or stop the USV), and a switch servo (used to select the ground station or the remote controller to control the USV). The four servos receive the control data from the on-board control computer sub-system or remote controller and take corresponding actions.

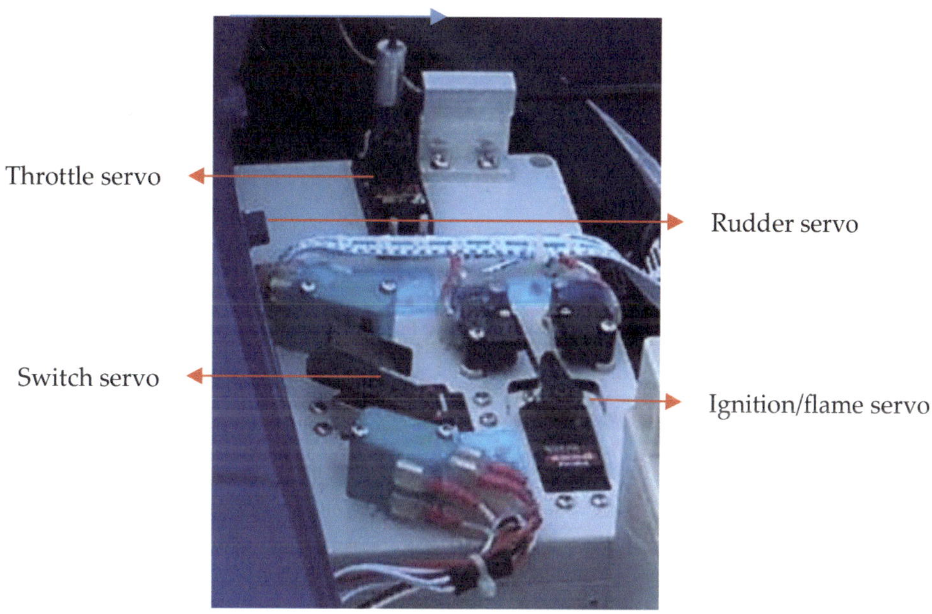

Figure 4. Execution sub-system.

2.2. Software design

Modular design is adopted in software structure. The software structure of computer control system is implemented using QNX SDP V6.5.0, a real-time operational system (RTOS), which is known as a real-time, high stable, and good portable OS. The software structure comprises six modules, the communicator module (TCP communicate and data transmit), GPS-IMU module (receive and process GPS-IMU signals), protocol module (parse the received data frame), tracker module (GPS tracking), controller module (including PID controller, ACC controller, and MPC controller), execute module (send PWM wave data to servos), and engine module (get the engine propeller speed) (**Figure 5**).

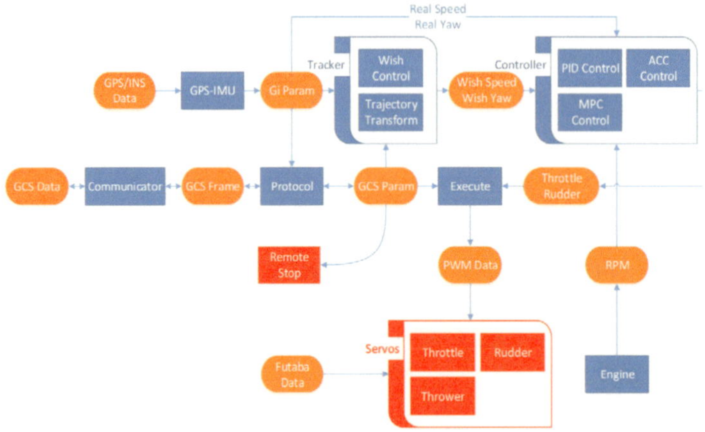

Figure 5. The software structure of USV system.

In execute module, there is an emergency sub-module that is used to stop the engine if no new data from ground station sub-system is received over 120 s. When new data from ground station sub-system are transferred to the on-board control sub-system, the execute module can generate PWM wave data to control the servos again.

In the software architecture, the key module is the controller module, which contains course keeping control, speed control, and GPS tracking control by using the measurements data from GPS-IMU module. The control period is 0.01 s.

3. USV dynamic modeling

Originally, linear models are often used owing to their simple structure, and mainly include first-order Nomoto model [12], first-order sideslip model [13], and linearized speed and steering model [14]. The Nomoto model describes the yaw dynamics between turn rate and the rudder angle input. However, when the USV's heading undergoes a small perturbation, the sideslip angle is no longer zero. To eliminate this defect in Nomoto model without considering slipping motion, one can incorporate a first-order lag model for sideslip. The sideslip model describes the sway dynamics that relates sideslip angle to turn rate. In addition to sideslip model, the linearized speed model relates the speed increment to throttle increment input. The linearized steering model relates the sideslip velocity and turn rate to rudder angle input. These models are usually effective under the assumptions: (1) the forward speed changes very slowly [15]; (2) the USV is in *displacement area* [16]. The deficiencies of the linear models are clear: (1) it presents poor performance when describing the details of the ship dynamics [17]; (2) the parameters change greatly for different speeds of the USV. This deteriorates the control performance greatly.

In addition to linear models, there are two kinds of nonlinear models, *maneuvring models* and *sea keeping models*. Maneuvring theory assumes that the USV is moving in restricted calm water.

Hence, the maneuvring models are derived for USV moving at positive speed U under a zero-frequency wave excitation assumption. The most famous model in maneuvring theory is the Abkowitz model [18] which adopts a nonlinear third-order truncated Taylor series expansion to approach the hydrodynamics at a nominal condition. Sea keeping theory, on the other hand, is the study of motion when there is wave excitation and the craft keeps its heading ψ and its speed U constant [18]. This introduces a dissipative force [19] known as fluid-memory effects. To estimate and identify the hydrodynamic derivatives of these nonlinear models, the forces and torque exerted on the USV need to be accurately measured. Unfortunately, this requires some strict experimental conditions that are often impossible [20].

In this section, we propose an active qLPV model to approach the hydrodynamics of USV. The advantage of the LPV model is that it is significantly simpler to be analyzed and allows the application of many linear control methods. The qLPV model is obtained by making the varying parameter of the LPV system a function of the state.

3.1. Active quasi-LPV modeling

The LPV model usually refers to the linear time-varying models that possess linear model structure but have exogenous variable $w(t)$ dependent system matrix $\mathbf{A}$ and input matrix $\mathbf{B}$ as shown in Eq. (1) [7]. While qLPV models means that the system matrix $\mathbf{A}$ and (or) input matrix $\mathbf{B}$ depend (or depends) on some state variable x of the system itself [8], that is, the system as Eq. (1).

$$\dot{x} = \mathbf{A}\left(w\left(t\right)\right)x + \mathbf{B}\left(w\left(t\right)\right)u \tag{1}$$

$$\dot{x} = \mathbf{A}\left(w\left(t\right), x\left(t\right)\right)x + \mathbf{B}\left(w\left(t\right), x\left(t\right)\right)u \tag{2}$$

That is, qLPV models are some kinds of generation of LPV models. The reason why we select the qLPV model structure to present the USV dynamics is that the nonlinearities of an USV cannot be ignored, especially when the USV is maneuvring with quickly varying velocity and course. Therefore, to denote the strong nonlinearities, the USV system can only be transferred into the qLPV form [21]:

$$\dot{v} = \mathbf{A}\left(v\right)v + \mathbf{B}u \tag{3}$$

where $\mathbf{A}(v) = \mathbf{M}^{-1}\mathbf{N}(v)$; $\mathbf{u} = T\left[\sin\delta \quad \cos\delta\right]^{\mathrm{T}}$; $\mathbf{B} = \mathbf{M}^{-1}\begin{bmatrix} 0 & -1 & x_\delta \\ 1 & 0 & 0 \end{bmatrix}^{\mathrm{T}}$.

The function $\mathbf{A}(v)$ in Eq. (3) can be denoted as linear combinations of the USV's velocity, that is,

$$\mathbf{A}(v) = \mathbf{M}^{-1}\left(\mathbf{N}_0 + \mathbf{N}_1|u| + \mathbf{N}_2|v| + \mathbf{N}_3 r + \mathbf{N}_4|r|\right) \tag{4}$$

where

$$\mathbf{N}_0 = \begin{bmatrix} X_u & 0 & 0 \\ 0 & Y_v & Y_r \\ 0 & N_v & N_r \end{bmatrix};\ \mathbf{N}_1 = \begin{bmatrix} X_{|u|u} & 0 & 0 \\ 0 & 0 & 0 \\ 0 & 0 & 0 \end{bmatrix};\ \mathbf{N}_2 = \begin{bmatrix} 0 & 0 & 0 \\ 0 & Y_{|v|v} & 0 \\ 0 & N_{|v|v} & 0 \end{bmatrix};$$

$$\mathbf{N}_3 = \begin{bmatrix} 0 & m + X_{vr} & mx_G + X_{rr} \\ -m & 0 & 0 \\ -mx_G & 0 & 0 \end{bmatrix};\ \mathbf{N}_4 = \begin{bmatrix} 0 & 0 & 0 \\ 0 & 0 & Y_{|r|r} \\ 0 & 0 & N_{|r|r} \end{bmatrix}.$$

The qLPV model structure provides two advantages: (1) its parameters can be identified by some linear algorithm resulting in a nonlinear mathematical model; and (2) mature linear control synthesis schemes can be generalized to obtain adequate performance.

To further eliminate the unstructured model error, the active modeling technique is used to account for the unstructured factors. In the preceding work, we have proposed a control architecture as following **Figure 6**.

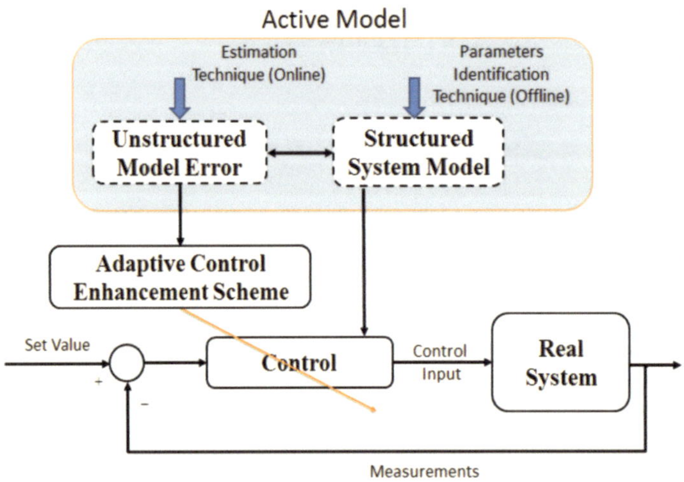

Figure 6. Active model-based control scheme.

In this architecture, the structured model can be used to design some nominal controller, while the online estimated model error is used to improve the closed loop performance of the nominal controller. To obtain the model error, we first rewrite the system Eq. (3) to follow the state-space form:

$$\dot{\boldsymbol{v}} = \mathbf{A}(\boldsymbol{v})\boldsymbol{v} + \mathbf{B}\boldsymbol{u} + \Delta\boldsymbol{v}$$
$$\boldsymbol{z} = \boldsymbol{v} + \boldsymbol{v}_m \tag{5}$$

where v is the system state vector; $\mathbf{A}(v)v + \mathbf{B}u$ is the structured system dynamics function which is shown in Eqs. (3) and (4); Δv is the model error; z contains the measurements; and v_m contains the measurement noise.

Then, UKF algorithm is used to estimate the model error Δv using the available measurements. The UKF is an often-used nonlinear Kalman filter owing to its good performance with respect to strong nonlinearities. The main idea of the UKF algorithm is that it uses the unscented transformation (UT) to handle the nonlinear part and compute the influence of the nonlinear function on some stochastic variables. The UT provides an approach for approximating the statistics of a nonlinear transformation through a finite set of "*sigma points*":

$$z = f(x) \tag{6}$$

where f is a nonlinear function and x denotes an $n \times 1$ stochastic variable with a mean of $\bar{\mathbf{x}}$ and a covariance of P_x.

To calculate the propagation statistics of x through f, that is, the mean $\bar{\mathbf{z}}$ and covariance P_z of the output z, the UT uses the following steps. The complete process is illustrated in **Figure 7**.

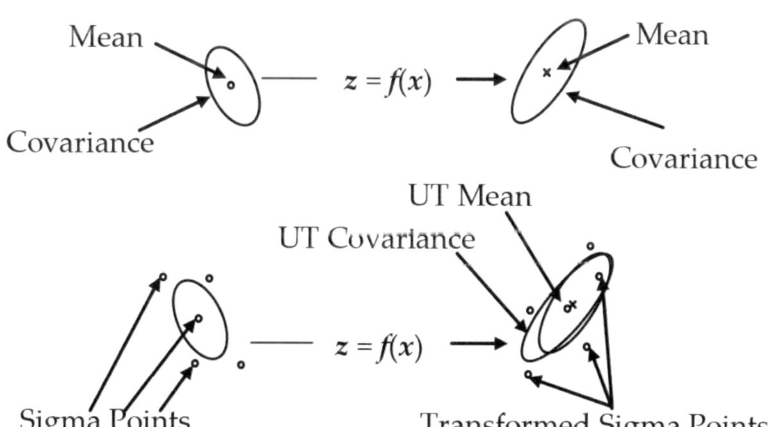

Figure 7. Unscented transformation.

The normal UKF can be deduced as follows.

UKF-I: Initialization. The measurement noise vector statistics are calculated from the initial measurement:

$$\hat{\boldsymbol{v}}_0^e = E\left(\boldsymbol{v}_0^e\right),$$
$$\boldsymbol{P}_0 = Var\left(\boldsymbol{v}_0^e\right) = E\left[\left(\boldsymbol{v}_0^e - \hat{\boldsymbol{v}}_0^e\right)\left(\boldsymbol{v}_0^e - \hat{\boldsymbol{v}}_0^e\right)^{\mathrm{T}}\right], \tag{7}$$

where $E[*]$ denotes the mean (i.e., expectation) of [*].

UKF-II: Sigma points calculation. At the kth time instant, using the kth time instant mean $\hat{\mathbf{v}}_k^e$ and covariance $\boldsymbol{P}_k$, the sigma points are defined using a Gaussian distribution.

$$\begin{aligned} \xi_{0,k}^e &= \hat{\boldsymbol{v}}_k^e, \\ \xi_{i,k}^e &= \hat{\boldsymbol{v}}_k^e + \left(\sqrt{(n+\lambda)\boldsymbol{P}_k}\right)_i, \quad (\forall i \in \{1,2,\cdots,n\}), \\ \xi_{i+n,k}^e &= \hat{\boldsymbol{v}}_k^e - \left(\sqrt{(n+\lambda)\boldsymbol{P}_k}\right)_i. \end{aligned} \tag{8}$$

UKF-III: Time update. Using the sigma points, the new state is acquired:

$$\begin{aligned} \gamma_{i,k|k-1}^e &= \boldsymbol{f}_{k-1}^e\left(\xi_{i,k-1}^e, \boldsymbol{u}_{k-1}\right) + \boldsymbol{q}_{k-1}, \quad (\forall i \in \{0,1,\cdots,L\}), \\ \hat{\boldsymbol{v}}_{k|k-1}^e &= \sum_{i=0}^{L} W_i^m \gamma_{i,k|k-1}^e = \sum_{i=0}^{L} W_i^m \boldsymbol{f}_{k-1}^e\left(\xi_{i,k-1}^e, \boldsymbol{u}_{k-1}\right) + \boldsymbol{q}_{k-1}, \\ \boldsymbol{P}_{k|k-1} &= \sum_{i=0}^{L} W_i^c \left(\gamma_{i,k|k-1}^e - \hat{\boldsymbol{v}}_{k|k-1}^e\right)\left(\gamma_{i,k|k-1}^e - \hat{\boldsymbol{v}}_{k|k-1}^e\right)^{\mathrm{T}} + \boldsymbol{Q}_{k-1}. \end{aligned} \tag{9}$$

UKF-IV: Measurements update. Similarly, using the updated state $\hat{\mathbf{v}}_{k|k-1}^e$ and $\mathbf{P}_{k|k-1}$ in ***UKF-III***, new sigma points $\xi_{i,k|k-1}^e$ can be calculated to update the measurements.

$$\begin{aligned} \chi_{i,k|k-1} &= \boldsymbol{h}_k^e\left(\xi_{i,k|k-1}^e\right) + \boldsymbol{r}_k, \quad (\forall i \in \{0,1,\cdots,L\}), \\ \hat{\boldsymbol{z}}_{k|k-1} &= \sum_{\mathrm{i}=0}^{L} W_i^m \chi_{i,k|k-1} = \sum_{\mathrm{i}=0}^{L} W_i^m \boldsymbol{h}_k^e\left(\xi_{i,k|k-1}^e\right) + \boldsymbol{r}_k, \\ \boldsymbol{P}_{\tilde{z}_k} &= \sum_{\mathrm{i}=0}^{L} W_i^c \left(\chi_{i,k|k-1} - \hat{\boldsymbol{z}}_{k|k-1}\right)\left(\chi_{i,k|k-1} - \hat{\boldsymbol{z}}_{k|k-1}\right)^{\mathrm{T}} + \boldsymbol{R}_k, \\ \boldsymbol{P}_{\tilde{v}_k \tilde{z}_k} &= \sum_{\mathrm{i}=0}^{L} W_i^c \left(\xi_{i,k|k-1}^e - \hat{\boldsymbol{v}}_{i,k|k-1}^e\right)\left(\chi_{i,k|k-1} - \hat{\boldsymbol{z}}_{k|k-1}\right)^{\mathrm{T}}. \end{aligned} \tag{10}$$

When the new measurement z_k is obtained, the state $\mathbf{v}_k^e$ is updated:

$$\begin{aligned} \boldsymbol{K}_k &= \boldsymbol{P}_{\tilde{v}_k \tilde{z}_k} \boldsymbol{P}_{\tilde{z}_k}^{-1}, \\ \hat{\boldsymbol{v}}_k^e &= \hat{\boldsymbol{v}}_{k|k-1}^e + \boldsymbol{K}_k \left(\boldsymbol{z}_k - \hat{\boldsymbol{z}}_{k|k-1} \right), \\ \boldsymbol{P}_k &= \boldsymbol{P}_{k|k-1} - \boldsymbol{K}_k \boldsymbol{P}_{\tilde{z}_k} \boldsymbol{K}_k^{\mathrm{T}}, \end{aligned} \tag{11}$$

where $\boldsymbol{Q}_k$ and $\boldsymbol{R}_k$ are the process and measurement noise covariance, respectively, which are both assumed to be known a priori. The parameter α is usually set within [0.0001, 1].

3.2. Identification results

In this section, we describe the results of two experiments: (1) full state model identification experiment; (2) active modeling enhanced qLPV model experiment. In the final, the results are analyzed and compared.

To quantitatively compare the identification errors, an index function is defined to evaluate the identification errors:

$$J_{index} = \sqrt{\frac{1}{N}\sum_{i=1}^{N}\left[x_{est}(i) - x_{ral}(i)\right]^2} \Bigg/ \sqrt{\frac{1}{N}\sum_{i=1}^{N}x_{ral}^2(i)} \tag{12}$$

To verify the model accuracy for different motion, experiments were conducted using a sine wave rudder angle input as shown in **Figure 8**. The quantitative comparisons are given in **Table 3**.

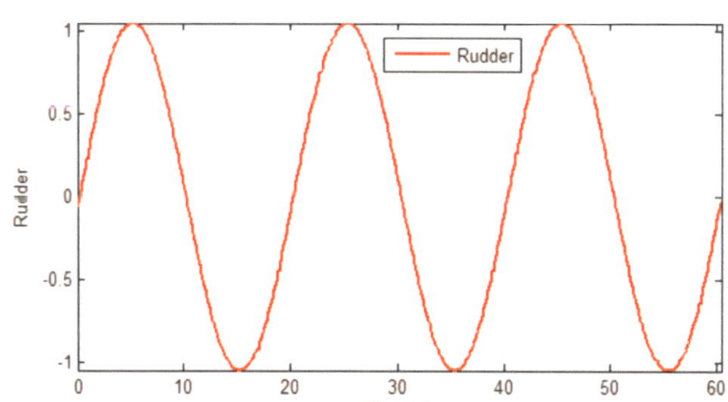

Figure 8. Sine wave rudder angle: input of the USV.

From these results, one can see that the qLPV model presents better performance for the USV system, especially for surge dynamics, for which the prediction error is <3% (no more than half of that for the two linear models). For sway dynamics, the qLPV model also provides better accuracy, but it is not as good as that for surge dynamics. This means that sway dynamics and yaw dynamics possess stronger nonlinearity since the sway and yaw speed are usually smaller and easily influenced by external disturbances compared to surge dynamics.

	Linearized	Nomoto with sideslip	qLPV
(a) Surge dynamics	4.6%	4.6%	2.3%
(b) Sway dynamics	19.7%	17.9%	16.4%
(c) Yaw dynamics	13.7%	13.4%	13.5%

Table 3. Modeling error for sine input.

To further reduce the model mismatch's influence and improve the estimation accuracy, the active modeling scheme is used. Using an UKF algorithm, the model error of the qLPV structured model is estimated online.

The parameters of the UKF algorithm are

$$\begin{aligned} &Q = 10^{-2}\, diag(0,0,0,16,2,1), R = 10^{-3}\, diag(16,2,1), \\ &\alpha = 1\times 10^{-3}, \beta = 2, \end{aligned} \tag{13}$$

and we use the same data as that of the full state model identification experiment, that is, the throttle was set to 30%, and the rudder angle follows a sine wave with amplitude $\pi/3$. The prediction error computed by using Eq. (12) is shown in **Table 4**. Compared to the results from the qLPV model without active modeling in **Table 3**, the model accuracy improvement is significant. The active modeling enhanced qLPV model significantly reduces the prediction errors (only one-third of that for the qLPV model). This is intended to make the USV autonomously adaptive to its internal and external uncertainties, that is, to achieve a robust tracking performance for time-varying unknown disturbances in the vehicle dynamics.

Surge	Sway	Yaw
0.8%	4.8%	4.5%

Table 4. Modeling error for active modeling enhanced qLPV model.

4. Application experiments

There are some application experiments to show the performance and application prospect of the USV system, such as rescuing rope throwing, rescuing of people, air–surface robots' cooperation, environment data collection, and water sampling. In this section, two typical

http://dx.doi.org/10.5772/64305

experiments, that is, rescuing experiment and water sampling experiment, are introduced in detail.

4.1. Rescuing experiment

The rescuing throwing experiment is aiming at searching and rescuing the trapped people (**Figure 9**). When trapped people were detected, the USV's direction would be adjusted by the ground station sub-system to launch a lifebuoy by using an automatic pneumatic. After the lifebuoy was exposed to water, it would be inflated automatically in 5 s.

a) rescuing rope throwing   b) rescuing of people

Figure 9. Rescuing experiment. (a) Rescuing rope throwing and (b) rescuing of people.

The launch power is from the high-pressure (30 MPa) gas in pneumatic cylinders. The trigger servo (**Figure 10**) of the pneumatic is controlled by the PWM wave from the on-board control computer or the remote controller. The farthest distance of dumping is 150 m. Using the kayak carried by the USV, we can achieve trapped people dragging. After the trapped people reached the kayak, the ground station sub-system would send commands to USV to drag the kayak to the safe place automatically.

Figure 10. Automatic pneumatic.

4.2. Water sampling experiment

The sampling equipment is composed of four sampling bottles, a water pump and some assist mechanical devices (**Figure 11**).

Figure 11. Water sampling equipment.

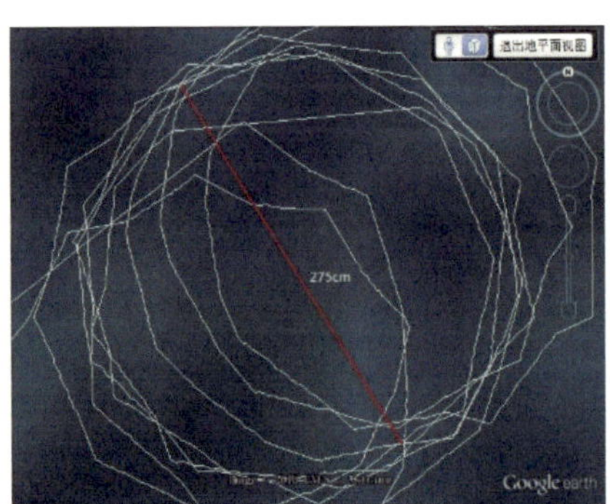

Figure 12. The minimum turning circle (radius 1.375 m).

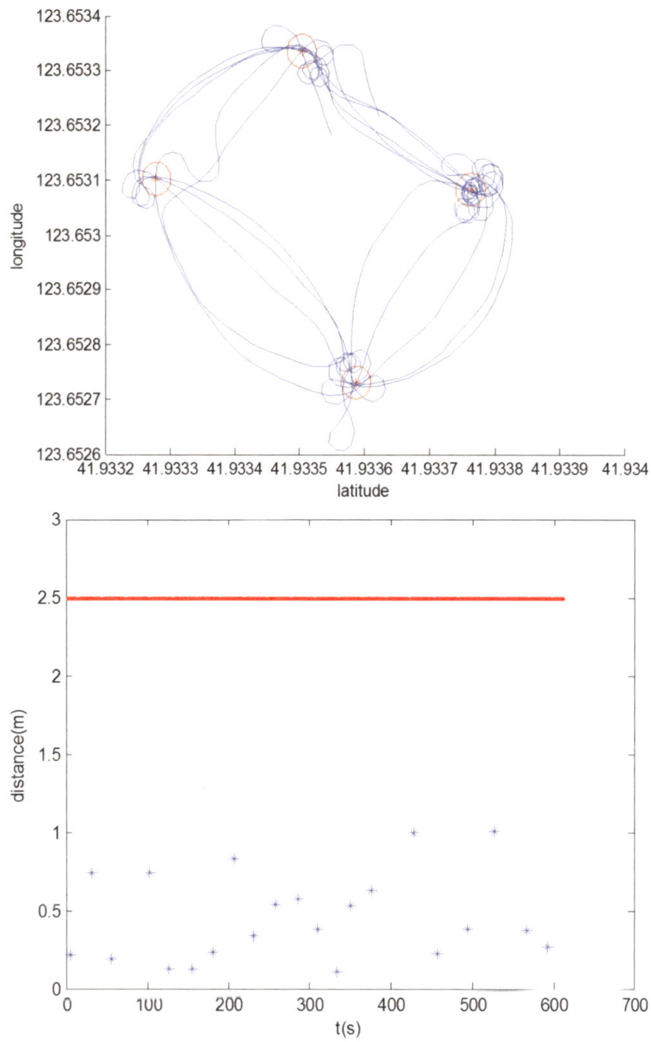

Figure 13. Four water sampling points and distance errors around the sampling points when water sampling.

First, the USV carried two sets of sampling equipment into a designed area (a 2.5 m radius circle around an GPS point). While the USV arriving at the interested point, it started to circle around the interested point at the minimum turning radius by using an GPS point tracking control algorithm (**Figure 12**). Then, the water pump protruded from the equipment and started drawing water into the bottle (each point for one bottle). After the work was done, the USV would return back automatically by using a course keeping control algorithm.

Figure 13 shows the trajectory of USV and the GPS points tracking error at four sampling points when water sampling. In the left figure, the green line represents the trajectory of USV, and the red circles are the 2.5 m radius circles around the designed sampling points. From **Figure 13**, we can see that the maximum tracking error 1.2 m, far <2.5 m.

5. Conclusions

The main concerns in this chapter is to design an real-time flooding disaster-oriented USV and construct a dynamics model for the USV system which is simple enough for model parameters identification and rich enough for motion control.

This work produced three contributions:

1. A real-time USV system adopting modular design by using different kinds of sensors and ground control system was introduced;
2. To improve model accuracy, an active qLPV nonlinear model was constructed and identified by considering both the 3-DOF rigid body dynamics and the hydrodynamics;
3. Extensive experiments were done to show the performance of the hardware and software systems.

Author details

Junfeng Xiong[1,2], Feng Gu[1], Decai Li[1], Yuqing He[1*] and Jianda Han[1]

*Address all correspondence to: heyuqing@sia.cn

1 State Key Laboratory of Robotics of Shenyang Institute of Automation, Chinese Academy of Sciences, Shenyang, China

2 University of Chinese Academy of Sciences, Beijing, China

References

[1] Hirabayashi, Y., et al., *Global flood risk under climate change.* Nature Climate Change, 2013. 3(9): p. 816–821.

[2] Murphy, R.R., et al., *Cooperative use of unmanned sea surface and micro aerial vehicles at Hurricane Wilma.* Journal of Field Robotics, 2008. 25(3): p. 164–180.

[3] Murphy, R.R., et al., *Marine heterogeneous multirobot systems at the great Eastern Japan Tsunami recovery.* Journal of Field Robotics, 2012. 29(5): p. 819–831.

[4] Qi, J., et al. *Design and implement of a trimaran unmanned surface vehicle system.* In: International Conference on Information Acquisition, 2007. ICIA'07. 2007. IEEE.

[5] Li, M., et al. *Design and implementation of a new jet-boat based unmanned surface vehicle*. In: International Conference on Automatic Control and Artificial Intelligence (ACAI 2012), 2012. IET.

[6] Faltinsen, O.M., *Hydrodynamics of high-speed marine vehicles*. 2005, Cambridge, UK: Cambridge University Press.

[7] Shamma, J.S., *An overview of LPV systems, in control of linear parameter varying systems with applications*. 2012, Springer: Berlin. p. 3–26.

[8] Raïssi, T., G. Videau, and A. Zolghadri, *Interval observer design for consistency checks of nonlinear continuous-time systems*. Automatica, 2010. 46(3): p. 518–527.

[9] Hwang, W.-Y., *Application of system identification to ship maneuvering, in Ocean Engineering*. 1980, MIT: Cambridge, MA.

[10] Zhang, X.-G. and Z.-J. Zou, *Identification of Abkowitz model for ship manoeuvring motion using ε-support vector regression*. Journal of Hydrodynamics, Series B, 2011. 23(3): p. 353–360.

[11] Peng, Q., F. Peng, and L. Ye, *Experimental study on GFRP pipes under axial compression*. Frontiers of Architecture & Civil Engineering in China, 2008. 2(1): p. 73–78.

[12] Nomoto, K., *On the steering qualities of ships*. International Shipbuilding Progress, 1957. 4(35): p. 75–82.

[13] Yu, Z., X. Bao, and K. Nonami, *Course keeping control of an autonomous boat using low cost sensors*. Journal of System Design and Dynamics, 2008. 2(1): p. 389–400.

[14] Fossen, T.I., *Guidance and control of ocean vehicles*. Vol. 199. 1994, New York: Wiley.

[15] Sarker, M.M.H., et al., *Modelling and analysis of control system for a multi-robotic system*. International Journal of Intelligent Control and Systems, 2009. 14(4): p. 221–227.

[16] Fossen, T.I., *Handbook of marine craft hydrodynamics and motion control*. 2011, New York: John Wiley & Sons.

[17] Kahveci, N.E. and P.A. Ioannou, *Adaptive steering control for uncertain ship dynamics and stability analysis*. Automatica, 2013. 49(3): p. 685–697.

[18] Abkowitz, M.A., *Lectures on ship hydrodynamics--Steering and manoeuvrability*. 1964, Hydroand Aerodynamic's Laboratory: Lyngby.

[19] Cummins, W., *The impulse response function and ship motions*. 1962, MIT: Cambridge, MA.

[20] Padilla, A., J.I. Yuz, and B. Herzer, *Continuous-time system identification of the steering dynamics of a ship on a river*. International Journal of Control, 2014. 87(7): p. 1387–1405.

[21] Xiong, J., et al., *Active quasi-LPV modeling and identification for a Water-Jet Propulsion USV: an experimental study*. IFAC-PapersOnLine, 2015. 48(28): p. 1359–1364.

Chapter 3

Muscle-Like Compliance in Knee Articulations Improves Biped Robot Walkings

Hayssan Serhan and Patrick Henaff

Additional information is available at the end of the chapter

http://dx.doi.org/10.5772/63746

Abstract

This chapter focuses on the compliance effect of dynamic humanoid robot walking. This compliance is generated with an articular muscle emulator system, which is designed using two neural networks (NNs). One NN models a muscle and a second learns to tune the proportional integral derivative (PID) of the articulation DC motor, allowing it to behave analogously to the muscle model. Muscle emulators are implemented in the knees of a three-dimensional (3D) simulated biped robot. The simulation results show that the muscle emulator creates compliance in articulations and that the dynamic walk, even in walk-halt-stop transitions, improves. If an external thrust unbalances the biped during the walk, the muscle emulator improves the control and prevents the robot from falling. The total power consumption is significantly reduced, and the articular trajectories approach human trajectories.

Keywords: humanoids, articular compliance, muscle modeling, neural networks, biped walking

1. Introduction

Robots are currently being designed based on human morphologies. Therefore, the most recent humanoid robots are technologically complex systems, with an extremely high level of mechanical and electronic integration. They are equipped with complete perceptive systems, which enable them to interact with the human beings and to move in human environments. However, controlling humanoid robots is difficult, particularly walking motions and balance when walking, during transitions from walking to stopping, or when the robot undergoes external thrusts. The walking and running abilities of humans have caused researchers to focus on muscular skeletal system properties, with a focus on improving the walking and interaction capabilities of robots.

One of the most significant properties of the mammalian muscle is compliance, that is, the capacity to adapt the muscle stiffness to various movements and interactions. When walking, the muscular compliance is implicated at a low level of control because it occurs in the muscle (the actuator) according to the required movement of the limb. Muscle compliance allows a muscle to optimize energy consumption by storing energy in passive phases and restoring energy for active phases. This intrinsic muscle stiffness control allows humans to run, walk, control walk-halt-walk transitions, and control posture related to external perturbations (other systems, such as the vestibular apparatus, are also involved in balancing, walking, and running processes).

1.1. Recent progress in articular compliance for legged robots

Creating an artificial system that can mimic mammalian muscle behavior would represent significant progress in the field of humanoid robots [1]. This goal can be achieved via numerous advancements [2, 3].

1.1.1. Compliant actuators

Special mechanical systems can be built that store and restore energy based on the walking phases. This approach, which is known as passive dynamic walking, was pioneered by McGeer more than a decade ago [4] and has been analyzed by several studies [5–8]. Passive dynamic walking is attractive due to its elegance and simplicity. However, active feedback control is necessary to achieve walking on the ground and varying slopes, for robustness related to uncertainties and disturbances, and to regulate the walking speed. The first active feedback control that exploits passive walking appeared for planar bipeds [6, 9–11]. Three-dimensional (3-D) passive walking was studied [12, 13], the results [11] of which were extended to the general case of 3-D walking [8]. An interesting extension of these concepts uses geometric reduction methods to generate stable 3-D walking from two-dimensional (2-D) gaits [14]. Robustness issues were addressed by using total energy as a storage function in the hybrid passivity framework [15].

Another manner in which to achieve compliance is to design actuators that reproduce the properties of mammalian muscles. These novel elastic actuators can be regarded as an artificial "muscle-tendon" [16]. The McKibben pneumatic muscle actuator [17–19] produces a high force at low speeds, but these actuators are difficult to control because of their nonlinearity depending on air temperature and pressure. However, artificial pneumatic muscles or elastic actuators can be used to actuate joints of bipedal robot and their natural compliance improves robustness to postural and motion in jumping experiments [20, 21].

Alternatives, such as piezoelectric actuator, electroactive polymers, or shape memory alloys, also possess energy disadvantages. Emerging technologies, such as those based on electroactive polymers, can provide high power density with reasonable energy efficiency [22]. Dielectric elastomer-based linear actuators are interesting; however, their thrust profile can be improved in terms of stiffness characteristics [23]. These techniques are very promising, but no electroactive polymer has yet yielded a walk within a biped.

1.1.2. Compliant joint with mechanical elasticities

Other approaches consist of developing new mechanisms in the robot actuators. Springs can be attached across the knee joints in parallel with the knee actuators [24], or a spring can be inserted in series with the actuator and thus reduce the energy consumption [25–27, 45]. This has been demonstrated for running, where tendons may be responsible for half or more of the overall work of the musculo-tendinous system [28]. However, the tendons could yield the required leg speed with only a small active work production, such as that from the transfer phase [29, 30] or from slow walking to fast running [31]. The necessary "push-off" phase, which allows humans to walk at all speeds, also benefits from elastic energy storage in the Achilles tendon [32, 33]. The springs could also be arranged in parallel with the actuators so that no active force is required to initiate these springs. The Delft biped [36] applies this principle with springs (and MacKibben muscles) at the ankles, which provides extended support for the leg and helps reduce collision losses. Although these principles can reduce the energy consumption, they tend to be poorly suited for theoretical control law designs. Mechanical systems (stops, clutches, latches, etc.) introduce nonlinearities, which are generally incompatible with traditional control approaches. The linear springs that store and return the energy introduce natural oscillation modes, which must be further identified and controlled. Adding elastic knees to biped robots offers more elastic gaits [34]. Knee prosthesis including elastic actuators positioned in parallel to reproduce agonist-antagonist muscle actions increases enhanced comfort of the human walking [35].

1.1.3. Compliant joint based on special control algorithms

In contrast to the insert compliance of the mechanical limb or actuator design, the compliance effect can be achieved via a DC motor control algorithm [44]. Online controllers can be used for maintaining dynamic stability of humanoid robots. Mixing damping joint and damping controller increases the balance of the robot [37].

One such approach consists of adapting the proportional integral derivative (PID) gains of the DC motor control loop in real time, according to a specific control law that explains the compliance behavior. For example, if this law represents a muscle model, then it can theoretically reproduce responses that are similar to those observed in the musculo-skeletal system. This approach then possesses the benefits of electric motors and those of human muscle without adding any physical elements to the robot. Nevertheless, this approach requires a muscle model.

Studies have mimicked muscle behavior using a DC motor and PID controller [38]. This muscle emulator (**Figure 1**) uses a muscle model developed based on a neural network autoregressive exogenous (NNARX) input structure. This NNARX was trained to learn data issued from experimentation described by Gollee and Donaldson et al. in [30, 39]. The PID parameters are tuned using an multi-layer perceptron MLP network with a special indirect online learning algorithm. The learning algorithm calculation is performed based on a model of the DC motor. The NNARX muscle model output is used as a reference for the DC motor control loop, in a model following control loop. The results show that the physical system was successfully forced to behave like the muscle model within acceptable error margins. This technique was able to physically emulate a nonlinear muscle model based on a DC motor with a PID controller tuned by a neural network (NN), enabling a robot to walk like a human. Using neural actuator, identification is possible when the actuator model is uncertain.

The rest of this chapter analyses the walking efficiency of a 3D simulated biped robot when the muscle emulator is implemented or not implemented in the robot's knees. The work focuses on the robustness of the dynamic walk during walk-halt-stop transitions and when external unknown forces unbalance the walking robot. The simulation results show that articular trajectories with the muscle emulator approach human trajectories and that the total motor power consumption is significantly reduced.

This chapter is organized as follows. After this introduction, Section 2 presents the robot and simulator, and summarizes the walking control approach presented in [40, 41]. The control algorithm implementation is described and the results are analyzed in Section 3. In addition, the results are compared and discussed. Section 4 provides the conclusions.

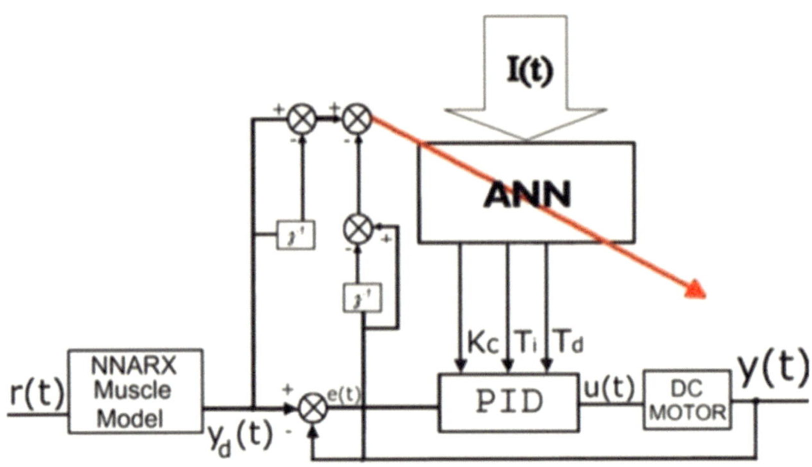

Figure 1. Block diagrams of the muscle emulator. *y(t)* is the torque output of the DC motor, $y_d(t)$ is the output of the muscle model, *r(t)* is the articular command, and *I(t)* is the input signals vector based on *e(t),e(t-1),e(t-2)* and *ρ(t)* which provides information on the fast variations of *r(t)*.

2. Walk control approach

This section integrates the muscle emulator into the knee joint control laws. The model of our biped robot consisted of six active degrees of freedom within each leg. Each knee is driven using a 90-W DC motor (Maxon RE-35) with 1/90 gear. The biped robot is simulated using the

http://dx.doi.org/10.5772/63746

Open Architecture Humanoid Robotics Platform (OpenHRP) simulator. The muscle emulator is implemented to control the DC motor in the low-level control mode without altering the high-level robot controls. The high-level controls include a walking algorithm based on a state machine that provides a stable 3D dynamic biped walk. The walking algorithm and muscle emulator are coded using C++ in the OpenHRP simulator.

2.1. Control levels description

The proposed robot actuator control diagram consists of two levels. The low-level control is designated by the PID controller. The high-level control is designated by the dynamic walking algorithm, which tunes the PD controllers and imposes the reference signals following the walking phases (**Figure 2**).

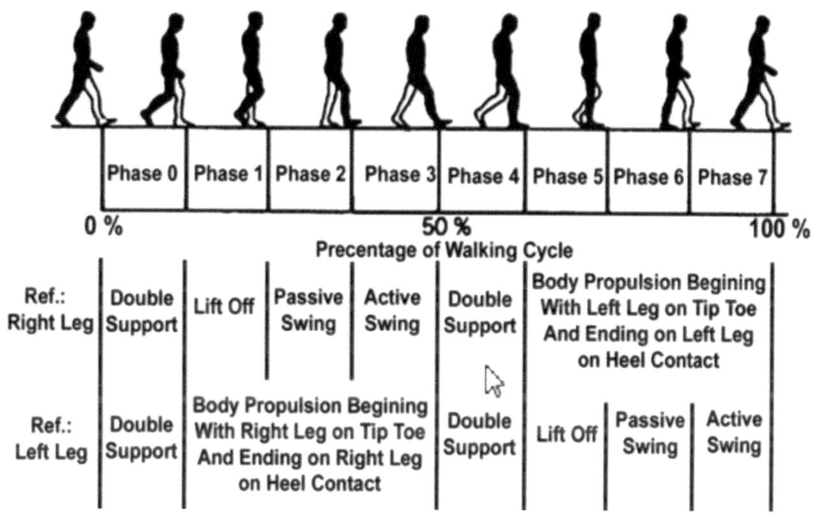

Figure 2. Full human walking cycle (ref. right leg).

Two possibilities exist for the lower level, which are based on using the fixed PID gains for the 12 motors (gains are calculated to have the same adjustments as the real robot) or using the fixed PID gains for all motors except those in the knees, for which PID gains are adapted using the muscle emulator (**Figure 3**).

The high-level control adapts the walking PD controller parameters to each walking phase and introduces the corresponding reference signal to the low-level loop. Note that all of the motors in the low level are driven by PID controllers with desired articular angles.

A general equation of each actuator input Y_d PD controller is given by:

$$Y_d(t) = \tau(t)^{Leg}_{Actuator} = K^{Phase}_{c\,Actuator}\left(\theta_d - \theta(t)^{Leg}_{Actuator}\right) + K^{Phase}_{d\,Actuator}\left(\dot{\theta}_d - \dot{\theta}(t)^{Leg}_{Actuator}\right) \quad (1)$$

where τ(t) is the desired torque, θ_d is the desired angle, θ_{d} is the desired velocity, "Leg" corresponds to the stance or swing leg, and "Phase" corresponds to one of the eight phases of the human walking cycle presented in **Figure 3** and "Actuator" corresponds to the hip, knee, and ankle. The values of K_c, and K_d change during each phase of the walk, as determined by an empirical method when the swing leg approaches the ground [40].

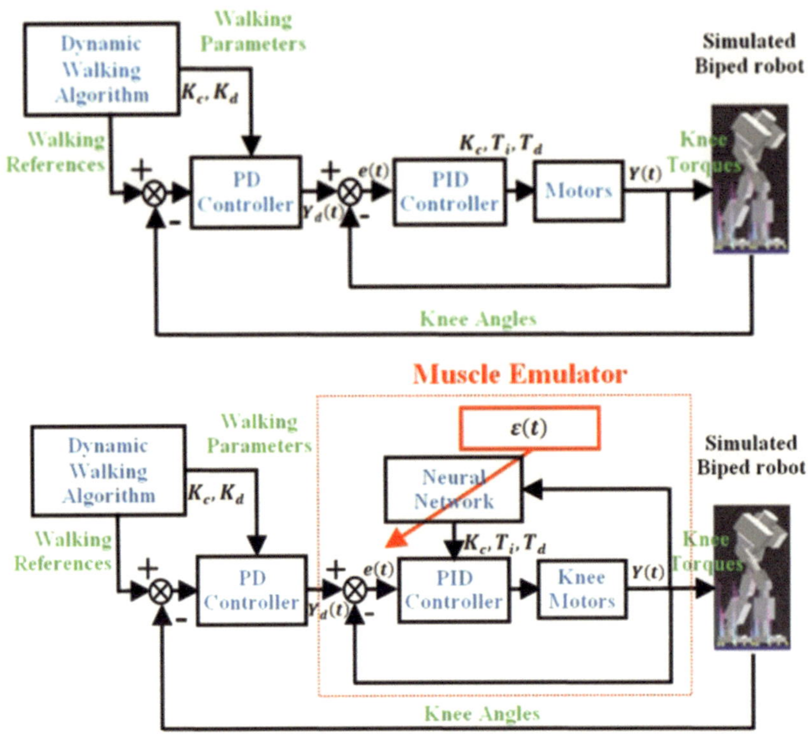

Figure 3. Proposed dynamic walk control diagram. Top: all articular control loop with fixed PID gains K_c, T_i, T_d. Bottom: knee adaptive control loop with muscle-like control loop.

2.2. Dynamic walking algorithm

The proposed walking control approach is based on a sequential analysis of the human walking cycle, the properties of which have been previously determined [37, 42, 43]. This cycle can be divided in eight phases, with one leg acting first as a swing leg and then as a stance leg (**Figure 3**). Phase 0 is the double support (DS) phase, during which the two legs are touching the ground. Phase 1 is the lift-off phase, in which hip muscle contractions accelerate the swing leg. Phase 2 is the passive swing, which begins when the thigh is sufficiently accelerated. Phase 3 is the active swing phase. Phase 4 is the DS phase in the next half of the walking cycle. The swing leg assumes the role of the stance leg (ST) in phase 5. Phases 6 and 7 are similar to the fifth phase in terms of functionality related to the stance leg.

We analyzed the human walking cycle [40] and determined that a robot's dynamic walk can be modeled using a Petri net algorithm (**Figure 4**). The cycle is divided into 12 states. States 1, 3, 4, 5, 6, and 10 correspond to the swing leg and states 2, 7, 8, 9, and 11 correspond to the stance leg. The initial position is defined when the two legs are in contact with ground (DS) and the robot is standing, but one of the legs is behind the other. All of the transitions in this state machine are defined based on the extreme values of essential leg angles in sagittal or frontal planes. All of the articulations are driven by PD controllers (Eq. (1)), for which each gain changes based on the corresponding phase of the Petri net algorithm. Finally, we can control the robot's velocity with respect to its body inclination.

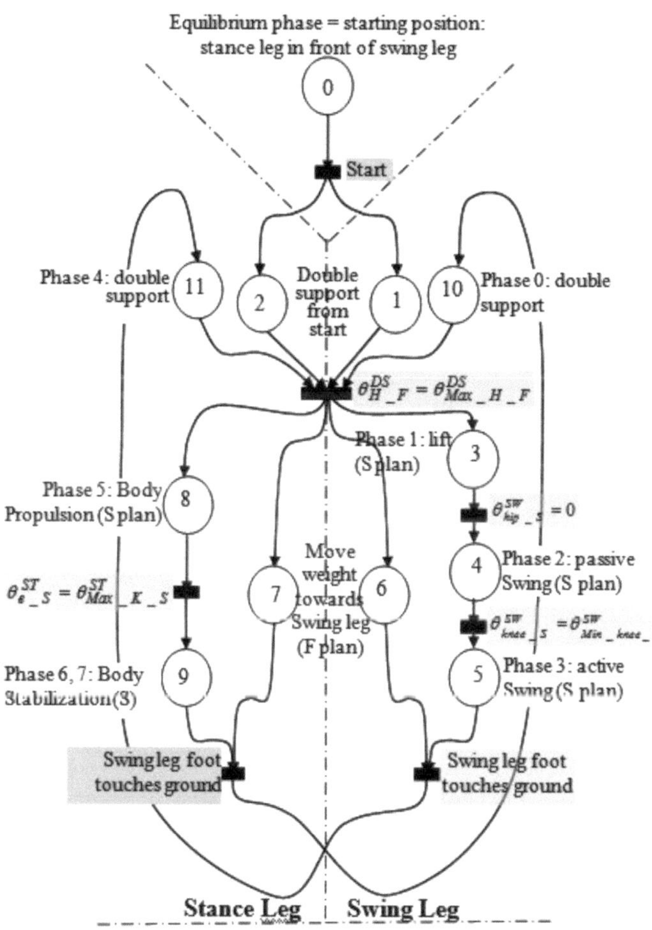

Figure 4. Petri net algorithm for a ROBIAN dynamic walk (S_Plane is sagittal **Figure 5**. plane, F_Plane is frontal plane).

To stop the robot walking cycle and keep it in a standup stable position, an extension of the Petri net **Figure 4** is applied to change our algorithm to introduce a stop phase (**Figure 5**). The change will start working after swing leg touches ground (State 9, **Figure 4**, or phases 3 or 7 in **Figure 3**). We controlled the hip motors to maintain the body in vertical position. By controlling stance and swing leg, the robot will be in stable standup position where we keep little distance between right and left legs foot. From this position we can control the robot to walk again by simple walking algorithm by reapplying our walking Petri net algorithm.

We applied the overall Petri net algorithm using a simulated model of our ROBIAN biped robot (**Figure 6**). The OpenHRP (Open Architecture Humanoid Robotics Platform) simulator was used, which is a dynamic humanoid robot simulation platform that was developed by AIST, the University of Tokyo, and MSTC. The robot model is programmed in VRML. All of the robot specifications were taken into account. Many control algorithms were tested on the real robot and on the simulated model to improve the similarity.

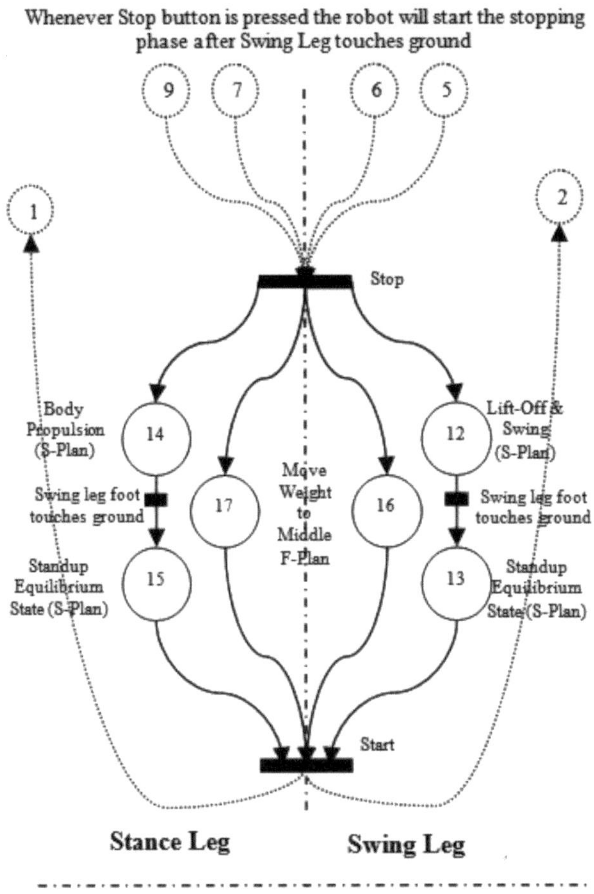

Figure 5. Petri net algorithm for a ROBIAN stop walk.

In the simulations, the walk begins from an initial position where the two feet are touching the ground. The swing leg is behind the stance leg (double support) and the walking speed is 0.65 m/s. The step length and walking speed are controlled by varying the four reference angles.

The duration of one complete walking cycle is approximately 1 s. The proposed approach allows us to achieve dynamic ROBIAN walking that is similar to human walking [40]. Simulations show that the walk is stable on a long distance, that is, more than 10 walking cycle [40]. Moreover, the approach also controls the robot's walk-standup-walk cycle, including transitions and stops [41] in a stable limit cycle.

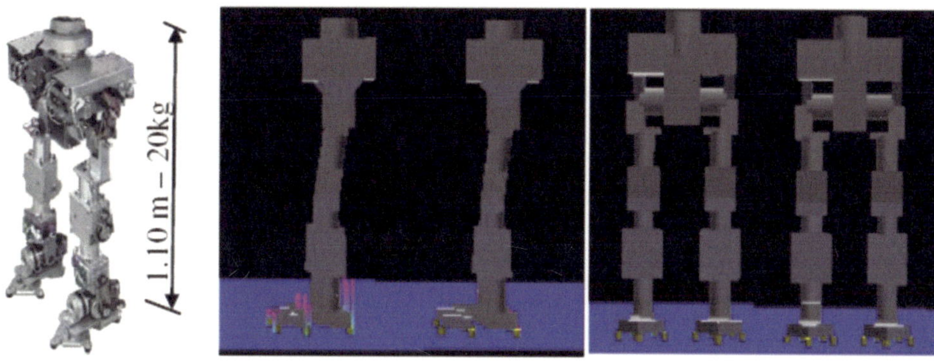

Figure 6. Modeling ROBIAN using OpenHRP.

Nevertheless, the adjustment of PID gains depending on the corresponding phase of the walk algorithm is not sufficient enough to have fluidity in the walk and to adapt the walking gait especially when the robot is unbalanced due to unexpected external forces. It has been seen in introduction that mixing damping joint and damping controller increases the balance [37]. Moreover, the muscle emulator presented in **Figure 1** acts simultaneously like a damping joint and a damping controller avoiding any mechanical changes in the robot y. The next section shows that implementation of this muscle emulator in the legs greatly improves the fluidity and stability of the walking algorithm (**Figures 4** and **5**) against external perturbation and reduces significantly the total power consumption of the robot.

3. Implementation of the muscle model and comparative results

This section implements the muscle emulator in the walking algorithm presented in the previous section. Only the knee joints are modified to not complicate the algorithm. The simulation results are compared with those obtained without the muscle model.

3.1. Continuous walking gait

The results presented in **Figure 7** illustrate that the robot's articular angles are close to those of humans and that trajectories generated with the muscle emulator are closer than those simulated without it. Stance phase ankle motions are significantly improved. However, the difference between the robot and human ankle angles remains the same in the swing phase, and the robot's foot control is dissimilar to a human foot in the landing phase. This is because the human foot exhibits significant flexibility compared to the rigid robot foot. This assumption has been confirmed by several experiments conducted by medical teams (Dr. B. Bussel and Dr. D. Pradon of the APHP Poincaré Hospital, Garches, France). These experiments recorded and compared human walking gaits based on rigid and non-rigid human foot soles. **Figure 8** depicts the hip, knee, and ankle during one of these experiments. The results show that the rigid sole changes the other articulation movements, especially those in the ankle, in which the extension movement is significantly reduced.

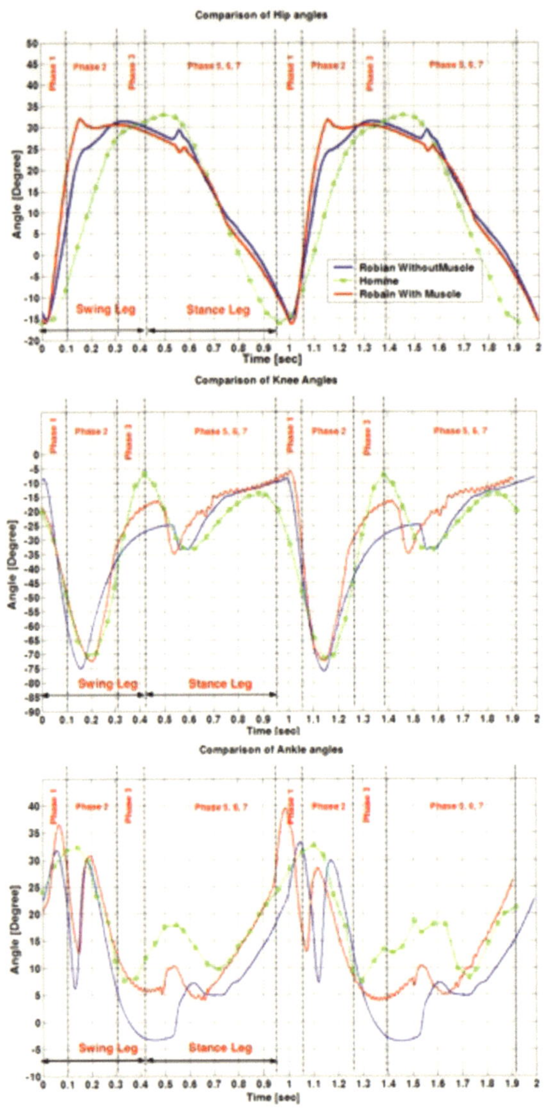

Figure 7. Comparison of the joint angles between an average 13-year-old and ROBIAN (right leg) over two walking cycles for the hip, knee, and ankle. The phases correspond to the human phases in **Figure 3**. The speed is 0.6 m/s.

The robot foot then lands parallel to ground and the swinging leg foot begins the swing phase. In the passive swing phase, ankle motors were controlled to keep the foot parallel to the ground. The foot was stabilized in this position in the active swing phase, allowing the foot to hit the ground in a manner that distributed the ground reaction forces equally throughout the foot.

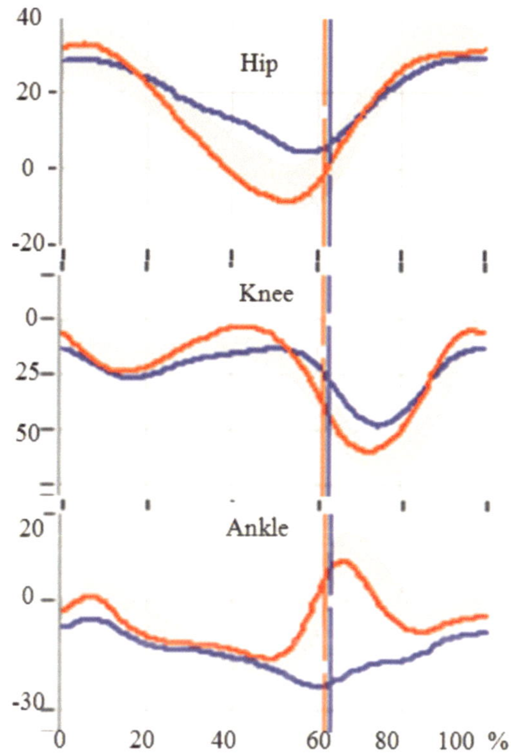

Figure 8. Comparison of the articular angles (in degrees) of the hip (flexion/extension), knee, and ankle (plantarflexion/dorsiflexion) in humans, with a rigid and non-rigid human foot soles. Time is expressed as a percentage of the human walk cycle duration. The experiments were conducted by Dr. D. Pradon and Dr. B. Bussel of the APHP R. Poincaré Hospital, Garches, France.

Figure 9 compares the articular joint power consumptions in the three articulations for one walking cycle with and without the muscle emulator in the knees. One can see that the muscle emulator reduces the power consumed at each articulation, even if it is only implemented in the knees. Other simulations illustrate that the emulator reduces the working duration of the DC motor in the overload zone defined by the constructor.

Figure 10 depicts the phase plane of three joint angles over 10 walking cycles with and without a muscle emulator in the knees. The muscle emulator induces changes in each articular movement, especially in the knees, for which the amplitude or flexion/extension is reduced and the extension reaches 0°. The convergence of the trajectories between the transient and stable walking cycles illustrates the stability of the walk.

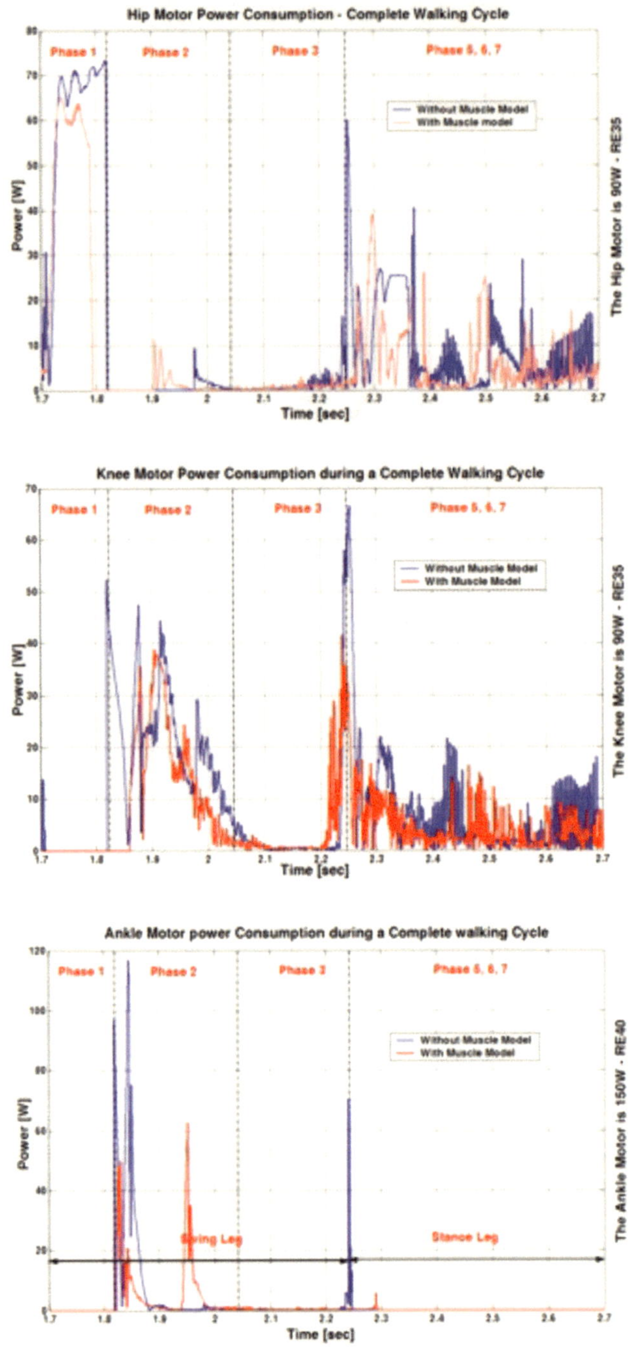

Figure 9. Comparison of the joint power consumption (right leg) of the hip, knee and ankle of ROBIAN over one walking cycles with and without a muscle emulator. The speed is 0.6 m/s.

http://dx.doi.org/10.5772/63746

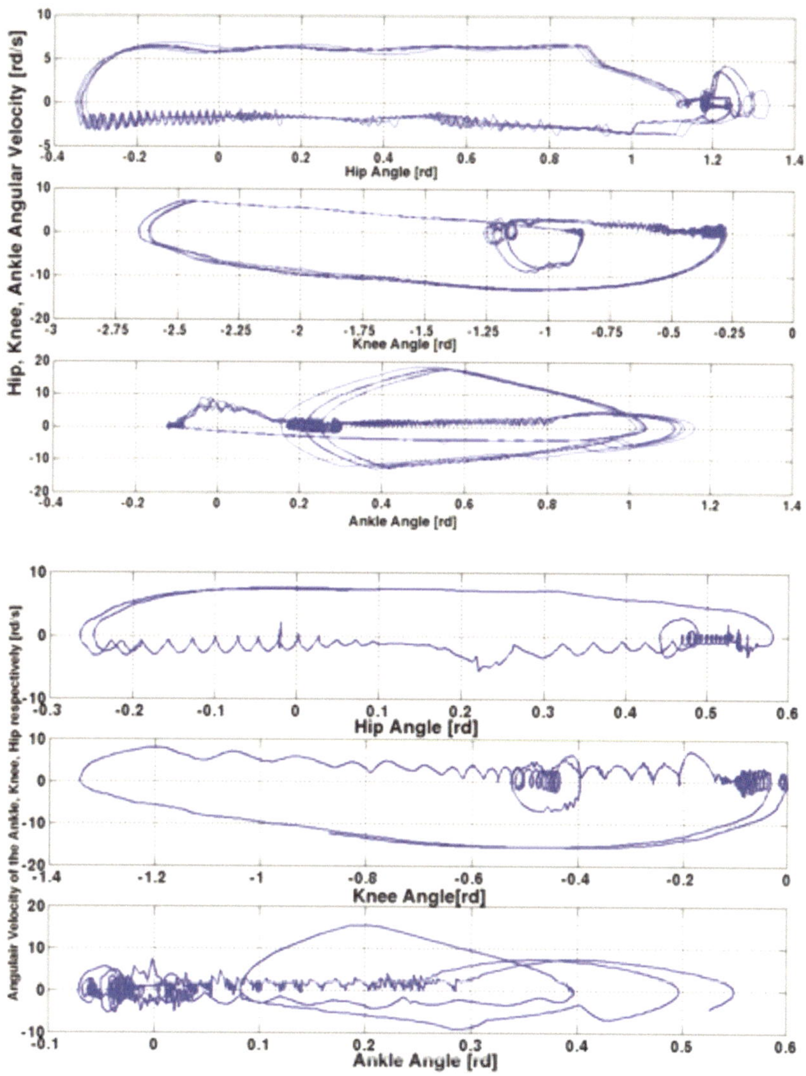

Figure 10. Phase plan of the three joints for 10 walking cycles. Top: without the muscle emulator. Bottom: with the muscle emulator.

3.2. Walk-halt-walk transitions

Figure 11 presents the different angular variations of both legs during a walk-halt-walk cycle that takes approximately 15 s. The curves show the differences between the two models during walking and stop processes. When walking, the curves are nearly identical, suggesting that the emulator does not change the robot's velocity. However, it is clear that the cycle induced with the muscle emulator displays more reduction than the simulation without the emulator. In addition, the emulator allows the robot to stop in a stand-up position with legs extended (hip and knee angles close to zero). The speed oscillations observed during the stop, which are reduced by the emulator, are due to the balance control instabilities, especially in the ankles. The muscle emulator acts as a low-frequency pass filter.

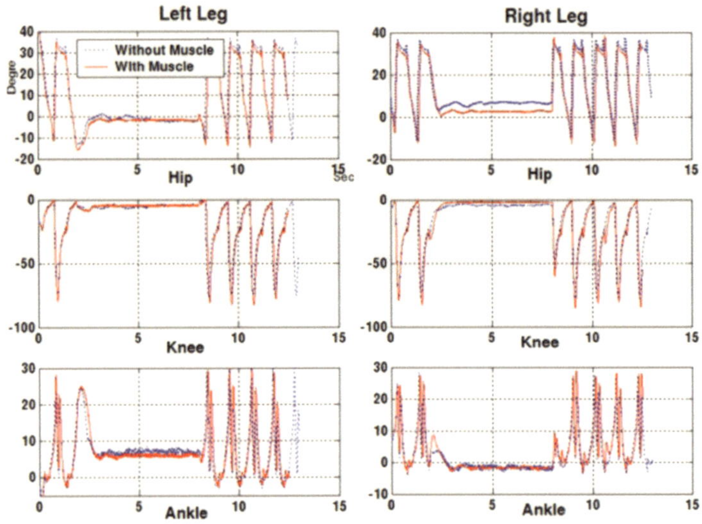

Figure 11. Angular variations of the hips, knees, and ankles of ROBIAN during a complete "walk-halt-walk" cycle (2 s of walking, 6 s of stopping and 8 s of walking).

Figure 12 shows the robot's trunk speed variations, which correspond to the curves in **Figure 10**. Transitions occur between the walking and stopping and stopping and walking processes. During these transitions, the balance of the robot is controlled based on the states and transitions of the Petri net. The shapes are nearly identical. However, the transient speed at the beginning of the stop is slightly amortized with the muscle.

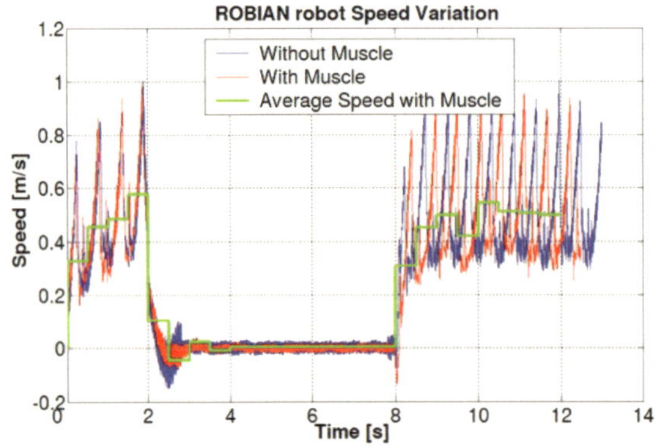

Figure 12. Trunk speed during a complete walk-stop-walk cycle.

3.3. Stability versus external perturbation forces

We simulated the external forces that affect the robot in the sagittal and frontal planes to assess the robustness of the walking algorithm with the muscle emulator. **Figure 13** shows the instantaneous sagittal plane horizontal velocity variation due to a 35 N and 100 ms thrust force applied forward and a 50 N force (100 ms duration) applied backward. Note that the speed of the robot initially decreases. The velocity then maintains the same average speed with muscle emulator, but a larger average speed without. This suggests that the robot is close to falling after being pushed without the muscle emulator.

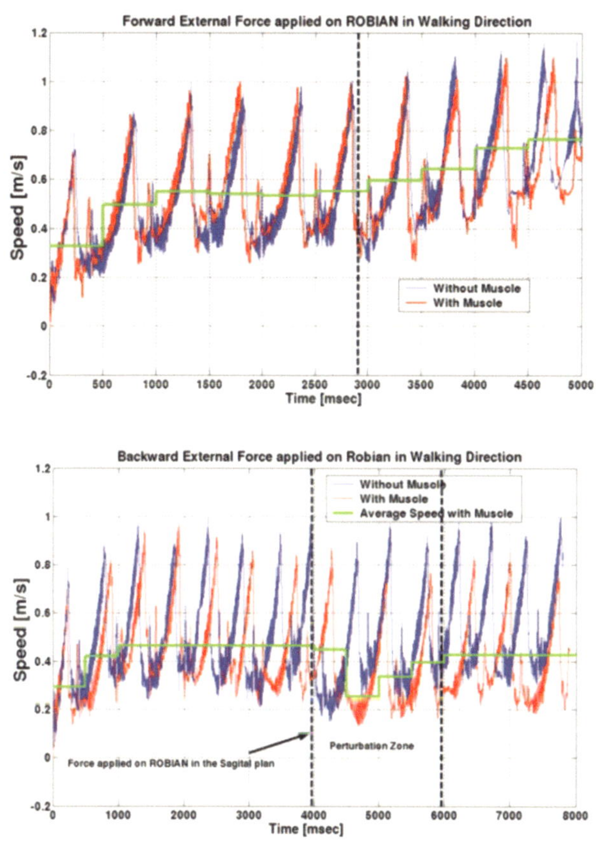

Figure 13. Trunk speed of ROBIAN in the sagittal plane, after applying a force. Top: in the direction of the walk. Bottom: in the direction opposite the walk.

Figure 14 shows the effect on the instantaneous frontal plane horizontal velocity due to a 120 N and 100 ms thrust force applied in the direction perpendicular to the walking robot. Note that movement of the robot in the plane is less than for the case with the emulator. It is clear that the stability of the robot is improved by the emulator, as the displacement of the robot is much smaller compared to the results without the emulator. The emulator allows the robot to withstand a 30% larger force than the simulation without the emulator. In fact, the robot falls when the same force is applied without the emulator.

Although the simulations with the emulator indicate that the robot can withstand a 30% larger applied force than those without the emulator, a sufficiently large sagittal plane force (approximately 85 N for 100 ms forward or 65 N backward) cannot be counteracted, and the robot falls in the direction of the force.

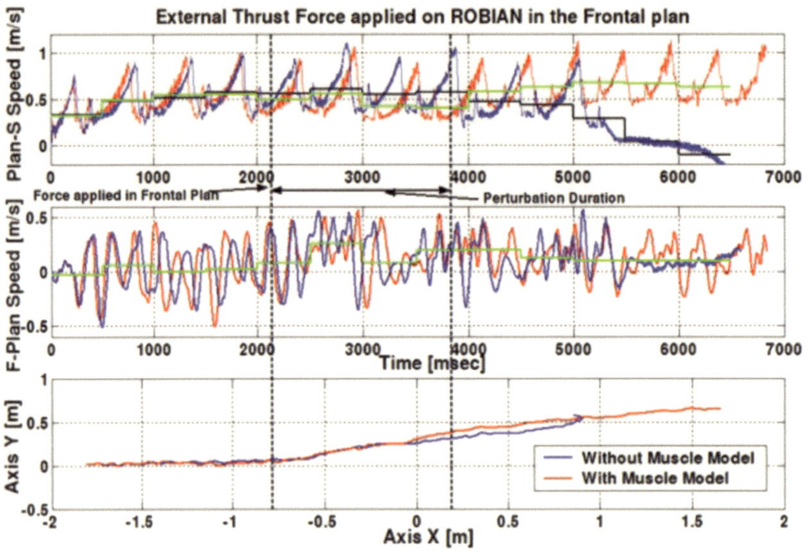

Figure 14. Trunk speed of ROBIAN (in sagittal plane) and trunk position (in the frontal plane), after applying a thrust force perpendicular to the direction of walking (frontal plane).

4. Conclusion

The contribution of this chapter is to compare and analyze the effects a muscle emulator on the walking efficiency of a biped robot. The results indicate that the use of a muscle emulator in the DC motor control loops of humanoid robots provides a compliance property in articular joints. Thus, the walking performance is improved, as follows:

- the muscle emulator significantly reduced the total power consumption.
- the work required by the electric motors decreased, and the motors work less in the authorized overload zone.
- the equilibrium of the robot can withstand 25–30% larger external forces with the muscle emulator.
- the algorithm allows the robot to stand with the knees fully extended in the stop position.

This model benefits from the advantages and simplicity of electric motor implementation, as well as from the benefits of human muscle compliance, which allow for dynamic walking without adding a physical element to the robot.

Author details

Hayssan Serhan[1,2] and Patrick Henaff[1,2*]

*Address all correspondence to: patrick.henaff@loria.fr

1 Faculty of Engineering I, Lebanese University, Lebanon

2 LORIA UMR 7503, University of Lorraine-INRIA-CNRS, Nancy, France

References

[1] Kim B.-H. Work analysis of compliant leg mechanisms for bipedal walking robots. *International Journal of Advanced Robotic Systems*. 2013. vol. 10, p. 334. DOI: 10.5772/56926

[2] Van der Kooij H., Veneman J.F., Ekkelenkamp R. Compliant actuation of exoskeletons. In Aleksandar Lazinica (Ed.), ISBN 978-3-86611-314-5*Mobile robots: Towards new applications*. InTech. 2006. DOI: 10.5772/4688

[3] Laffranchi M., Tsagarakis N.G., Caldwell D.G. Improving safety of human-robot interaction through energy regulation control and passive compliant design. In Maurtua Inaki (Ed.). *Human machine interaction – getting closer*. 2012. pp. 155–170. Intech. DOI: 10.5772/27781

[4] McGeer T. Passive dynamic walking. *The International Journal of Robotic Research*. 1990. vol. 9, no. 2, pp. 62–82.

[5] Garcia M., Chatterjee A., Ruina A., Coleman M. The simplest walking model: Stability, complexity, and scaling. *ASME Journal of Biomechanical Engineering*. 1998. vol. 120, no. 2, pp. 281–288.

[6] Goswami A., Espiau B., Keramane A. Limit cycles in a passive compass gait and passivity-mimicking control laws. *Autonomous Robots*. 1997. vol. 4, no. 3, pp. 273–286.

[7] Goswami A., Thuilot B., Espiau B. A study of the passive gait of a compass-like biped robot: symmetry and chaos. *The International Journal of Robotic Research*. 1998. vol. 17, no. 12, pp. 1282–1301.

[8] Spong M.W., Bullo F. Controlled symmetric and passive walking. *IEEE Transaction on Automatic and Control*. 2005. vol. 50, no. 7, pp.1025–1031.

[9] Ohta H., Yamakita M., Furuta K. From passive to active dynamic walking. *In IEEE Conference on Decision Control*, Phoenix, AZ. 1999. pp. 3883–3885.

[10] Spong M.W. Bipedal locomotion, robot gymnastics, and motor air hockey: a rapprochement. In *TITech COE/Super Mechano-Systems Workshop*, Tokyo, Japan. 1999. pp. 34–41.

[11] Spong M.W. Passivity based control of the compass gait biped. In *IFAC Triennial World Congress*, Beijing, China. 1999. vol. 3, pp. 19–23.

[12] Kuo A.D. Stabilization of lateral motion in passive dynamic walking. *The International Journal of Robotic Research*. 1999. vol. 18, no. 9, pp. 917–930,

[13] Collins S.H., Wisse M., Ruina A. A three-dimensional passive dynamic walking robot with two legs and knees. *The International Journal of Robotics Research*. 2001. vol. 20, no. 7, pp. 607–615.

[14] Ames A.D., Gregg R.D., Wendel E.D.B., Sastry S. Towards the geometric reduction of controlled three-dimensional robotic bipedal walkers. In *Workshop Lagrangian Hamiltonian Methods Nonlinear Control*, Nagoya, Japan. 2006. pp. 117–124.

[15] Spong M.W., Bhatia G. Further results on control pf the compass gait biped. In *IEEE International Conference on Intelligent Robots and Systems (IROS)*, Las Vegas, Nevada. 2003, pp. 1933–1938.

[16] He F., Liang Y., Zhang H., Pagello E. Dynamics and control of an extended elastic actuator in musculoskeletal robot system. In *Proceedings of the 12th International Conference IAS*. June 26–29, 2012; Jeju Island, Korea. 2012. pp. 671–681. DOI: 10.1007/978-3-642-33932-5_63

[17] Klute G.K., Hannaford B. Accounting for elastic energy storage in McKibben artificial muscle actuators. *Journal of Dynamic Systems, Measurement and Control*. 2000. vol. 122, pp. 386–388, DOI: 10.1115/1.482478

[18] Tondu B., Lopez, P. Modeling and control of McKibben artificial muscle robot actuators. *Control Systems. IEEE*. 2000. vol. 20, no. 2, pp. 15–38, DOI: 10.1109/37.833638

[19] Villegas D., Van Damme M., Vanderborght B., Beyl P., Lefeber D. Third-generation pleated pneumatic artificial muscles for robotic applications: development and comparison with McKibben Muscle Daniel Villegas, *Advanced Robotics*. 2012. vol. 26, no. 11–12, pp 1205-1227, DOI:10.1080/01691864.2012.689722

[20] Gollee H., Murray-Smith D.J., Jarvis J.C. A nonlinear approach to modeling of electrically stimulated skeletal muscle. *IEEE Transactions on Biomedical Engineering*. 2001. vol. 48, no. 4, pp. 406–415. DOI: 10.1109/10.915705

[21] Nishikawaab S., Tanakaa K., Shidaa K., Fukushimac T., Niiyamad R., Kuniyoshia Y. A musculoskeletal bipedal robot designed with angle-dependent moment arm for dynamic motion from multiple states. *Advanced Robotics*. 2014. vol. 28, no. 7, pp. 487–496. DOI: 10.1080/01691864.2013.876936

[22] Carpi F., De Rossi D., Kornbluh R.-, Pelrine R., Sommer-Larsen P. *Dielectric Elastomers as Electromechanical Transducers*. Amsterdam: Elsevier. 2008.

[23] Berselli B., Vassura G., Parenti Castelli V., Vertechy R. On designing compliant actuators based on dielectric elastomers for robotic applications, robot manipulators new achievements. In: Aleksandar Lazinica and Hiroyuki Kawai (Ed.), ISBN: 978-953-307-090-2, InTech. 2010. DOI: 10.5772/9311

[24] Kim J.-H., Oh J.-H. Walking control of the humanoid platform KHR-1 based on torque feedback control. ICRA '04. In *Proceedings 2004 IEEE International Conference on Robotics and Automation*. 26 April–1 May 2004. 2004. vol. 1, pp. 623, 628. DOI: 10.1109/ROBOT.2004.1307218

[25] Radkhah K. et al. Concept and design of biobiped1 robot for human-like walking and running. *International Journal of Humanoid Robotics*. 2011. vol. 8, no. 3, pp. 439–458. DOI: 10.1142/S0219843611002587

[26] Ghorbani R., Qiong W. Environmental-interaction robotic systems: Compliant actuation approach. *International Journal of Advanced Robotic Systems*. 2007. vol. 4, no. 1, pp. 81–92. DOI: 10.5772/5705

[27] Potkonjak V., Svetozarevic B., Jovanovic K., Holland O. The puller-follower control of compliant and noncompliant antagonistic tendon drives in robotic systems. *International Journal of Advanced Robotic Systems*. 2011. Vedran Kordic, Aleksandar Lazinica, Munir Merdan (Ed.), ISBN: 1729-8806, InTech. DOI: 10.5772/10690.

[28] Roberts T.J., Marsh R.L., Weyand P.G., Taylor C.R. Muscular force in running turkeys: The economy of minimizing work. *Science*. 1997. vol. 275, no. 5303, pp. 1113–1115.

[29] Kuo A.D. Energetics of actively powered locomotion using the simplest walking model. *Journal of Biomechanical Engineering*. 2002. vol. 124, pp. 113–120,

[30] Martinez-Villalpando E.C., Herr H. Agonist antagonist active knee prosthesis: A preliminary study in level-ground walking. *Journal of Rehabilitation Research and Development* 2009. vol. 46, no. 3, pp. 361–374.

[31] Kwon O.-S., Choi R.-H., Lee D.-H. Locomotion control of a compliant legged robot from slow walking to fast running. *International Journal of Advanced Robotic Systems*. 2012. vol. 9, p. 240. DOI: 10.5772/54469

[32] Lakatos D., Rode C., Seyfarth A., Albu-Schaffer A. Design and control of compliantly actuated bipedal running robots: Concepts to exploit natural system dynamics. *In 14th IEEE-RAS International Conference on Humanoid Robots (Humanoids)*. 18–20 Nov. 2014. 2014. pp. 930–937. DOI: 10.1109/HUMANOIDS.2014.7041475

[33] Sato R., Miyamoto I., Sato K., Aiguo Ming, Shimojo M. Development of robot legs inspired by bi-articular muscle-tendon complex of cats. *In 2015 IEEE/RSJ International Conference on Intelligent Robots and Systems* (IROS). Sept. 28–Oct. 2 2015. 2015. pp. 1552–1557. DOI: 10.1109/IROS.2015.7353574

[34] Monasterio-Huelin F., Gutierrez A., Berenguer F., Zappa J. A compliant quasi-passive biped robot with a tail and elastic knees. In Behnam Miripour (Ed.). *Climbing and Walking Robots*. Intech. 2010. DOI: 10.5772/8834.

[35] Yang T., Westervelt E., Schmiedeler J., Bockbrader R. Design and control of a planar bipedal robot ERNIE with parallel knee compliance. *Autonomous Robots*, Springer, USA. 2008. vol. 25, pp. 317–330.

[36] Wisse M., Schwab A.L., van der Helm F.C.T. Passive dynamic walking model with upper body. *Robotica*. Nov. 2004. vol. 22, no. 6, pp. 681–688.

[37] Winter D. *Biomechanics and Motor Control of Human Movement*, Third Edition. Wiley. 2005.

[38] Serhan H., Nasr C., Hénaff P. Muscle emulation with DC motor and neural networks. *International Journal of Neural Systems*. 2010. vol. 20, no. 4, 341–353.

[39] Donaldson N.de N., Gollee H., Hunt K.J., Jarvis J.C., Kwende M.K.N. A radial basis function model of muscle stimulated with irregular inter-pulse intervals. *Medical Engineering Physics*. 1995. vol. 17, no. 6, pp. 431–441. DOI: 10.1016/1350-4533(94)00013

[40] Serhan H., Nasr C., Henaff P., Ouezdou F. A new control strategy for ROBIAN biped robot inspired from human walking. *In IEEE/RSJ International Conference on Intelligent Robots and Systems* (IROS). 2008. pp. 2479–2485.

[41] Serhan H., Hénaff P., Nasr C., Ouezdou F. State machine-based controller for walk-halt-walk transitions on a biped robot. *In IEEE-RAS International Conference on Humanoid Robots (Humanoids 2008)*, Korea. 2008. pp. 535–540.

[42] Viel E. La marche humaine, la course et le saut. Masson. 2000.

[43] Bouisset S., Maton B. *Muscles, posture et mouvement*. Hermann. 1995.

[44] Kim M.-S., Oh J.H. Posture control of a humanoid robot with a compliant ankle joint. *International Journal of Humanoid Robotics*. 2010. vol. 7, no. 1, pp. 5–29. DOI: 10.1142/S0219843610001988

[45] Jonathon W., Sensinger, Lawrence E. Burkart, Gill A. Pratt, Richard F. ff. Weir. Effect of compliance location in series elastic actuators. *Robotica*. 2013. vol. 31, no. 8, pp. 1469–8668. DOI: 10.1017/S0263574713000532

Chapter 4

Kinematic Analysis of the Triangle-Star Robot with Telescopic Arm and Three Kinematics Chains as T-S Robot (3-PRP)

Ahmad Zahedi, Hadi Behzadnia,
Hassan Ghanbari and Seyed Hamed Tabatabaei

Additional information is available at the end of the chapter

http://dx.doi.org/10.5772/64556

Abstract

In this chapter, the limitations and weaknesses of the motion geometry and the workspace of Triangle-Star Robot {T-S (3-PRP)} are diagnosed after research and consideration of the issues at hand. In addition, they are offered in index form. To remove the problems with the abovementioned cases, at first, a robot with telescopic arms and a similar kinematics chain is rendered to give a kinematics analysis approach like Hartenberg-Denavit. Furthermore, in order to increase the workspace, Reuleaux Triangle-Star Robot {RT-S (3-PRP)} with kinematics chains 3-PRP and Circle-Star Robot{C-S (3-PRP)} with kinematics chains 3-PRP and a new improved structure are introduced.

Keywords: automation, robot, kinematics, telescopic arms, stiffness, workspace

1. Introduction

In recent decades, the use of industrial robots in the production and development of nano technological processes, industrial automation, chemically contaminated environments, medical & biological industrial, the calibration of measurement system, etc., are seriously in circulation. So the scientific research centers are conducting some research in this regard to meet customers' demands. To compete with international markets, parameters such as precision, high repeatability in production, control management, and emphasis on standards have proved to be necessary in the usage of robot in modern technology. This has caused the companies to move toward a practical system to produce the maximum production variety at minimum time with the lowest expenses and the highest quality. Thus, the industrial automation with the help of robots replaces human force in production and in assembly lines.

Today, robots are divided into several groups: Serial robots, Parallel robots, Synthetic robots, and Mobile robots [2, 5]. Synthetic robots are to incorporate the serial and parallel manipulators by connecting them in serial. The serial connections of serial and parallel manipulators can be categorized into the following four types: Parallel-Parallel, Serial-Parallel, Parallel-Serial, and Serial-Serial [8]. Characteristics such as precision, speed, stiffness, and a workspace without singularity points have differentiated the Parallel robot from the Serial robots [4, 5]. The Parallel manipulator robots are used in making flight simulators, helicopters, machinery tools, precise robots, etc. The reverse kinematics solution of these robots as compared with the simple Serial robots and direct kinematics solution is hard with complicated equations [3, 8]. In this paper, first, the geometric attributes of the Triangle-Star Robot are offered and then based on motion geometry and robotic workspace, the limitations and weaknesses are recognized and indexically represented. In addition, to removing the abovementioned shortcomings, a new Triangle-Star Robot with telescopic arms and two new robots with improved structures are represented. Finally, the kinematics analysis of the robots similar to Denavit–Hartenberg approach is carried out.

2. Geometric characteristics of the Triangle-Star Robot

The geometric model of the Triangle-Star Robot (T-S Robot), shown in **Figure 1** (**a**, **b**), is composed of a $A_1\ A_2\ A_3$ fixed triangle and a moving star with $B_1\ B_2\ B_3$ arms in which the B star can move on triangle A with three kinematics chains 3-PRP.

(a)

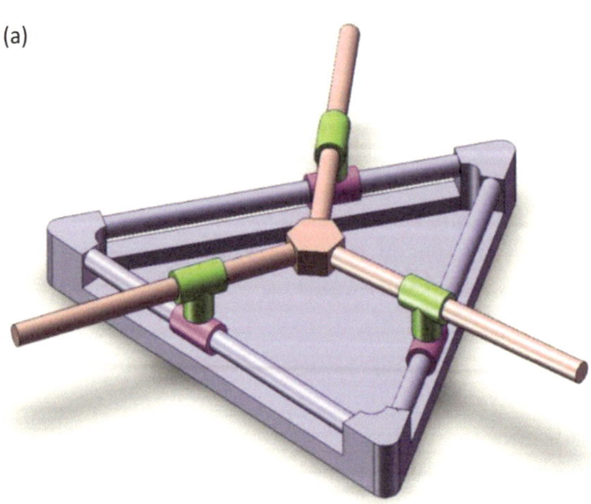

http://dx.doi.org/10.5772/64556

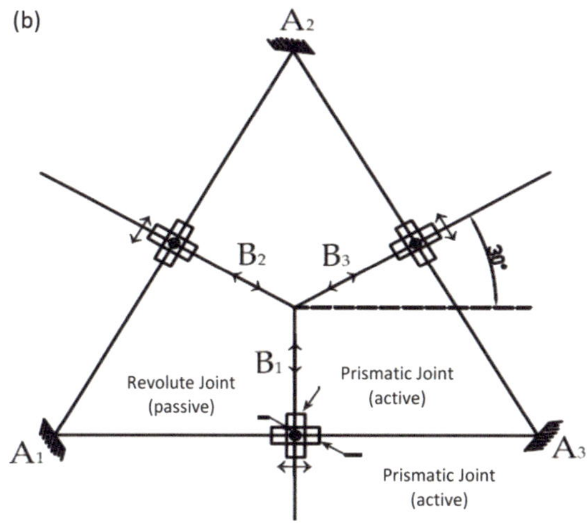

Figure 1. (a) The Triangle-Star Robot (T-S Robot) with rigid arms and (b) the geometric model of T-S Robot with rigid arms.

The motion geometry of each of the kinematics chains 3-PRP, which has been applied to T-S Robot, is achieved through relative movement of the Prismatic joint, Revolute joint, and again the Prismatic joint as shown in **Figure 2**.

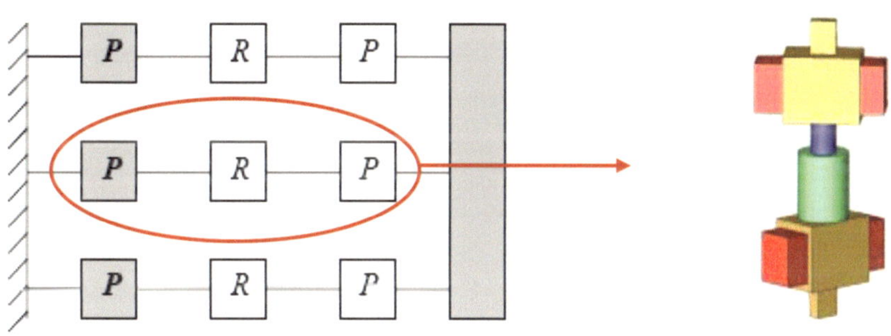

Figure 2. The motion geometry of the kinematics chains (3-PRP) for T-S Robot.

The lower active Prismatic joint movement joined to the stimulator in the direction of the angle of the triangle causes movement in the upper passive Prismatic joint connected to lower active Prismatic joint by the Revolute joint. As a result, star motion geometry, which is sited (fixed) on the upper Prismatic joints, serves as a function of the lower Prismatic movement joined to the stimulators.

3. The number of degree-of-freedom for T-S Robot

The most famous and prevalent method to calculate the degree of freedom (DOF) is using formula (1) and according **Table 1** [9].

$$F = 6(N - L - 1) + \sum_{i=1}^{n} f_i \tag{1}$$

N = The number of arms, L = The number of robotic joints, f_i = Indicated of DOF related i joint as shown in the above mentioned table. On the other hand, in the parallel manipulator robot, $N = 8$, $L = 9$ and f_i for $i = 1,\ldots,9$ equals 1, thus the robot's DOF is equal to 3.

Joint	Diagram	Symbol	DOF
Revolute	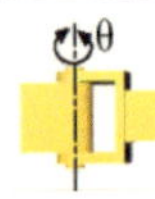	R	1
Prismatic		P	1
Cylindric		C	2
Universal	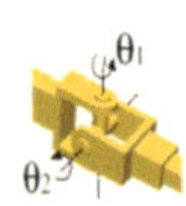	T	2
Spherical		S	3

Table 1. DOF for type of joint [9].

4. The weaknesses of the motion geometry for T-S Robot

Although T-S Robot has so many advantages such as high precision, speed, stiffness, and lack of singularity in workspace, it has two main weaknesses:

1. At the time of Robotic motion, the star arms occupy a large space around the triangle
2. The abovementioned robot has a rather limited workspace

Therefore, to remove the first problem, a robot with telescopic arms, **Figure 3** (**a**, **b**), is introduced whose kinematics are conducted without getting involved in the whole discussion.

To remove the second problem, new robots with improved structures are introduced. The schematic design of these robots, **Figure 4** (**a**, **b**) and their three-dimensional designs in the kinematics analysis section of each robot in **Figure 14a** and **Figure 14b** along with robot motion geometry structure are shown.

http://dx.doi.org/10.5772/64556

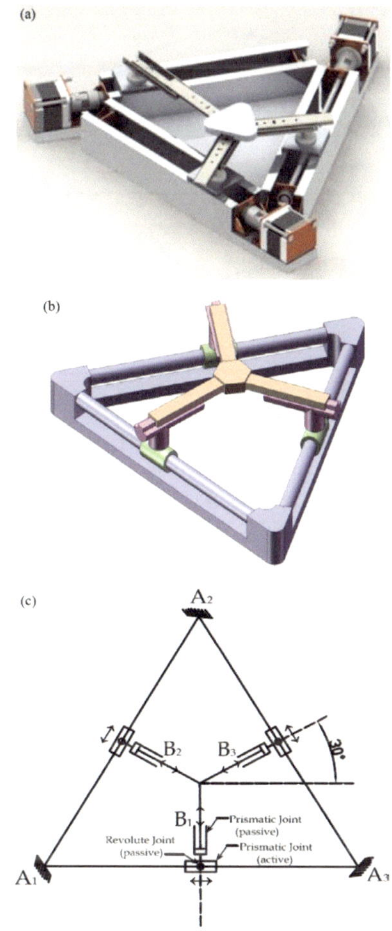

Figure 3. (a) Real model of the Triangle-Star Robot' components with telescopic arms. (b) The Triangle-Star Robot (T-S Robot) with telescopic arms. (c) The geometric model of T-S Robot with telescopic arms.

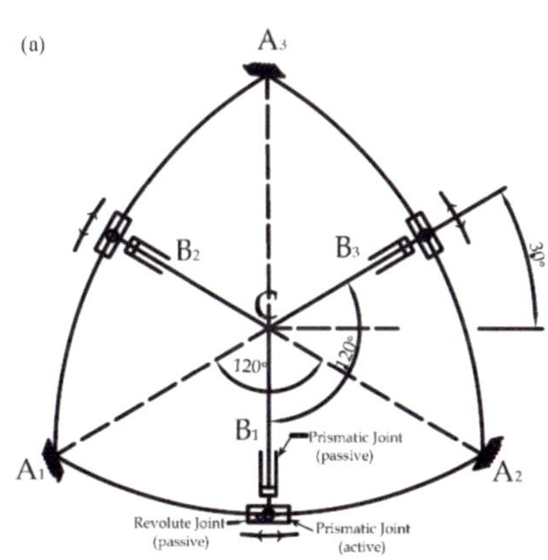

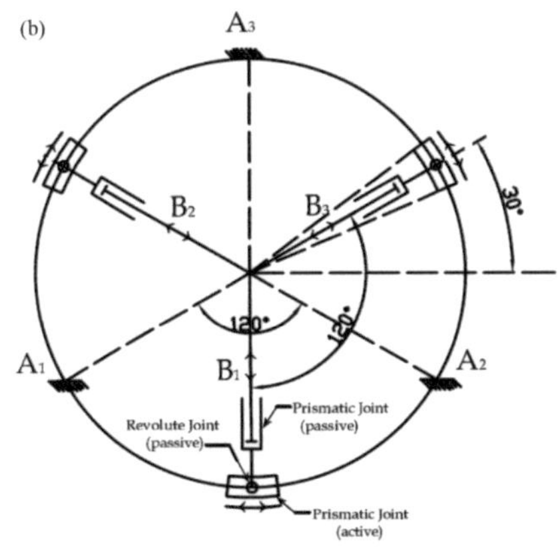

Figure 4. (a) Introduction a new geometric model as Reuleaux Triangle-Star Robot with telescopic arms. (b) Introduction of a new geometric model as Circle-Star robot with telescopic arms.

These robots are as follows:

a. Reuleaux Triangle-Star Robot with kinematics structure {RT-S (3-PRP)}

b. Circle-Star Robot with kinematics structure {C-S (3-PRP)}

5. Kinematic analysis of the T-S Robot

In the boundary of kinematics science, place, speed, acceleration, and all the derivations higher than local variable (in proportion to time) are examined. As defined [2, 8], the robotic kinematics is the study of the robotic movement without the consideration of the forces and torques applied to it. As a matter of fact, the examination of robotic motion geometry is regarded as an invariable coordinate frame work proportional to the intended time.

The kinematics problem includes the following sections [2, 8]:

1. Direct kinematics: local calculation and the central manipulator orientation in relation to basic frame work.

2. Reverse kinematics: if the location and central manipulator orientation are given, the calculation of all the possible joint angles involved in directing the robot toward a desired location and orientation is called reversed kinematics.

A systematic, practical approach has been represented by Denavit-Hartenberg to determine rotation and transformation between the two adjacent links in a robot [2].

In this paper, direct kinematics and reverse kinematics are examined through this approach and then position analysis, speed, input, and output acceleration are closely studied. The interesting point is that the direct kinematics of Parallel arms is as complicated as the reverse kinematics of Serial arms and the simplicity of the reverse kinematics of parallel mechanisms is as much as the simplicity of the direct kinematics of Serial arms [2, 5].

http://dx.doi.org/10.5772/64556

6. The extraction of Denavit-Hartenberg parameters for T-S Robot

To achieve a_i^k, α_i^k, S_i^k, θ_i^k parameters, first, the position of coordinate axis according to Denavit-Hartenberg model as shown in **Figure 6** (**a**, **b**) is determined. Then by placing these parameters in Denavit-Hartenbary matrix, which is obtained by relation (2) according to **Figure 5**. Consequently, the achieved transmitted matrixes and robotic kinematics analysis are carried out.

$$q = [q_1, q_2, q_3,]^T,\ r = [r_1, r_2, r_3,]^T,\ r = f(q) \tag{2}$$

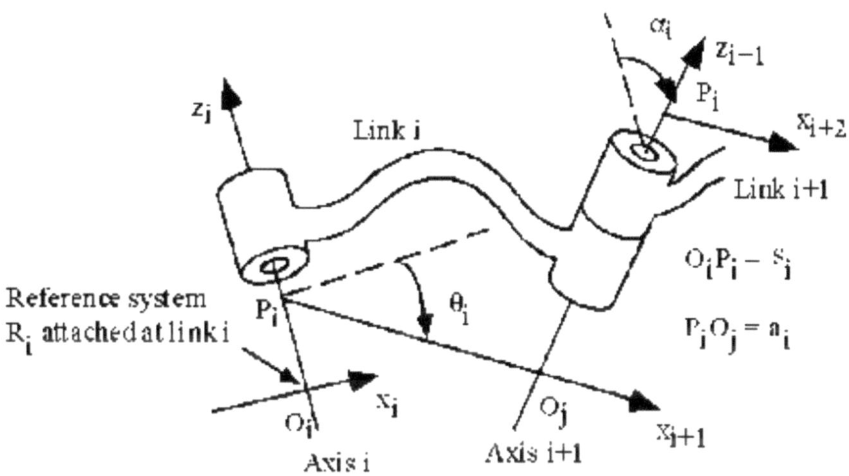

Figure 5. Kinematic chain and parameters representation of the Hartenbary–Denavit model for two adjacent links [1].

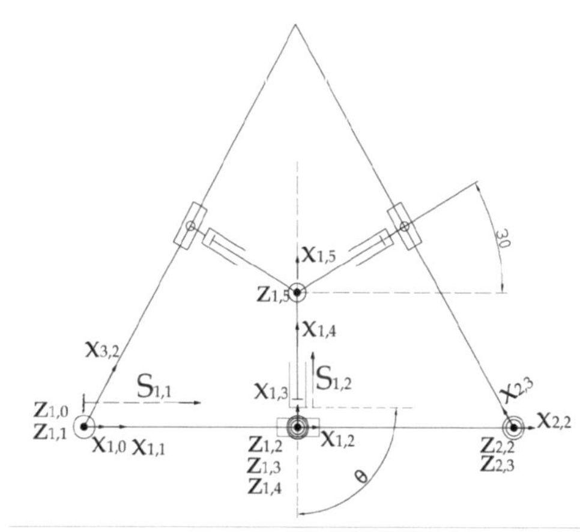

(a)

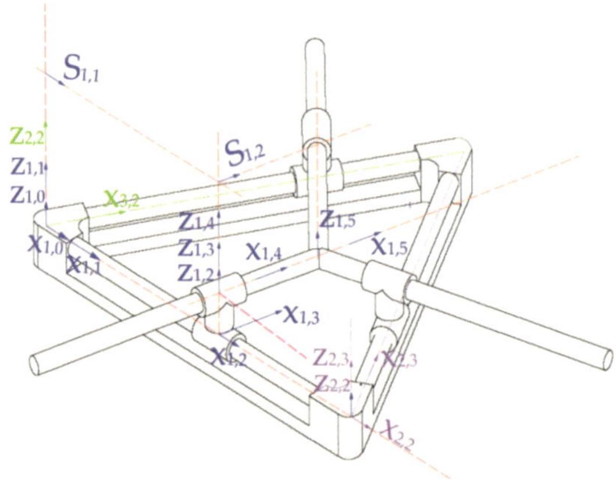

(b)

Figure 6. (a) Kinematic chain and parameters of the proposed T-S (3-PRP) robot with telescopic arms (b) Kinematic chain and parameters of the T-S robot with rigid arms.

$$H^k_{i,i+1} = \begin{bmatrix} R_{i,i+1} & P_{i,i+1} \\ [0] & 1 \end{bmatrix}, \quad i = 0,1,2,.(joint) \qquad (3)$$

$$k = 1,2,3,4,...(\text{N.chain})$$

$$H_{i,i+1} = \begin{bmatrix} \cos\theta_i & -\sin\theta_i\cos\alpha_i & \sin\theta_i\sin\alpha_i & a_i\cos\theta_i \\ \sin\theta_i & \cos\theta_i\cos\alpha_i & -\cos\theta_i\sin\alpha_i & a_i\sin\theta_i \\ 0 & \sin\alpha_i & \cos\alpha_i & S_i \\ 0 & 0 & 0 & 1 \end{bmatrix} \qquad (4)$$

a_i: offset distance between two adjacent joint axes, where $a_i = | p_i - o_j |$.

α_i: twist angle between two adjacent joint axes. It is the angle required to rotate the z_i axis into alignment with the z_{i+1}–axis about the positive x_{i+1}–axis according to the right-hand rule.

θ_i: joint angle between two incident normals of a joint axis. It is the angle required to rotate the x_i–axis into alignment with the x_{i+1}–axis about the positive z_i–axis according to the right-hand rule.

S_i: translational distance between two incident normals of a joint axis. $S_i = | p_i - o_i |$ is positive if the vector $p_i - o_i$ points in the positive z_i–direction; otherwise, it is negative (**Figure 6**).

To analyze the robot kinematically, first, the central manipulator point is transferred to the corner of the triangle to which the reference coordinate device is joined and through the extraction of a_i^k, α_i^k, S_i^k, θ_i^k parameters, **Table 1**, related to Denavit–Hartenberg parameters, is achieved. In this case, k = 1, 2, 3 is the number of central manipulator point transmission,

Figure 7. These transmissions are as follows: Path 1: C, P_1, A_1, Path 2: C, P_3, A_3, A_1, Path 3 : C, P_2,, A_1

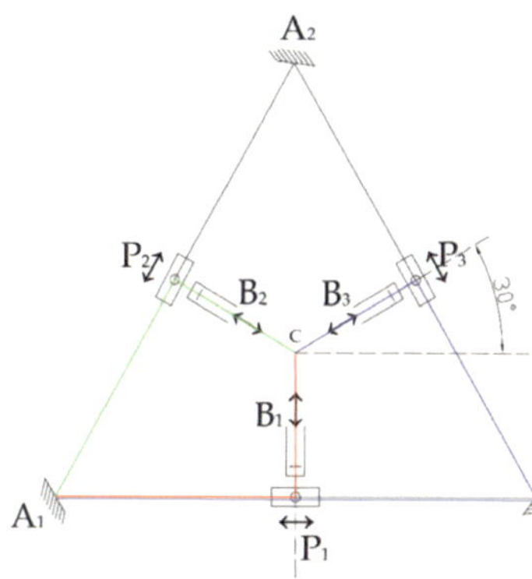

Figure 7. Transmission paths from point C to A_1.

	Joint no.	0	1	2	3	4	5	6
Path 1	$a_{1,i}$	0	0	S_{11}	0	0	S_{13}	x
	$\alpha_{1,i}$	0	0	0	0	0	0	0
	$S_{1,i}$	0	0	0	0	b	0	0
	$\theta_{1,i}$	0	0	0	$\frac{\pi}{2}+\theta$	0	0	0
Path 2	$a_{2,i}$	e	0	S_{21}	0	0	S_{23}	x
	$\alpha_{2,i}$	0	0	0	0	0	0	0
	$S_{2,i}$	0	0	0	0	b	0	0
	$\theta_{2,i}$	0	$\frac{2\pi}{3}$	0	$\frac{\pi}{2}+\theta$	0	0	0
Path 3	$a_{3,i}$	0	0	S_{31}	0	0	S_{33}	x
	$\alpha_{3,i}$	0	0	0	0	0	0	0
	$S_{3,i}$	0	0	0	0	b	0	0
	$\theta_{3,s}$	$\frac{\pi}{3}$	0	0	$-\frac{\pi}{2}-\theta$	0	0	0

Table 2. Kinematics parameters of T-S robot.

In the above indexes, the first shows the number of the direction, and the second shows the number of the parameter. To avoid making mistakes, the number of direction indexes (first index) is shown to the power of k.

By placing **Table 2** parameters, in matrix 2, the amount of $H_{i,i+1}^{k}$ matrix is achieved in which, i =1, 2, 3, 4 is the number of coordinate systems joined to the links and k = 1, 2, 3 is the number of the central manipulator reference point directions joined to C point. The speed-acceleration points are obtained through placing the position analysis matrixes two by two in an equal way.

7. Direct kinematics for T-S robot

For the direct kinematics analysis, point o, located in the origin of the coordinate of A_1, reference system is regarded as the primary point and o', point in the origin of the coordinate joined to the center of C point, is considered to be the final point.

The number of unknown parameters in each path is two. As a result, the total number of unknown parameter for each path is 6, but since the number of degree of freedom for each kinematics chains 3-PRP equals 9, the number of each path of the unknown parameters is 3, that is, the three parameters are related to other parameters that are geometrically obtained. In this case, the robotic degree of freedom is 3. If S_{11}, S_{21}, S_{31} parameters are considered to be the inputs, the unknown parameters will be S_{13}, S_{23}, S_{33} and θ_{11}, θ_{21}, θ_{31}.

What is stated is for the simplification and avoidance of mistakes in the number of direction index to the power of k. Therefore, to obtain the unknown parameters, final point transformation matrix o' is obtained in proportion to the primary point o and then relation (5, 11) is used.

$$[P]_i = H^k_{o,o'}[P]_{i+1} \tag{5}$$

$$H^k_{o,o'} = H^K_{0,1} \times H^K_{1,2} \times H^K_{2,3} \times H^K_{3,4} \ldots\ldots H^K_{i,i+1} \tag{6}$$

$$H^1_{o,o'} = H^1_{o,o'} = H^3_{o,o'} \tag{7}$$

$$[P_{O'}]_{(O)(for.path1)} = [P_{O'}]_{(O)(for.path2)} = [P_{O'}]_{(O)(for.path3)} \tag{8}$$

$$\begin{aligned} [P_i] &= [H^k_{i,i+1}] \cdot [P_{i+1}] \\ [P_i] &= [x_i; y_i; z_i; 1] \end{aligned} \tag{9}$$

In the above relation, $[P]_i$ is the position in relation to coordinate axes joined to member i and $[P]_{i+1}$ is the position in relation to coordinate axes joined to member $i+1$. The robot has three kinematics chains, and three different transformation matrixes are achieved. In this way, the three matrixes should be equalized according to relation (5). The unknown parameters are obtained through the solution of these equations.

Point o' coordinate (P_3 origin of coordinate) located in the center of the star in P_2 coordinate is as follows. Point o' coordinates in P_0 coordinate is as follows:

http://dx.doi.org/10.5772/64556

$$[P_{O'}]_{(2)} = [P_i] = [x_O; y_O; z_O;\ 1]_{(2)} = [0; 0; S_2^k; 1]_{(2)} \tag{10}$$

$$[P_{O'}]_{(0)} = \begin{bmatrix} x_{O'} \\ y_{O'} \\ z_{O'} \\ 1 \end{bmatrix}_{(0)} \begin{bmatrix} x_{O'} \\ y_{O'} \\ z_{O'} \\ 1 \end{bmatrix}_{(0)} = \left[H_{0,1}^K\right] \times \left[H_{1,2}^K\right] \times \left[H_{2,3}^K\right] \times \begin{bmatrix} 0 \\ 0 \\ S_2^K \\ 1 \end{bmatrix}_{(2)} \tag{11}$$

$$H_{O,O'}^1 = \begin{bmatrix} -Sin\theta & -Cos\theta & 0 & -Sin\theta \times x - Sin\theta \times S_3^1 + S_1^1 \\ Cos\theta & -Sin\theta & 0 & Cos\theta \times x + Cos\theta \times S_3^1 \\ 0 & 0 & 1 & B \\ 0 & 0 & 0 & 1 \end{bmatrix} \tag{12a}$$

$$H_{O,O'}^2 = \begin{bmatrix} \wp & -\Re & 0 & \wp \times x + \wp \times S_3^2 - 0.5 \times S_1^2 + e \\ \Re & \wp & 0 & \Re \times x + \Re \times S_3^2 + 0.866 \times S_1^2 \\ 0 & 0 & 1 & b \\ 0 & 0 & 0 & 1 \end{bmatrix}$$

$$\Re = -.0.866Sin\theta - 0.5Cos\theta$$

$$\wp = 0.5Sin\theta - 0.866Cos\theta, \tag{12b}$$

$$H_{O,O'}^2 = \begin{bmatrix} \wp' & -\Re' & 0 & \wp' \times x + \wp' \times S_3^3 + 0.5 \times S_1^3 \\ \Re' & \wp' & 0 & \Re' \times x + \Re' \times S_3^3 + 0.866 \times S_1^3 \\ 0 & 0 & 1 & b \\ 0 & 0 & 0 & 1 \end{bmatrix}$$

$$\wp' = -0.5Sin\theta + 0.866Cos\theta$$

$$\Re' = -0.866Sin\theta - 0.5Cos\theta$$

$$path(1): \begin{bmatrix} x_{O'} \\ y_{O'} \\ z_{O'} \\ 1 \end{bmatrix}_{(0)} = \begin{bmatrix} -x \times Sin\theta - S_3^1 \times Sin\theta + S_1^1 \\ x \times Cos\theta + S_3^1 \times Cos\theta \\ b \\ 1 \end{bmatrix}$$

$$path(2): \begin{bmatrix} x_{O'} \\ y_{O'} \\ z_{O'} \\ 1 \end{bmatrix}_{(0)} = \begin{bmatrix} \wp \times x + \wp \times S_3^2 - 0.5 \times S_1^2 + e \\ \Re \times x + \Re \times S_3^2 + 0.866 \times S_1^2 \\ b \\ 1 \end{bmatrix} \tag{13ab}$$

$$path.(3):\begin{bmatrix} x_{O'} \\ y_{O'} \\ z_{O'} \\ 1 \end{bmatrix}_{(0)} = \begin{bmatrix} \wp' \times x + \wp' \times S_3^3 + 0.5 \times S_1^3 \\ \Re' \times x + \Re' \times S_3^3 + 0.866 \times S_1^3 \\ b \\ 1 \end{bmatrix} \quad (13c)$$

8. The motion geometry of T-S Robot

$\phi_3 + \phi_5 = 180,\ \phi_5 = \frac{\pi}{2} + \theta_3,\ \phi_3 = \frac{\pi}{2} + \theta_2 \Rightarrow \theta_2 = -\theta_3, \phi_1 + \phi_2 = 180,\ \phi_1 = \frac{\pi}{2} + \theta_1,\ \phi_2 = \frac{\pi}{2} - \theta_2 \Rightarrow \theta_1 = \theta_2,$
$\theta_1^1,\ \theta_1^2,\ \theta_1^3 = \theta,\ Sin\theta = \frac{2t}{1+t^2},\ Cos\theta = \frac{1-t^2}{1+t^2},\ t = \tan\frac{\theta}{2}$

From the motion geometry of the T-S robot and then relations (5,13), the motion equation are achieved.

$$-0.75x \times Sin\theta - S_3^1 \times Sin\theta + S_1^1 + 0.866x \times Cos\theta - 0.5S_3^2 \times Sin\theta + 0.866S_3^2 Cos\theta + 0.5S_1^2 - e = 0 \quad (14a)$$

$$0.75x \times Cos\theta + S_3^1 \times Cos\theta + 0.866x \times Sin\theta + 0.866S_3^2 \times Sin\theta + 0.5S_3^2 Cos\theta + 0.866S_1^2 = 0 \quad (14b)$$

$$-0.75x \times Sin\theta - S_3^1 \times Sin\theta + S_1^1 - 0.866x \times Cos\theta - 0.5S_3^3 \times Sin\theta - 0.866S_3^3 Cos\theta - 0.5S_1^3 = 0 \quad (14c)$$

$$0.75x \times Cos\theta + S_3^1 \times Cos\theta - 0.866x \times Sin\theta - 0.866S_3^3 \times Sin\theta + 0.5S_3^3 Cos\theta - 0.866S_1^3 = 0 \quad (14d)$$

$$-1.7x \times Cos\theta + 0.5S_3^2 \times Sin\theta - 0.866S_3^2 \times Cos\theta - 0.5S_1^2 - 0.5S_3^3 Sin\theta - 0.866S_3^3 Cos\theta - 0.5S_{13}^3 = 0 \quad (14e)$$

$$-1.7x \times Sin\theta - 0.866S_3^2 \times Sin\theta - 0.5S_3^2 \times Cos\theta + 0.866S_1^2 - 0.866S_3^3 Sin\theta + 05S_3^3 Cos\theta - 0.866S_1^3 = 0 \quad (14f)$$

9. Case analysis

If we consider the fixed angle of the triangle $A_1A_2A_3$ in **Figure 8**, $e = 1000$ mm, $b = 100$ mm and each of the S_{11}, S_{21}, S_{31} inputs are applied to the direction of the triangle, then **Figure 9** (**a**, **b**) are obtained.

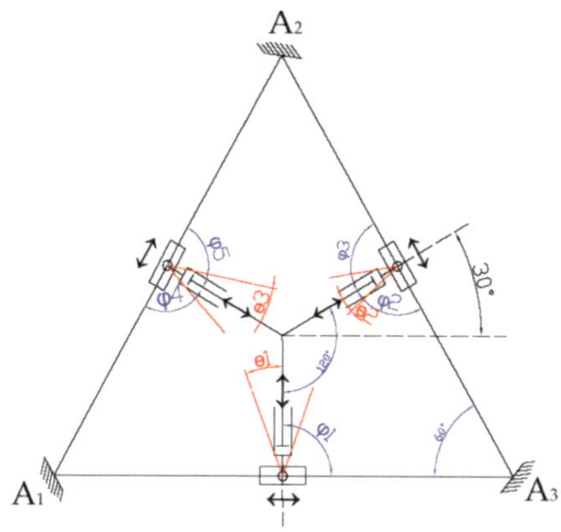

Figure 8. Geometric relation between triangle and star sides.

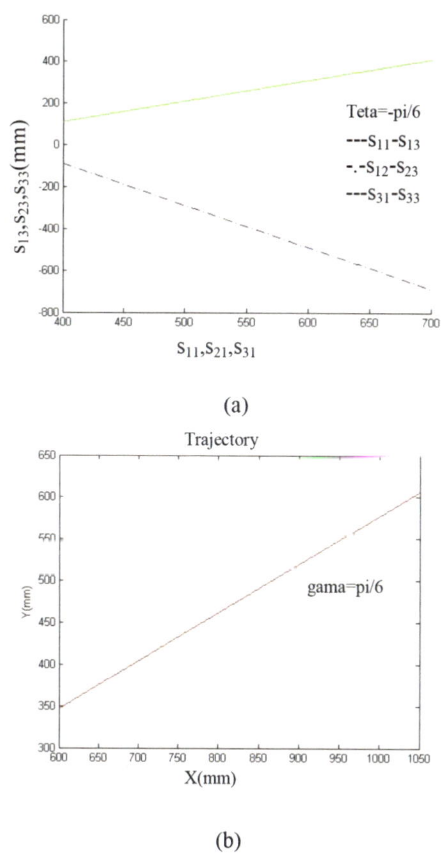

Figure 9. (a) Variation S_{13}, S_{23}, S_{33} (output) versus actuator S_{11}, S_{21}, S_{31} (input) and (b) X–Y plots of the trajectory of the motion point c.

10. Inverse kinematics for T-S Robot

The representation of the location and orientation of the robotic final manipulator can lead to the estimation of all the possible joint collections, which are used to transfer the robot and to obtain the assumed orientation. This process is called kinematics reverse estimation [2]. Therefore, in the reverse kinematics analysis, we have access to the coordinate components of the central point of the moving star (point C located in the reference system O′) which are relative to the primary reference system and which are the goal for obtaining the unknown parameters S_{11}, S_{21}, S_{31}, (S_1^1, S_1^2, S_1^3). The independent linear equations are obtained through equalizing the transformative matrixes related to the central point of the moving star.

Solving these equations in which the inputs are θ_1^1, θ_1^2, $\theta_1^3=\theta$ and S_{13}, S_{23}, S_{33} (S_3^1, S_3^2, S_3^3), then S_{11}, S_{21}, S_{31} (S_1^1, S_1^2, S_1^3) are obtained.

$$-X \times Sin\theta - Sin\theta \times S_3^1 + S_1^1 - q_x = 0 \tag{15a}$$

$$X \times Cos\theta + Cos\theta \times S_3^1 - q_y = 0 \tag{15b}$$

$$0.5X \times Sin\theta - 0.866X \times Cos\theta + 0.5S_3^2 \times Sin\theta - 0.866S_3^2 \times Cos\theta - 0.5S_2^1 + e - q_x = 0 \tag{15c}$$

$$-0.866X \times Sin\theta - 0.5X \times Cos\theta - 0.866S_3^2 \times Sin\theta - 0.5S_3^2 \times Cos\theta + 0.866S_2^1 - q_y = 0 \tag{15d}$$

$$-0.5X \times Sin\theta + 0.866X \times Cos\theta - 0.5S_3^3 \times Sin\theta + 0.866S_3^3 \times Cos\theta + 0.5S_1^3 - q_x = 0 \tag{15e}$$

$$-0.866X \times Sin\theta - 0.5X \times Cos\theta - 0.866S_3^3 \times Sin\theta - 0.5S_3^3 \times Cos\theta + 0.5S_1^3 - q_y = 0 \tag{15f}$$

To conduct the reverse kinematics analysis, we should assume two directions: (1) liner, (2) circular, in which the center of the star (point located in O′ reference system) goes through the two mentioned directions as shown in **Figure 10** (**a**) and **Figure 11** (**a**, **e**). The outcome is respectively represented in **Figure 10** (**b**) and **Figure 11** (**b**, **c**, **d**, **e**).

http://dx.doi.org/10.5772/64556

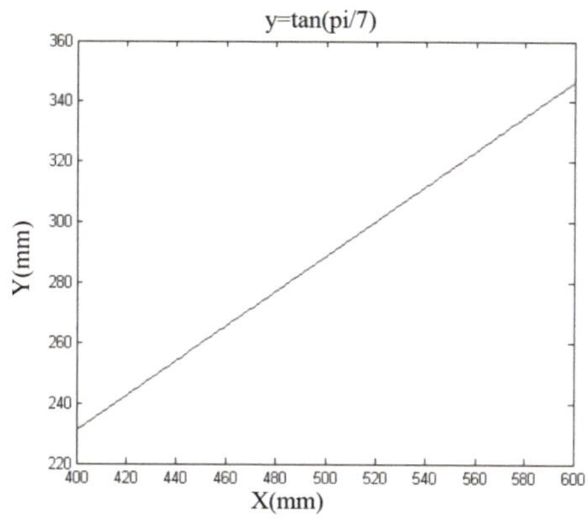

(a)

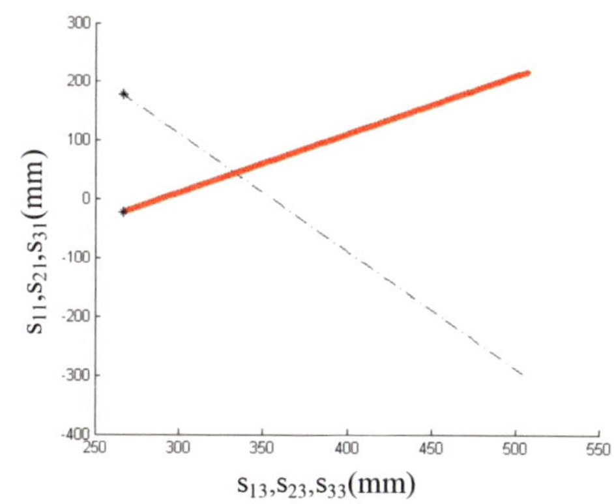

(b)

Figure 10. (a) *X-Y* plots of the path of the motion point c and (b) variation S_{11}, S_{21}, S_{31} (output) versus S_{13}, S_{23}, S_{33} S_{13}, S_{23}, S_{33} (mm).

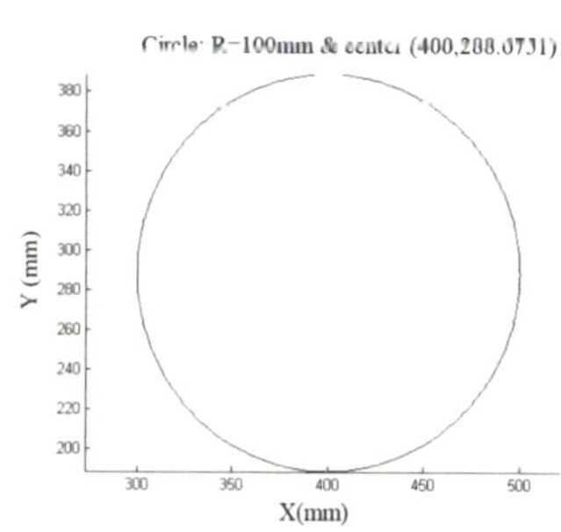

(a)

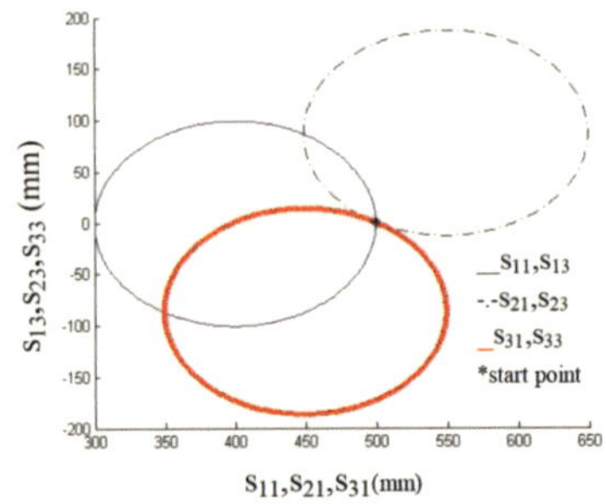

(b)

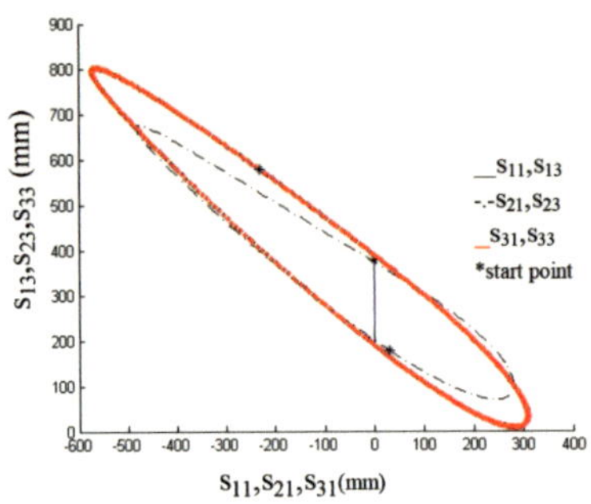

(c)

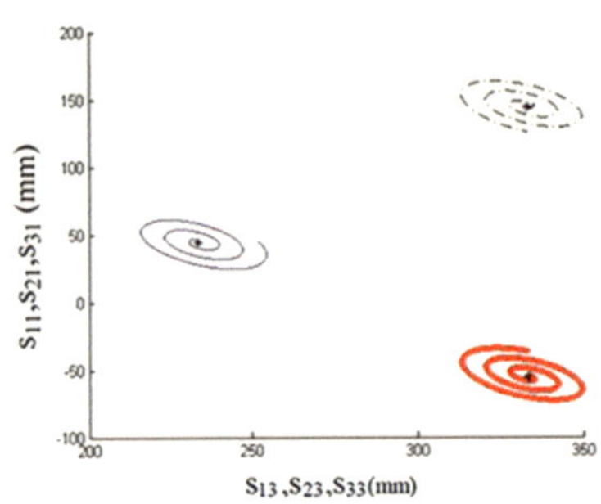

(d)

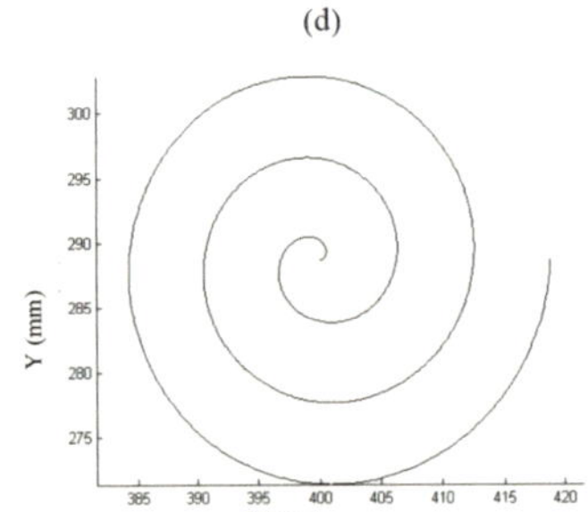

(e)

Figure 11. (a, e) *X-Y* plots of the path of the motion point c; (b, d) variation S_{11}, S_{21}, S_{31} (outputs) versus S_{13}, S_{23}, S_{33} for $\theta = 0$; (c) variation S_{11}, S_{21}, S_{31} (outputs) versus S_{13}, S_{23}, S_{33} (inputs) for $\theta = pi/6$.

http://dx.doi.org/10.5772/64556

11. Workspace analysis

The existence or nonexistence of the kinematics solution determines the robotic workspace. The lack of solution means that the robot is not able to obtain optimal orientation because it is located out of the workspace. These conditions are called robotic singularity states. Almost all the robots have singularity points in either the border of their workspace or in their workspace. The singularity point in the border of workspace denotes a state that occurs when the arm is fully stretched or folded on itself when the final manipulator is almost or precisely located in the border of the workspace.

On the other hand, the singularity states in the workspace signify the conditions that occur in the mechanism workspace or in general when two or some joint axes are located in one direction. When the robot is located in the singularity position, it loses all or some of its degrees of freedom in the deicardean space. It is obvious that this process is done in the border of the robotic workspace. The examination of the T-S(3-PRP) robotic workspace has shown that it has no singularity in its workspace; the robotic workspace is presented in **Figure 12** (**Figure 13**).

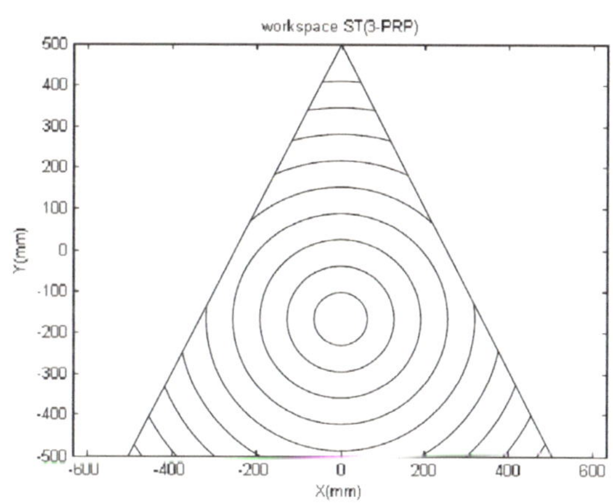

Figure 12. X-Y plots of the Achievable workspace of the T-S robot.

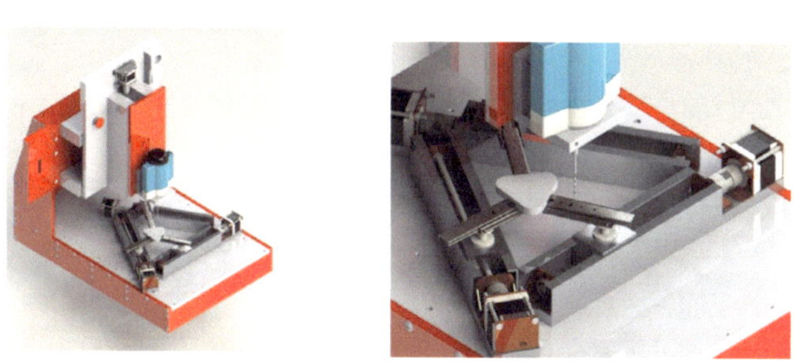

Figure 13. T-S Robot as milling machine' table.

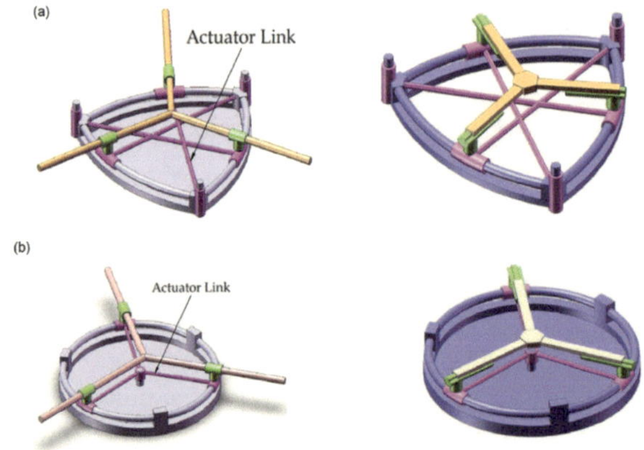

Figure 14. Introduction a new geometric model as (a) Reuleaux Triangle –Star Robot with telescopic and (b) arms Circle-Star robot with telescopic arms.

12. Conclusion

Features such as precision, speed, stiffness, and workspace with singularity point distinguish the parallel robots from the Serial robots. But the weak points of these robots compared with Serial robot are the move limited workspace. In this paper, first, the limitations and weaknesses of the Triangle-Star Robot {T-S 3(PRP)} were recognized. To remove these disadvantages, a robot with telescopic arms were presented and its kinematics analysis done through Hartenbeg–Denavit instruction. In addition, there is no disturbance in the totality of the discussion.

Afterwards, to increase the workspace,

a. Reuleaux Triangle-Star Robot with the kinematics structure {RT-S 3(PRP)}

b. Circle-Star Robot with kinematics structure{C-S 3(PRP)} are introduced and kinematically analyzed.

Author details

Ahmad Zahedi[1], Hadi Behzadnia[2*], Hassan Ghanbari[1] and Seyed Hamed Tabatabaei[3]

*Address all correspondence to: mashaykhan@yahoo.com

1 Islamic Azad University, Firoozkooh Campus, Firoozkooh, Iran

2 Machinery Unit, The Ministry of Roads and Urban Development (MRUD), Iran

3 Islamic Azad University, South Tehran Campus, Tehran, Iran

References

[1] Ibarreche JI, Altuzarra O, Petuya V, Hernandez A, Pinto C. Structural Synthesis of the families of parallel manipulators with 3 degrees of freedom. Romansy 19-Robot Design, Dynamics and Control. 2013;544: 35-42. DOI: 10.1007/978-3-7091-1379-0_5.

[2] Chennakesava-Reddy A. Difference between Denavit - Hartenberg (D-H) Classical and Modified Conventions for Forward Kinematics of Robotics with case study, International Conference on Advanced Materials and manufacturing Technologies (AMMT); 18-20 December 2014; JNTUH College of Engineering Hyderabad. 2014. P. 267-286.

[3] Alexander Yu, A.Bonev I, Zsombor-Murray P. Geometric Approach to the Accuracy Analysis of a Class of 3-DOF Planar Parallel Robots. Mechanism And Machine Theory. 2007; 43: 364-375. DOI: 10.1016/J.mechmachtheory.2007.03.002

[4] Pandilov Z, Dukovski V. Comparison of the characteristics between serial and parallel Robots. ACTA TEHNICA CORVINIENSIS – Bulletin of Engineering Tome VII. 2014; 5: 143-160. DOI: 10.3311/pp.ch.2014-5.01

[5] Lovasz E-CH, Grigorescu S, Mărgineanu D, Pop C, Gruescu C, Maniu I. Kinematics of the planar parallel Manipulator using Geared Linkages with linear Actuation as kinematic Chains 3-R(RPRGR)RR. The 14th IFToMM World Congress; 25-30 October 2015; Taipei, Taiwan; 2015.P. 1-6

[6] Pandilov Z, V.Dukovski V. Several open problems in parallel robotics. ACTA TECHNICA CORVINIENSIS-Bulletin of Engineering, Tome IV. 2011; 3:77-84. DOI: 10.3311/pp.ch.2011-3.01

[7] Chablat D, Staicu S. KINEMATICS OF A 3-PRP PlANAR PARALLEL ROBOT. U.P.B. Sci. Bull., Series D. 2009; 71:1-15. DOI: 0904/0904.0058

[8] Chablat D, Wenger P. Self Motions of a Special 3-RPR Planar Parallel Robot. Advances in Robot Kinematics. 2006; 1: 221 – 228. DOI: 10.1177/0278364908092466

[9] Haijun Su,Peter Dietmaier, J. M. McCarthy" Trajectory Planning for Constrained Parallel Manipulators "Robotics and Automation Laboratory, UCI,August 1, 2002

Chapter 5

Recent Developments in Monocular SLAM within the HRI Framework

Edmundo Guerra, Yolanda Bolea,
Rodrigo Munguia and Antoni Grau

Additional information is available at the end of the chapter

http://dx.doi.org/10.5772/63820

Abstract

This chapter describes an approach to improve the feature initialization process in the delayed inverse-depth feature initialization monocular Simultaneous Localisation and Mapping (SLAM), using data provided by a robot's camera plus an additional monocular sensor deployed in the headwear of the human component in a human-robot collaborative exploratory team. The robot and the human deploy a set of sensors that once combined provides the data required to localize the secondary camera worn by the human. The approach and its implementation are described along with experimental results demonstrating its performance. A discussion on the usual sensors within the robotics field, especially in SLAM, provides background to the advantages and capabilities of the system implemented in this research.

Keywords: mapping and localization, sensors, visual odometry, HRI, features initialization

1. Introduction

A great deal of the investigation done in the field of robotics is addressed to the Simultaneous Localisation and Mapping (SLAM) problem [1, 2]. The SLAM problem is generally described as that of a robot—or robotic device with exteroceptive sensor/s—which explores an unknown environment, performing two different tasks at the same time: It builds a map with the observations obtained through the exteroceptive sensor/s [3] and localizes itself into the map during the exploration, thus knowing the position and trajectory.

The works defining the origin of the field can be traced to Smith and Cheeseman [4], Smith et al. [5], and Durrant-Whyte [6], which established how to describe the relationships between landmarks while accounting for the geometric uncertainty through statistical methods. These eventually led to the breakthrough represented in Smith's work. In such a research, the problem was presented for the first time as a combined problem with a joint state composed of the robot pose and the landmark estimations. These landmarks were considered correlated due to the common estimation error on the robot pose. That work would lead to several works and studies, being [7] the first work to popularize the structure and acronym of SLAM as known today.

The problem related with SLAM techniques is considered of capital importance given that a solution to it is required to allow an autonomous robot to be deployed in an unknown environment and operate without human assistance. But there is a growing field of robotics research that deals with the interaction of human and robotic devices [8]. Thus, there are several applications of robotic mapping and navigation that include the human as an actor. The basic application would be the exploration of an environment by a human, but mapped through a robotic platform [9]. Other works deal with more complex applications, such as mapping the trajectory of a group of humans and robots during the exploration of an environment and coordinating them with the help of radio frequency identification (RFID) tags [10]. Another application gaining weight is the use of SLAM to allow assistance robots to learn environments, improving the usability of the device [11].

All these approaches solve some kind of SLAM problem variant where the human factor is present: to assist, to track, to navigate, etc. But none uses data captured by human senses. There are works that deal with the mapping of human-produced data into map generated by a robot, but these data are not used in the map estimation process, but 'tagged' to it. So currently, no approach uses the data from human into the solution to the SLAM problem. This is a waste of useful resources, given the power of the human sight, still superior in terms of image processing to the most advanced techniques which are increasingly adopting the strategies discovered by scientists, but designed and adopted by human evolution millennia ago.

So, in this chapter, we will discuss about the monocular SLAM problem in the context of human-robot interaction (HRI), with comments on available sensors and technologies, and different SLAM techniques. To conclude the chapter, a SLAM methodology where a human is part of a virtual sensor is described. His/her exploration of the environment will provide data to be fused with that of a conventional monocular sensor. These fused data will be used to solve several challenges in a given delayed monocular SLAM framework [12, 13], employing the human as part of a sensor in a robot–human collaborative entity, as was first described in authors' previous work [14].

2. Sensors in the SLAM problem

In robotic systems, all relations between the system and the physical environment are performed through transducers. Transducers are the devices responsible for converting one kind of energy into another. There are basically 2 broad types of transducers: sensors and actuators. Actuators use energy from the robotic system to produce physical effects, such as forces and displacements, sound, and lightning. Sensors are the transducers responsible for sensing and measuring by way of the energy conversion they perform: turning the energy received into signals (usually of electrical nature), which can be coded into useful information.

The sensors used in SLAM, just like in any other fields of robotics, can be classified according to several criteria. From a theoretical point of view, one of the most meaningful classifications is that if the sensor is of proprioceptive or exteroceptive nature. Proprioceptive (i.e., '*sense of self*') sensors are generally responsible for measuring values internal to the robot system, like the position of a joint, the remaining battery charge, or a given internal temperature. On the other side, exteroceptive sensors measure different characteristics and aspects of the environment, normally with respect to the sensor itself.

The encoders are proprioceptive sensors, responsible for measuring the position or movement of a given joint. Although there are linear encoders, only the rotary encoders are frequently used in the SLAM problem [15]. These encoders can measure directly the position of the rotary axis, in terms of position if they are 'absolute encoders' or in terms of movement for the 'incremental encoders'. Their great accuracy when measuring rotation allows computing the exact distance traveled by a wheel, assuming that its radius is known. Still they present several problems related to the nature of how they measure: The derived odometers assume that all the movement against the wheel surface is transformed into rotation at a constant and exact rate, which is false in many circumstances. This makes them vulnerable to irregular and dirty surfaces. As a proprioceptive sensor, with no exterior feedback, the error of a pure odometry-based SLAM approach will grow unbound, suffering the drift due to dead reckoning.

Range finders are exteroceptive sensors which measure distances between them and any point in the environment. They use a variety of active methods to measure distance, sending out sound, light, or radio waves and listening to the receiving waves. Generally, these are known as sonar, laser range finders (LRF), or radar systems. The devices destined to robotics applications generally perform scans, where a set of measurements is performed concurrently or over such a short time that they are considered all simultaneously. When scans are performed, each sub-measurement in a set is usually paired with bearing data, to note the relation between the different simultaneous measurements, generally performed in and arc.

Sonar systems use sound propagation through the medium to determine distances [16]. Active sonar creates a pulse of sound (a ping) and listens to its reflections (echoes). The time of the transmission of the pulse to its reception is measured and converted to distance by knowing the speed of sound being a time-of-flight measurement. Laser rangefinders (LRF) [17] can work on different principles, using time-of-flight measurements, interferometers, or the phase shift method. As the laser rays are generally more focused compared to the other types of waves, they tend to provide higher accuracy measurements. Radars [18] also employ electromagnetic waves, using time-of-flight measures, frequency modulation, and the phased array method between others to produce the measurements. As they usually produce a repeated pulse at a given frequency (RPF), they present both a maximum and minimum range of operation.

These sensors can have great accuracy given enough time (the trade-off between data density and frequency is generally punishing), and as they capture the environment, they do not suffer from dead reckoning effects. On the other side, the data they provide are just a set of distance at given angles, so these data need to be interpreted and associated, requiring cloud matching methodology (like iterative closest point, ICP, and other similar and derived ones), which is computationally expensive. Besides, they have all their specific weaknesses: Sonar has limited usefulness outside of the water given how sound works on the air; LRF are vulnerable to ambient pollutants (dust, vapors) that may distort the lightning processes of the measurement; radar has very good range but tends to be lacking in accuracy compared to the other range-finders.

The Global Positioning System (GPS) [19] is a proprioceptive sensor based on synchronizing radio signal received from multiple satellites. With that information, it can compute the coordinates and height position of the sensor on any point of the world with up to 10 m margin. This 10 m margin grows rapidly if fewer satellites are visible (direct line of sight is required), making it useless on closed environments, urban canyons, etc. Besides, the weakness to satellite occlusion and wide error margin, the GPS presents other challenges, like a rather slow update rate for most of the commercial solutions.

The inertial measurement unit (IMU) is a proprioceptive sensor that combines several sensing components to produce estimations of the linear and angular velocities and the forces of the device. They have generally linear and angular accelerometers, and sometimes they include also gyroscopes and magnetometers, producing the sensory part of inertial navigation system (INS). The INS includes a computing system to estimate the pose and velocities without external references. The systems derived from the IMU have generally a good accuracy, but they are vulnerable to drift when used in dead reckoning strategies due to their own biases. The introduction on external reference can improve the accuracy, and thus, they are frequently combined with GPS. Introducing other external references leads to the development of the inertia-visual odometry field, which is closely related to the SLAM [20, 21]. Still, the accuracy gain is limited by the nature of the exteroceptive sensor added (which keeps its own weaknesses), and the IMU part of the system becomes unreliable in the presence of strong electromagnetic fields.

Vision-based sensors are exteroceptive sensors which measure the environment through the reflection of light on it, capturing a set of rays conformed as a matrix, thus producing images. The most common visual sensor is the camera, which captures images of the environment observed in a direction, similarly to the human eye. Still, there are many types of cameras, depending on the technology which they are based, which light spectrum they capture, how they convert measurement into information, etc. An standard camera can generally provide color or grayscale information as an output at 25 frames per second (fps) or more, being generally focused on the wavelength range visible by the human eye, and presenting that information in a way pleasant to the human eye. But specific cameras can work with different frameworks as target, thus capturing other spectra not seen by human eye (IR, UV...), producing vastly higher fps rates, etc.

One of the main weaknesses of cameras within the context of the SLAM problem is that they produce only visual angular data: Each element of the matrix which composes an image shows the visual appearance information about a projected point where a ray (which theoretically can reach the infinite) finds an object. Thus, cameras alone cannot produce depth estimation in a given instant. This can be solved by more specific sensors, like time-of-flight cameras. These sensors generally have poorer resolutions, frame rate, dynamic range, and performance overall, while being several orders of magnitude more expensive, which made them barely used until few years ago.

There are other types of visual sensors that while still being cameras, they result more divergent from the standard monocular cameras. A good example would be the works on multiple camera stereo vision. Stereo cameras generally include two or more cameras and based on epipolar geometry can find the depth of the elements on the environment. Omnidirectional cameras expand the field of view, so that they can see almost all their surroundings at any given time, vision, presenting several challenges of their own in terms of image mapping and representation.

3. Classic only-bearing monocular SLAM approaches

There are many approaches to solve the SLAM problem, depending on the sensors available and the mathematical models and procedures used. From particle filters [22] to sums of Gaussian distributions, passing through the use of graph-based approaches [23] and RANSAC methods [24], the SLAM problem has been treated using many different mathematical techniques. Latest trends rely on bundle adjustment and other optimization methods [25, 26]. Still, one of the most commonly found approaches to the problem is using the extended Kalman filter (EKF) [27, 28], treating it as an incremental estimation problem.

The general monocular EKF-SLAM procedure is based on detecting points of interest which can be detected and distinguished from other robustly, introduce them into the map representation which is being built inside the filter, and track them through the sequence of frames, estimating both their pose and the camera odometry. For each landmark, a patch of the image around it, describing its 'appearance', is stored and will be used to identify it, and the landmark itself is generally modeled through unified inverse depth parametrization [29], although other model exists [30].

The estimation process is based on probabilistic filtering, where an initial prediction step makes a prediction of the movement of the robot and so of the position of the camera. Data from any sensors can be used; although in pure monocular SLAM methodologies, due to lack of data, predicted motion is assumed to be described by Gaussian distributions [31]. Thus, a constant velocity movement model is used, with random impulses of angular and linear accelerations modeled as white noise. The prediction of the map is much simpler: As the landmarks or points of interest in the map are assumed to be part of the environment, the hypothesis used is that they will remain static and so their position does not change.

After the prediction step, a conventional EKF-SLAM would produce an actual measurement from the sensors and compare them with the predicted measurements obtained through the direct observation model. This step requires solving the data association problem, which consists in matching the predicted measurements with the actual measurements from the sensors. Given the computational cost of extracting all the possible points of interest at each frame and matching them with those predicted points, and the issues produced by the uncertainty of the prediction given that it is based on random movements, an active search strategy is used to deal with the problems [32]. Under this strategy, the features in the map are predicted into the pixel space using the direct observation model, and for each pixel, a search is performed looking for the most similar point (according to the stored patch), using zero-normalized cross-correlation (ZNCC). Ideally, each feature predicted will be matched to a new pixel that is the same feature in the latest frame. This process can fail due to visual artifacts, the geometry of the environment, the presence of dynamic objects, and several other causes. So these pairs of points (namely, the feature predicted and the match found in the latest frame) are checked through a data association validation methodology, in our case the HOHCT [13].

Once the predicted landmarks and its associated pairing are found in image space, in pixel coordinates, the innovation of the Kalman filter or residual can be computed following the usual EKF methodology.

Although the iterative estimation of the map as an EKF is pretty straightforward, there is still a critical process which defines many characteristics of any give monocular SLAM approach: the feature initialization process. When using conventional point detector and descriptors, it is frequent that several dozens or even hundreds of points will appear in an image, and most of them will be ignored for the SLAM process—based on spatial distribution, position on image, etc.—but still, no depth information is available in an instant way. Thus, two main strategies exist to deal with this issue: Undelayed approaches try to 'guess' the value to initialize the depth, normally relying on heuristics, while delayed approaches track a feature over a time, until they have a good estimation of its depth and only then proceed to initialize it.

These two types of strategies define many characteristics of the SLAM procedures. As undelayed approaches try to use point features as landmarks just after have been seen, the points are quickly introduced into the filter, accepting many outliers that have to be validated later or rejected at the data association validation step [28, 31]. On the other side, delayed approaches track and estimate the points before using them, so the used landmarks are generally more stable and reliable with delayed initialization [33].

The delayed inverse-depth (DI-D) monocular SLAM is a delayed feature initialization technique [12, 13]. The delay between a landmark being observed for the first time and being initialized allows estimating the parallax achieved through the estimated odometry. This in turn enables obtaining depth estimations for the landmarks through triangulation.

http://dx.doi.org/10.5772/63820

4. Introducing the human component into monocular SLAM

The DI-D procedure, although it was shown to be a strong monocular EKF-SLAM methodology, still presents several features that reduces its usability and scalability, mainly the need for an initialization process using synthetic or known a priori landmarks. These known landmarks would help initially to produce the odometry estimation and thus are critical to solve the scale problem of the map.

Shifting the monocular SLAM problem from an isolated sensor fusion point of view to a component into a bigger human-robot collaborative effort allows considering new options. Given the features of current exploratory robots, it is worth noting that an exploratory team composed of robots and humans will outperform any robotic device. If the desired tasks increase in complexity (emergency situations, those required management and decision under high uncertainty), the advantage of a human-robot collaborative team increases dramatically. Assuming that the human wears a headwear device with several sensors, the SLAM capabilities of the robot can be improved (**Figure 1**). Thus, the camera deployed in the helmet will be used to obtain 'instantaneous parallax', thus achieving complete measurement when the human is looking at the same direction as the robot, in a stereo-like situation, as it was initially proposed and described in authors previous work [14].

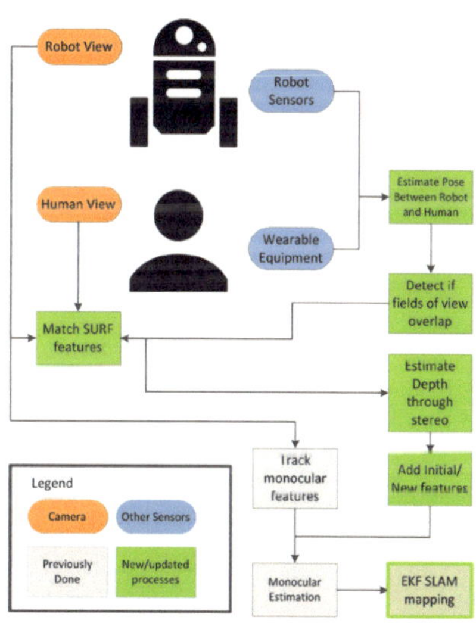

Figure 1. DI-D monocular EKF-SLAM components within human-robot collaborative.

To achieve this, in addition to the new camera on the human (C_h) which will perform the depth estimation with the robot camera (C_r), a combination of sensors and mechanisms able to estimate the pose between the cameras should be deployed. From a software point of view, the modules required to treat these data and estimate the pose, and those new which will deal with the new depth estimation process, must be implemented. From the EKF-SLAM methodology, the initialization of features is a local process that introduces the features into the EKF using the inverse-depth parametrization, which remains the same, but will require also using a new inverse observation model that treats the whole system to estimate features as a single complex sensory process.

4.1. Multiple monocular vision sensor array: pseudo-stereo vision

The weak points discussed before can be solved within a cooperative context exploiting data from another monocular camera. Assuming that the C_h with known pose is near to the robotic camera performing SLAM (C_r), joining observations from both cameras allow performing stereo-like estimation when their fields of view overlap. This way, a new non-constant stereo inverse-depth feature initialization approach will be used to address the issues.

Classical stereo approaches [34, 35] rely on epipolar geometry to create a calibrated camera rig with multiple constraints. These constraints typically include that both cameras' projection planes lie in the same plane in world coordinates; this allows optimizing the correspondence problem as the match on an image of another's image pixel will lie in the corresponding epipolar line, and rectification can turn them into straight lines parallel to the horizontal axis. Several works have dealt with rectification of stereo images for unrestricted pose cameras both calibrated [35] and uncalibrated [36].

Figure 2. Image pair sample captured at one experimental sequence.

Figure 3. Rectification of images left and right at **Figure 2**. Scale distortions are produced due to the multiple reprojection operations.

Fusiello et al. [35] detailed the first method to rectify stereo pairs with any given pairs of calibrated cameras. The method is based on rotating the cameras until they have one of their axis aligned to the baseline and forcing them to have their projective planes contained within the same plane to achieve horizontal epipolar lines. Other works have proposed similar approaches to rectifying stereo pairs assuming calibrated, uncalibrated, or even multiple view [37, 38] stereo rigs. These approaches need to warp both images according to the rectification found (see **Figure 2** left and right and **Figure 3**) and, in some cases, producing great variations in terms of orientation and scale (**Figure 3**), thus rendering them less attractive in terms of our approach.

At any case, dealing with stereo features without rectified images is not a big problem in the proposed approach. The process of stereo features search and matching will be done sparsely, only to introduce new features: during the initialization, or when the filter needs new features. For both cases, only a part of the image will be explored, and when adding new features in a system already initialized, additional data from the monocular phase can be used to simplify the process.

4.2. Scaled feature initialization with collaborative sensing

The requirement of metric scale initialization of the DI-D method can be avoided under the assumption of a cooperative framework. Classical DI-D required the presence of a set of known, easily identifiable features to estimate them initially through the PnP problem and initiate the EKF with scale. Assuming that at the start of the exploration a cooperating, free moving camera is near, the data from this camera can produce the features needed through pseudo-stereo estimation. This process is shown in **Figure 4**, where, after the pose between the robot camera and the human camera is known, the maximum distance from a camera where a point with a given minimum parallax (pl_{min}) could lie is found. This distance is employed to build a model of the field of view of each camera, as a pair of pyramids, with each apex in the optical center of a pinhole camera, and the base centered along the view axis. Then, it can be guaranteed that any point with parallax—between cameras—equal or greater than pl_{min} will lie in the space intersected by the two fields of view modeled as pyramids, as seen in **Figure 5**. So the intersection between the different polygons composing the pyramids is computed as a set of segments (two point tuples), as described by **Algorithm 1**. Once all the segments are known, they are projected into the 2D projective space of each camera, and a search region is adjusted around them, determining the regions of interest where the stereo correspondence may be useful and significant.

In the interest regions found, SURF-based feature descriptors [39] are matched to produce new stereo features to initialize in the EKF state vector when needed. SURF is chosen over SIFT and FAST [39] due to the more convenient trade-off offered in terms of matching accuracy and efficiency, and could be replaced by any other feature descriptor. Each pair of matched points between cameras allows estimating the world coordinates of the landmark feature seen through triangulation, back tracing the points on the images from the robot camera and the human camera. Then, the landmarks found and fully measured (with real depth estimation) are introduced in the monocular EKF according to the unified inverse depth parametrization.

To take advantage from the computational effort made during the non-overlapping frames, the landmarks that were being tracked to be initialized prior to the pseudo-stereo measurement are given priority to be introduced; these landmarks are robust because they were tracked for several frames previously.

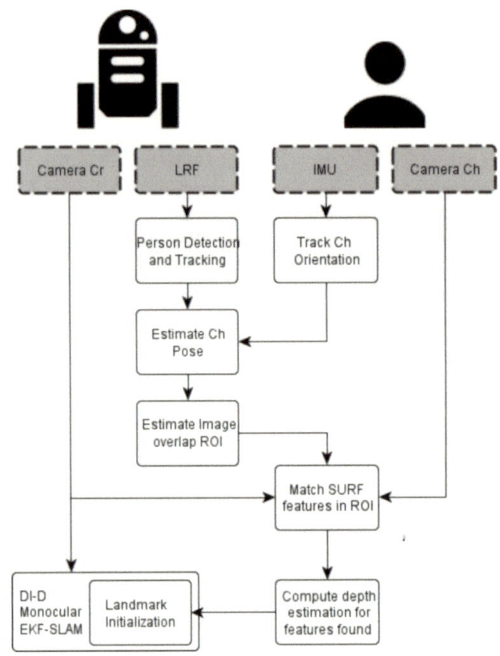

Figure 4. Block diagram of the final implementation, with sensors on gray boxes and software processes on clear blocks. The updated landmark initialization is one of the cornerstone processes in any feature-based EKF-SLAM technique.

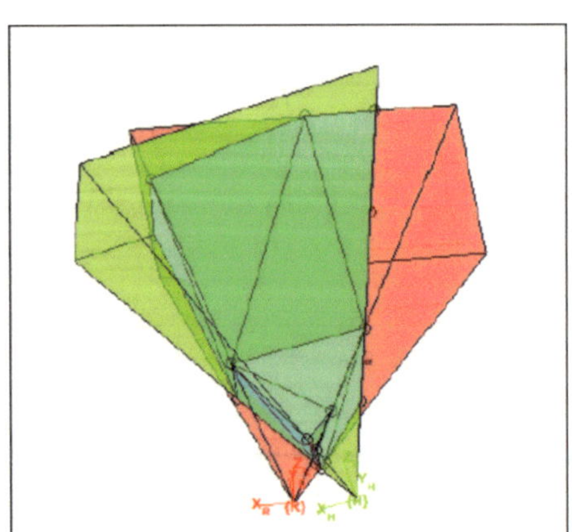

Figure 5. Graphical representation of the intersection of both cameras' fields of view.

```
ALGORITHM 1:
(ri_r, ri_h):= Find Stereo ROI (cam_r, cam_h, pl_min)
begin
ri_r:= Ø; ri_h:= Ø
distance:= FindDistance (cam_r.pose, cam_h.pose)
PyramidDepth:= FindMaxDepth (distance, pl_min)
Py1:= ModelFoV(cam_r,PyramidDepth)
Py2:= ModelFoV(cam_h,PyramidDepth)
intersection = Ø
for each polygon_i in Py1
    for each polygon_j in Py2
        segment := Intersect(polygon_i, polygon_j)
        intersection.add(segment)
    end for
end for
if ¬(intersection = Ø) then
    ri_r := Envelope (ProjectTo2D (cam_r.pose, intersection.points))
    ri_h := Envelope (ProjectTo2D (cam_h.pose, intersection.points))
end if
end
```

5. Experimentation and results

The approach described in this work was fully implemented and tested with real data. The DI-D SLAM with pseudo-stereo feature initialization was programmed in MATLAB® to test and evaluate it. Several sequences were captured in semi-structured environments using a robotic platform and wearable headgear.

5.1. Experimental system implementation

The sequences were reduced to a resolution of 720 × 480 pixels and grayscale color, shortening the computational effort for the image processing step. Each sequence corresponds to a collaborative exploration of the environment at low speed, including a human and a robotic platform, each one equipped with the monocular sensors assumed earlier, C_h and C_r, respectively. The data collected include monocular sequences, odometry from the robot, estimation of the human pose with respect to the robot, and the orientation of the camera C_h. During the sequences, the camera C_r whose sequence would be used for the SLAM process was deployed looking forward, towards the advance direction. This kind of movements produces singularities in estimation, as the visual axis of the cameras is aligned with the movement, producing 'tunnel blindness', where the elements near the centre of the captured images produce negligible parallax, and thus, only variations in scale are perceptible in short intervals.

The robot functions were performed by a robotic platform based on the Pioneer 3 AT (see **Figure 6**). The platform runs a ROS distribution over an Ubuntu 14.04 OS. The platform is equipped with a pair of laser range finders Leuze RS4-4 and a Logitech C920 webcam, able to work up to 30 frames per second (fps) at a resolution of 1080p. The sensors worn by the human are deployed on a helmet, including another Logitech webcam camera and an Xsens AHRS. All the data have been captured and synchronized through ROS in the robotic platform hardware. The ROS middleware provides the necessary tools to record and time-stamp the data from the sensors connected to the platform.

Figure 6. The robotic platform (based on the Pioneer 3 AT) used to capture the data sequences to test the described approach.

To estimate the pose of C_h, orientation data from the IMU are combined with the approximate pose of the human, estimated with the range finders [17]. The final position of the camera is computed geometrically as a translation from the estimated position of the Atlas and Axis vertebrae (which allow most of the freedom of movement of the head). These vertebrae are considered to be at a vertical axis over the person position estimated with the range finders, with height modeled individually for each person. In this work, it is assumed that the environment is a flat terrain, easing the estimation process.

The pose of the camera C_h with respect to the C_r is not assumed to be perfectly known. Instead, it is considered that a 'noisy' observation of the pose of C_h with respect to C_r is available by means of the methodology described above. The inherent error to the observation process is

http://dx.doi.org/10.5772/63820

modeled, assuming that the observation is corrupted by Gaussian noise. The value of the parameters used to model the inaccuracies for computing the pose of C_h was obtained statistically by comparing actual and estimated values. It is also important to note that an alternate method could be used for computing the relative pose of C_h, for instance, using different sensors.

5.2. Experiments and results

The introduction of an auxiliary monocular sensor which can provide non-constant stereo information was proven useful. One of the weaknesses discussed earlier of the DI-D was the need to set an initial metric scale through synthetic feature, which has been removed. This grants more autonomy to the system, exploiting the implicit human–robot interaction without enforcing utilization of artificial landmarks. Besides, as the metric scale initialization can introduce more features into the initial state because it is not limited to the artificial landmark, the scale propagates in a smoother way with reduced drift on the local scale.

5.2.1. Visual odometry accuracy and scaling

Figure 7 shows results for two sample trajectories, with and without the utilization of the proposed non-constant stereo DI-D feature initialization approach, in blue and orange lines, respectively. The trajectory on **Figure 7** (left) was captured in an inner courtyard, with several seats and trees. This trajectory ran for 19 m, with two 90° turns, on a semi-structured environment with plenty of objects within view that could be mapped. On the other side, the trajectory shown in **Figure 7** (right) was capture as a straight 10 m movement in a blind alley with painted walls with homogenously textured surfaces, reducing the chances to obtain robust features. These two sequences contained plenty the most disadvantageous characteristics for monocular SLAM: singular movements where parallax cannot be observed for the central region of the camera, quick changes in orientation and turning velocities, surfaces/environments showing low count of robust visual features, natural lightning and shadows, etc.

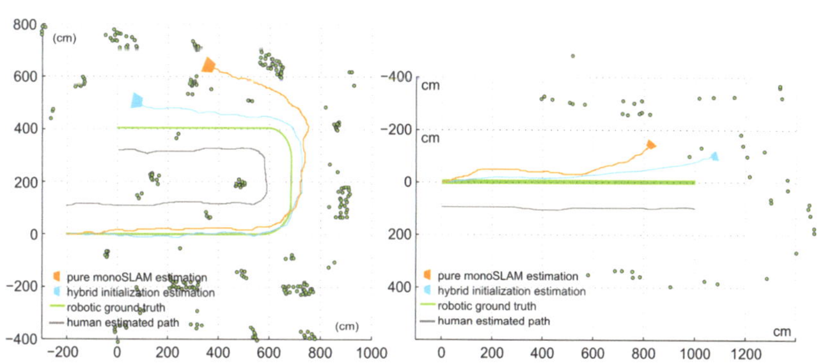

Figure 7. Trajectories estimated with classical DI-D monocular SLAM (orange plots) and with the new non-constant stereo DI-D approach for feature initialization (blue plots). Green line denotes robot ground truth; gray line denotes C_h ground truth.

The introduction of the pseudo-stereo initialization of features enables initialization of features with actual depth estimation instantly, without relying on heuristic or having a delay where data are being processed but not used in the estimation. Each of these situations has strong deterrents; for example, the heuristics used for depth initialization can vary between sequences, or even different SLAM 'runs' of the same video sequence, accounting for the uncertainty in the prediction model and the feature selection process. When a feature is initialized with a delayed method after it has been seen, computational power is spent on estimating a landmark that likely will not be used and never introduced into the EKF map.

At the end, the proposed approach made the system more resilient, especially to quick view changes, such as turning, and long singular movements—front advance. These movements can be seen in **Figure 7**, left and right, respectively. During close turns, delayed monocular SLAM approaches have very little time to initialize features because the environment changes quickly and the features are observed for short periods. This produces a decrease in the number of initialized features that decreases the odometry estimation accuracy. At the end of the run, the uncertainty becomes so big that errors cannot be corrected, the EKF loses convergences, and the estimation process results become useless. **Figure 7** (left) shows how the two turns can greatly degrade the orientation estimation for a classic delayed SLAM method, while the proposed approach can track the turns much more closely, with less than half the error. On the other side, **Figure 7** (right) illustrates the issues of singular movements: The odometry scale is very hard to estimate for pure monocular methods, because features present reduced parallax. Not only the length of the trajectory is affected by this phenomenon, but the accuracy of the orientation estimation also becomes compromised due to the inability of the EKF to reduce uncertainty quickly enough.

5.2.2. Computational costs analysis

The apparent increase in the computational effort that would suppose the utilization of the presented approach could be hard to justify within the field of filtering-based SLAM, which generally try to keep computational costs as low as possible.

In the considered sequence set, there were a total of 9527 frames for C_r. Although C_h had a paired frame for each one on the C_r sequences, the overlap only was found in 3380 of the pairs. This means increasing the visual processing and feature capturing costs on a 35.47% of the frames. Increasing the computational cost of the most demanding step in a third of iterations may look daunting, but there are few considerations. The technique rarely implies processing an additional full frame: The region where the overlap is interesting is predicted and modeled as a ROI into the C_h image, limiting the area to explore. Besides, the cost increase is bounded by the number of frames where it is applied, so, if there are enough features visible in the map, there is no need to execute the pseudo-stereo depth estimation.

Moreover, it is worth noting that newly proposed approach made less effort per feature to initialize it, as it can 'instantly' estimate landmarks on the process of being initially measured through parallax accumulation. This trades off with the fact that the pseudo-stereo initialization can initialize with more frequency weak features which the delayed initialization would not been able to handle, and must be rejected during the data association validation step.

http://dx.doi.org/10.5772/63820

Table 1 shows the features initialized by each approach and the tracking effort required until the initialization of the features is done. Note how the non-constant stereo DI-D feature initialization approach uses about 8% more features, but the effort used to initialize them is much lower, as seen by the number of frames which the feature is tracked prior to being introduced into the map. This is because many features that are being tracked are instantly initialized through stereo once they lay in the overlapped field of view. This is advantageous because it allows to introduce features known to be strong (enough to be tracked) directly without more tracking effort, compensating the effort used for the C_h processing and stereo-based initialization.

	Classic EKF monocular SLAM	Pseudo-stereo feature initialization
Features initialized (total)	1871	2032
Features per sequence (avg.)	267.28	290.26
Average frames to initialize a feature	19.63	8.53

Table 1. Statistics of initialized features and frames from first observation until initialization for DI-D SLAM and pseudo-stereo initialization.

Furthermore, in real-time applications employing this technique, the C_r sensor could be upgraded to an 'intelligent' sensor, with processing capabilities, using off-the-shelf technologies—low-cost microcomputers, FPGA, etc. This approach would integrate image processing in the C_h sensor, allowing parallel processing of features, and sending only extracted features, reducing required bandwidth and transmission time. This processing step could be done while the robotic camera C_r makes the general EKF-SLAM process, and thus, it would be possible to have the SURF landmarks' information after the EKF update, in time for the possible inclusion of new features.

6. Conclusions

A novel approach to monocular SLAM has been described, where the capabilities of additional hardware introduced in a human-robot collaborative context are exploited to deal with some of the hardest problems it presents. Results in quickly changing views and singular movements, the bane of most of the EKF-SLAM approaches, are greatly improved, proving the proposed approach.

A set of experiments on semi-structured scenarios, where a human wearing a custom robotic headwear explores the unknown environments with a robotic platform companion, were captured to validate the approach. The system proposed profits from the sensors carried out by the human to enhance the estimation process performed through monocular SLAM. As such, data from the human-carried sensors are fused during the measurement of the points of interest, or landmarks. To optimize the process and avoid unnecessary image processing, the usefulness of the images from the camera on the human is predicted with a geometrical model which estimates if the human was looking at the same places that the robot, and limits the search regions in the different images.

During the tests using real data, the MATLAB implementation of the approach proved itself to be more reliable and robust than the other feature initialization approaches. Besides, the main weakness of the DI-D approach, the need of a calibration process, was removed, thus producing a locally reliable technique able to benefit from more general map extension and loop closing techniques. While the model to estimate the pose between cameras has a given uncertainty very difficult to reduce (accumulated through the kinematic chain of the model), the measurement uncertainty is still lower than that of the purely monocular measurements, even with the parallax-based (in the delayed DI-D case) approach.

To conclude, the system proves the validity of a novel paradigm in human-robot collaboration, where the human can become part of the sensory system of the robot, lending its capacities in very significant ways with low-effort actions like wearing a device. This paradigm can open up the possibility of improving the capabilities of robotics systems (where a human is present) at a faster pace than what purely technical development would allow.

Acknowledgements

This research has been funded by Science Spanish ministry project reference DPI2013-42458-P.

Author details

Edmundo Guerra[1], Yolanda Bolea[1], Rodrigo Munguia[2] and Antoni Grau[1*]

*Address all correspondence to: antoni.grau@upc.edu

1 Automatic Control Dept., Techncial Univ of Catalonia, Spain

2 Computer Science Dept., University of Guadalajara, Guadalajara, Mexico

References

[1] Durrant-Whyte H, Bailey T. Simultaneous localization and mapping: part I. IEEE Robot Autom Mag. 2006;13(2):99–110.

[2] Bailey T, Durrant-Whyte H. Simultaneous localization and mapping (SLAM): part II. IEEE Robot Autom Mag. 2006;13(3):108–17.

[3] Thrun S, et al. Robotic mapping: a survey. Explor Artif Intell New Millenn. 2002;1:1–35.

[4] Smith RC, Cheeseman P. On the representation and estimation of spatial uncertainty. Int J Robot Res. 1986;5(4):56–68.

[5] Smith R, Self M, Cheeseman P. Estimating uncertain spatial relationships in robotics. In: IEEE International Conference on Robotics and Automation Proceedings. 1987. pp. 850–850.

[6] Durrant-Whyte HF. Uncertain geometry in robotics. IEEE J Robot Autom. 1988;4(1):23–31.

[7] Durrant-Whyte H, Rye D, Nebot E. Localization of Autonomous Guided Vehicles. In: Giralt G, Dr.-Ing GHP, editors. Robotics Research [Internet]. Springer London; 1996 [cited 2013 Jun 4]. pp. 613–25. Available from: http://link.springer.com/chapter/10.1007/978-1-4471-1021-7_69

[8] De Santis A, Siciliano B, De Luca A, Bicchi A. An atlas of physical human–robot interaction. Mech Mach Theory. 2008 Mar;43(3):253–70.

[9] Fallon MF, Johannsson H, Brookshire J, Teller S, Leonard JJ. Sensor fusion for flexible human-portable building-scale mapping. In: Intelligent Robots and Systems (IROS), 2012 IEEE/RSJ International Conference on [Internet]. 2012 [cited 2013 Jun 13]. pp. 4405–4412. Available from: http://ieeexplore.ieee.org/xpls/abs_all.jsp?arnumber=6385882

[10] Kleiner A, Prediger J, Nebel B. RFID Technology-based Exploration and SLAM for Search and Rescue. In: IEEE; 2006 [cited 2013 Jun 10]. p. 4054–9. Available from: http://ieeexplore.ieee.org/lpdocs/epic03/wrapper.htm?arnumber=4059043

[11] Cheein FAA, Lopez N, Soria CM, di Sciascio FA, Pereira FL, Carelli R. SLAM algorithm applied to robotics assistance for navigation in unknown environments. J Neuroeng Rehabil. 2010;7(1):10.

[12] Munguía R, Grau A. Monocular SLAM for visual odometry: a full approach to the delayed inverse depth feature initialization method. Math Probl Eng. 2012;2012:1–26.

[13] Guerra E, Munguia R, Bolea Y, Grau A. A highest order hypothesis compatibility test for monocular SLAM. Int J Adv Robot Syst. 2013, 10:311.

[14] Guerra E, Munguia R, Grau A. Monocular SLAM for autonomous robots with enhanced features initialization. Sensors. 2014;14(4):6317–37.

[15] Johannsson H, Kaess M, Fallon M, Leonard JJ. Temporally scalable visual SLAM using a reduced pose graph. In: Robotics and Automation (ICRA), 2013 IEEE International Conference on [Internet]. IEEE; 2013 [cited 2016 Feb 10]. pp. 54–61. Available from: http://ieeexplore.ieee.org/xpls/abs_all.jsp?arnumber=6630556

[16] Diosi A, Taylor G, Kleeman L. Interactive SLAM using laser and advanced sonar. In: Robotics and Automation, ICRA 2005 Proceedings of the IEEE International Conference

on [Internet]. 2005 [cited 2013 Jun 10]. pp. 1103–1108. Available from: http://ieeexplore.ieee.org/xpls/abs_all.jsp?arnumber=1570263

[17] Sanfeliu A, Andrade-Cetto J. Ubiquitous networking robotics in urban settings. In: Workshop on Network Robot Systems Toward Intelligent Robotic Systems Integrated with Environments Proceedings of [Internet]. 2006 [cited 2013 Dec 26]. pp. 10–13. doi: 10.1.1.75.2850&rep=rep1&type=pdf

[18] Checchin P, Gérossier F, Blanc C, Chapuis R, Trassoudaine L. Radar Scan Matching SLAM Using the Fourier-Mellin Transform. In: Howard A, Iagnemma K, Kelly A, editors. Field and Service Robotics [Internet]. Springer Berlin Heidelberg; 2010 [cited 2013 Jun 10]. pp. 151–61. (Springer Tracts in Advanced Robotics). doi: 10.1007/978-3-642-13408-1_14

[19] Joerger M, Pervan B. Autonomous ground vehicle navigation using integrated GPS and laser-scanner measurements. In: San Diego: Position, Location, and Navigation Symposium [Internet]. 2006 [cited 2013 Jun 10]. Available from: http://mmae.iit.edu/~gps/publications/open/mathieu%20plans%2006.pdf

[20] Li M, Mourikis AI. High-precision, consistent EKF-based visual–inertial odometry. Int J Robot Res. 2013;32(6):690–711.

[21] Lupton T, Sukkarieh S. Visual-inertial-aided navigation for high-dynamic motion in built environments without initial conditions. IEEE Trans Robot. 2012;28(1):61–76.

[22] Montemerlo M, Thrun S, Koller D, Wegbreit B. FastSLAM: a factored solution to the simultaneous localization and mapping problem. In: Proceedings of the National conference on Artificial Intelligence [Internet]. 2002 [cited 2013 Jun 4]. pp. 593–598. Available from: http://www.aaai.org/Papers/AAAI/2002/AAAI02-089.pdf

[23] Folkesson J, Christensen H. Graphical SLAM—a self-correcting map. In: 2004 IEEE International Conference on Robotics and Automation, 2004 Proceedings ICRA'04. 2004. pp. 383–390, Vol.1.

[24] Civera J, Grasa OG, Davison AJ, Montiel JMM. 1-Point RANSAC for extended Kalman filtering: application to real-time structure from motion and visual odometry. J Field Robot. 2010;27(5):609–631.

[25] Klein G, Murray D. Parallel tracking and mapping for small AR workspaces. In: Mixed and Augmented Reality, ISMAR 2007 6th IEEE and ACM International Symposium on [Internet]. 2007 [cited 2013 Dec 26]. pp. 225–234. Available from: http://ieeexplore.ieee.org/xpls/abs_all.jsp?arnumber=4538852

[26] Newcombe RA, Lovegrove SJ, Davison AJ. DTAM: Dense tracking and mapping in real-time. In: 2011 IEEE International Conference on Computer Vision (ICCV). 2011. pp. 2320–7.

[27] Abrate F, Bona B, Indri M. Experimental EKF-based SLAM for mini-rovers with IR sensors only. 3rd European Conference on Mobile Robots (ECMR 2007), Friburg, Germany, 2007.

[28] Grasa OG, Civera J, Montiel JMM. EKF monocular SLAM with relocalization for laparoscopic sequences. In: IEEE International Conference on Robotics and Automation (ICRA), [Internet]. 2011 [cited 2013 Jun 6]. pp. 4816–4821. Available from: http://ieeexplore.ieee.org/xpls/abs_all.jsp?arnumber=5980059

[29] Civera J, Davison AJ, Montiel JMM. Unified inverse depth parametrization for monocular slam. In: Proceedings of Robotics: Science and Systems. 2006.

[30] Sola J, Vidal-Calleja T, Civera J, Montiel JMM. Impact of landmark parametrization on monocular EKF-SLAM with points and lines. Int J Comput Vis. 2012;97(3):339–368.

[31] Davison AJ. Real-time simultaneous localisation and mapping with a single camera. In: IEEE International Conference on Computer Vision. 2003. p. 1403–1410.

[32] Davison AJ, Murray DW. Mobile robot localisation using active vision. Proc 5th European Conference on Computer Vision, Freiburg, Germany, 1998; Vol.2: 809–825.

[33] Munguia R, Grau A. Camera localization and mapping using delayed feature initialization and inverse depth parametrization. In: IEEE Conference on Emerging Technologies and Factory Automation, 2007 ETFA. 2007. pp. 981–8.

[34] Loop C, Zhang Z. Computing rectifying homographies for stereo vision. In: IEEE Computer Society Conference on Computer Vision and Pattern Recognition. 1999. p. 131, Vol. 1.

[35] Fusiello A, Trucco E, Verri A. A compact algorithm for rectification of stereo pairs. Mach Vis Appl. 2000;12(1):16–22.

[36] Fusiello A, Irsara L. Quasi-euclidean uncalibrated epipolar rectification. In: ICPR 2008 19th International Conference on Pattern Recognition, [Internet]. 2008 [cited 2013 Dec 26]. pp. 1–4. Available from: http://ieeexplore.ieee.org/xpls/abs_all.jsp?arnumber=4761561

[37] Kang SB, Webb JA, Zitnick CL, Kanade T. A multibaseline stereo system with active illumination and real-time image acquisition. In: Proceedings of the Fifth International Conference on Computer Vision, [Internet]. 1995 [cited 2013 Nov 18]. pp. 88–93. Available from: http://ieeexplore.ieee.org/xpls/abs_all.jsp?arnumber=466802

[38] Fanto PL. Automatic Positioning and Design of a Variable Baseline Stereo Boom [Internet]. Virginia Polytechnic Institute and State University; 2012 [cited 2013 Nov 14]. Available from: http://scholar.lib.vt.edu/theses/available/etd-07252012-081926/

[39] Juan L, Gwun O. A comparison of sift, pca-sift and surf. Int J Image Process IJIP. 2009;3(4):143–152.

Chapter 6

Validation and Experimental Testing of Observers for Robust GNSS-Aided Inertial Navigation

Jakob M. Hansen, Jan Roháč, Martin Šipoš,
Tor A. Johansen and Thor I. Fossen

Additional information is available at the end of the chapter

http://dx.doi.org/10.5772/63575

Abstract

This chapter is the study of state estimators for robust navigation. Navigation of vehicles is a vast field with multiple decades of research. The main aim is to estimate position, linear velocity, and attitude (PVA) under all dynamics, motions, and conditions via data fusion. The state estimation problem will be considered from two different perspectives using the same kinematic model. First, the extended Kalman filter (EKF) will be reviewed, as an example of a stochastic approach; second, a recent nonlinear observer will be considered as a deterministic case. A comparative study of strapdown inertial navigation methods for estimating PVA of aerial vehicles fusing inertial sensors with global navigation satellite system (GNSS)-based positioning will be presented. The focus will be on the loosely coupled integration methods and performance analysis to compare these methods in terms of their stability, robustness to vibrations, and disturbances in measurements.

Keywords: inertial navigation, INS/GNSS, integration, nonlinear observers, extended Kalman filter, UAV

1. Introduction

Inertial navigation systems (INSs) are widely used with price being a crucial factor predetermining the application. In case of unmanned vehicles, "low-cost" or "cost-effective" systems are preferred in general applications. As long as low-cost inertial measurement units (IMUs) use micro-electro-mechanical system (MEMS)-based inertial sensors, they are small in dimension and light and are low power consuming, and thus their presence can be found for instance in mobile phones, terrestrial vehicles, robots, stabilized platforms as well as in unmanned aerial vehicles (UAVs), small aircraft, and satellites. Even if the applications are cost-effective, the performance commonly requires data fusion from various sources due to the inertial sensors' imperfections, such as insufficient resolution for navigation purposes, bias instabilities, noise, etc. Therefore, special data treatment is required. In sense of aerial applications, the usage of UAVs has increased rapidly in recent years. UAVs can be used in many applications [1, 2] fulfilling a broad spectrum of assignments in fields of reconnaissance, surveillance, search and rescue, remote sensing for atmospheric measurements, traffic monitoring, natural disaster response, damage assessment, inspection of power lines, or for aerial photography [2, 3]. These applications generally require navigation to be carried out which includes the position, velocity, and attitude (PVA) estimation [4], and thus cost-effective solutions have been commonly studied and implemented with advantage.

Current research and development in the area of low-cost navigation systems are focused on small-scale and integrated solutions [5]. As mentioned, as long as MEMS-based IMUs are used, the evaluation process requires data fusion from other aiding sources available. These sources stabilize errors in navigation solutions and thus increase navigation accuracy. Over the last few years, a solution for vehicle navigation without absolute position measurements provided by global positioning system (GPS) or radio frequency beacons has become very popular. For indoor or low-altitude navigation, it can use for example cameras, laser scanners, or odometers in terrestrial navigation [6, 7]. However, the solutions fusing inertial and GPS measurements are still preferable for aerial vehicles operating outside in large areas simply because of unblocked GPS signals. The implementation of other aiding sensors, such as magnetometer or pressure sensors, can further enhance the overall accuracy, reliability, and robustness of a navigation system [8, 9]. Attention is also paid to data processing algorithms used for PVA estimation, so that many literature sources can be found dealing with filtering techniques used for instance complementary filters [10], particle filters [11], or Kalman filters (KFs) [12, 13]. In the last named case, the extended KF (EKF) is used most of the time since it provides an acceptable accuracy with a reasonable computational load. Therefore, KF represents one of the most used algorithms for UAV attitude estimation (see comprehensive survey of estimation techniques in [14]) and is often complemented by other algorithms and decision-based aiding [15]. Since the accuracy of navigation systems is always directly related with the choice of sensors, the chapter also includes a short introduction on sensors suitable for cost-effective navigation systems as well as topics concerning stochastic sensor parameter evaluation methods and data pre-processing.

The contribution of this chapter is dedicated to comparison of two approaches suitable for navigation solutions and thus provides a clear understanding of the differences in the studied approaches. These approaches are tuned to satisfy a certain level of accuracy and applied on real flight data. The results are compared to an accurate referential attitude obtained from a multi-antenna GPS receiver. Such comparison with an independent referential system provides a thorough evaluation of performances of the studied approaches and shows their capabilities to handle sensors' imperfections and vibration impacts of harsh environment on the accuracy of attitude estimation in aerial applications.

2. Theoretical background

This section will include an introduction to sensors suitable for cost-efficient navigation systems, as well as topics concerning deterministic and stochastic sensor parameters. Moreover, a review of the current state of the art on state estimation will be included, while also the kinematic vehicle model and general assumptions are presented.

2.1. Inertial sensors

Navigation systems providing the tracking of an object's attitude, position, and velocity play a key role in a wide range of applications, e.g., in aeronautics, astronautics, robotics, automotive industry, underwater vehicles, or human motion observation. A basic technique to do so is via dead reckoning. One technique for dead reckoning is using an initial position, velocity, and attitude and consecutively updating the estimates based on acceleration and angular rate measurements. These measurements are generally provided by three axial accelerometer (ACC) and three axial angular rate sensors (gyros) forming a so-called inertial measurement unit (IMU). The inertial sensors have to be chosen according to required accuracy and economical aspects. The sensors are a major source of errors in navigation systems. Therefore, the type of application should be considered as well. The required accuracy related to various applications is shown in **Figures 1** and **2**, see [16], for gyroscopes and accelerometers, respectively.

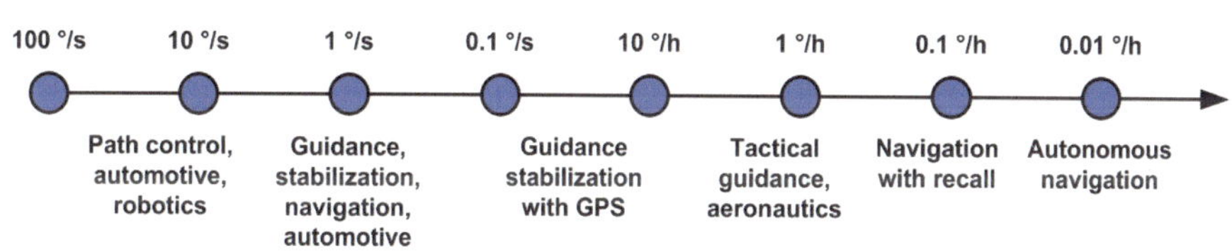

Figure 1. Bias instability of gyroscopes related to specified applications [16].

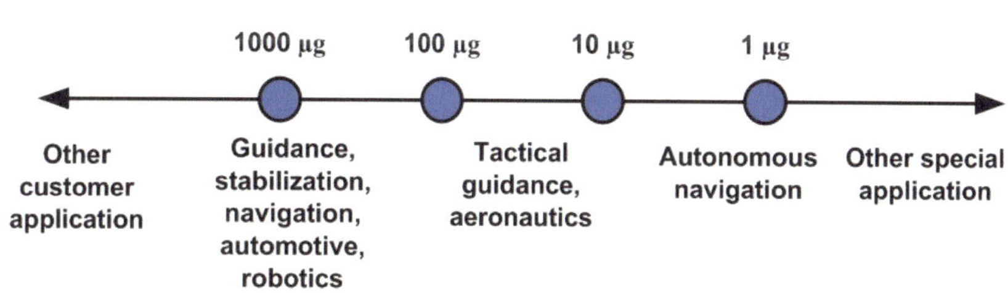

Figure 2. Bias instability of accelerometers related to specified applications [16].

Accuracy of performed navigation is related to inertial sensors' characteristics such as resolution and sensitivity and their imperfections in terms of bias instability, scale factor nonlinearity, and dependency of sensitive element on other quantities than just gyros or ACCs.

Unwanted deterministic behavior can be reduced by calibration, but stochastic parameters such as bias instability and initial offset can be described by statistical values.

An uncompensated accelerometer bias, b_{ACC}, contributes in position error based on $\Delta p = 1/2 b_{ACC} t^2$, where t is time. Thus, even small deviations in sensed acceleration will cause unbound error in position with time. For instance, when $b_{ACC} = 0.1$ mg is considered, it leads to position errors of 0.05 m after 10 s, and an error of 177 m after 600 s. Generally speaking, all navigation systems dedicated for aircraft need to fulfill the requirement of a maximum error of 1 nautical mile per hour. In navigation attitude, accuracy also plays a key role. When attitude is considered, the azimuth is the most difficult parameter to estimate, since ACCs can be used for pith and roll angle compensation. For azimuth compensation, other sources can be used, e.g., magnetometers and GNSS, but these are not inertial sensors and thus depend on environmental conditions.

For aircraft navigation it is, according to **Figures 1** and **2**, required to use gyros with the precision better than 1 deg/h and ACC not more than 10μg, see [17]. The higher precision, the more expensive the device is. The other aspect, which has to be taken into account, is to check if a particular device is in solid state or is using moving parts. **Figures 3** and **4** depict the current state of gyroscope and accelerometer technology. Recently, there has been a progress on MEMS-based sensors increasing the sensitivity of the gyroscopes to compete with the fiber optic gyros (FOGs). However, FOGs still provide better stability than the MEMS-based gyros. For ACCs, it has become very popular to use quartz resonator in applications with high-accuracy requirements due to its costs. If higher accuracy is still required, only mechanical pendulous rebalance (servo) ACCs have to be utilized. According to **Figures 3** and **4**, one can see that mechanical gyros and ACCs still satisfy the precision requirements best; nevertheless, there is a trend to replace them with solid-state devices for their better reliability, stability, and mean-time-before-failure parameter. Therefore, in the following paragraphs, solid-state devices will be considered.

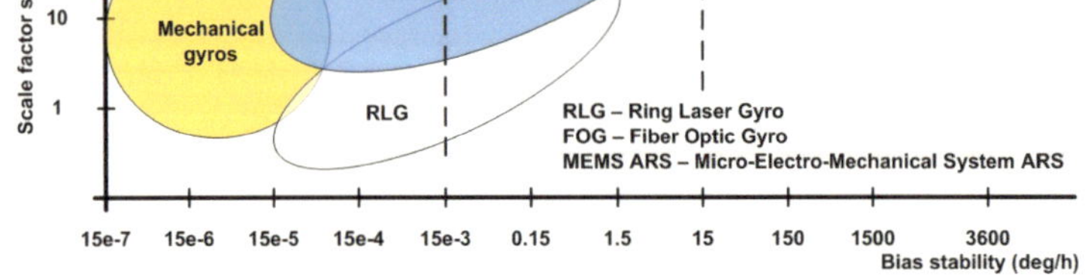

Figure 3. Gyro technology [17].

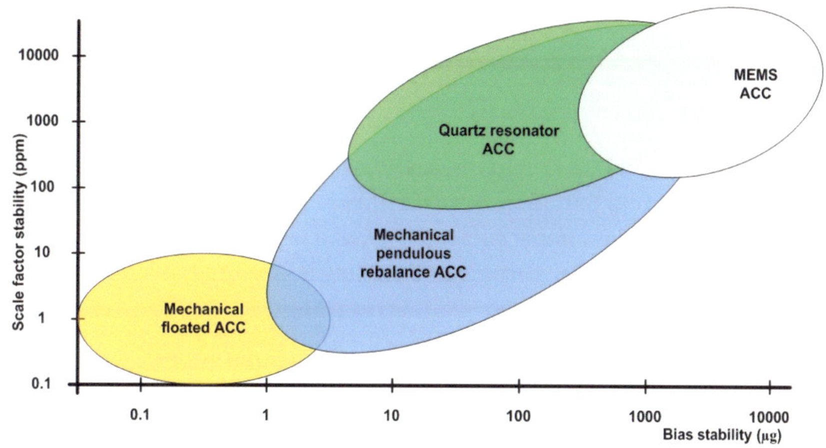

Figure 4. Accelerometer types and their performances [17].

The most precise device for angular rate measurements is a ring laser gyroscope (RLG), which has stability better than 0.1 deg/h and resolution better than 10^{-6} deg/s. In the case of ACC, the most precise existing device is a servo ACC with a resolution about 1 μg. These devices would have been ideal for all applications, if they were not so expensive. Due to this reason, other systems, such as micro-electro-mechanical system (MEMS)-based ones, have been used in cost-effective applications, such as navigation of small aircraft and unmanned vehicles both terrestrial and aerial. MEMS sensors offer reduced power consumption, weight, manufacturing and assembly costs, and increased system design flexibility. Reducing the size and weight of a device allows multiple MEMS components to be used to increase functionality, device capability, and reliability. In contrast, MEMS-based systems might suffer from low resolution, noisy output, bias instability, temperature dependence, etc. Nevertheless, their applicability in navigation is wide due to fast technology improvements, applied data processing algorithms, and aiding systems. In navigation, aiding systems are commonly used to provide corrections for position, velocity, or attitude. Those systems might be based for instance on GNSS, electrolytic tilt sensors, pressure-based altimeter, odometer, laser scan, or vision-based odometry usage.

In cost-effective applications, MEMS-based devices are preferred. Therefore, their usage has to be accompanied by modern methods of signal and data processing, algorithms for their calibration, parameters identification, and fusion.

2.1.1. Gyroscopes

The basic parameters generally provided in product datasheets include dynamic range, initial sensitivity, nonlinearity, alignment error, initial bias error, in-run bias instability, angular random walk, linear acceleration effect on bias, and rate noise density. According to these parameters, the following types of gyros can be defined: low-cost, moderate-cost, and high-performance gyros. When looking at datasheets, the in-run bias stability provides the information about the best sensor performance corresponding to the gyro resolution floor.

Unfortunately, there are other exhibiting error factors which affect a gyro performance. In all cases, the gyro noise and its frequency dependency have to be taken into account and handled. Stochastic sensor parameters can be generally estimated via power spectral density (PSD) analyses or via Allan variance analysis (AVAR).

In the case of low- and moderate-cost gyros, scale factor, alignment error, and null bias errors accompanied by parameters variation over a temperature range highly decrease the gyro performance. To minimize their impacts, it is required to perform the calibration within which a correction table or polynomial correction function is acquired. More details about the calibration methods and procedures can be found in [18].

The other perspectives of the gyro performance are produced by the fact that it does not measure just a rotational rate, but also its sensitive element has linear acceleration and g^2 sensitivities. It is caused by the asymmetry of a mechanical design and/or micromachining inaccuracies, and it can vary from design to design. Due to Earth's 1 g field of gravity, according to [19], it can suffer from large errors when uncompensated. In the case of low-cost gyros, the g and g^2 sensitivities are not specified because their design is not optimized for a vibration rejection. They can have g sensitivity about 0.3°/s/g. Therefore, looking at a bias instability in these cases is almost pointless due to the high effect of this vibration behavior. Higher performance gyros improve the vibration rectification by design so the g sensitivity can go down to about 0.01°/s/g. To further decrease this sensitivity, anti-vibration mounts might be applied. Nevertheless, these anti-vibration mounts are very difficult to design, because they do not have a flat response over a wide frequency range, and they work particularly poorly at low frequencies. Moreover, their vibration reduction characteristics change with temperature and life cycle.

2.1.2. Accelerometers

MEMS technology-based accelerometers (ACCs) in the navigation area measure specific force mainly on a capacitive principle and/or on a vibrating differential structure. The basic parameters include measurement range, nonlinearity, sensitivity, initial bias error, in-run bias instability, noise density, bandwidth of frequency response, alignment error, and cross-axis sensitivity. In the case of multi-axial ACCs, the *z*-axis often has a different noise and bias performance. Vibrations will affect ACCs; however, if the frequency spectrum is adequate for the application, no problem arises from this point of view. The current MEMS technology cannot compete with high-performance types and cannot be implemented to stand-alone inertial navigation systems due to their low resolution, bias instability, and insufficient noise level reduction. Generally, this type of ACC is used in navigation systems in which a GNSS receiver is also implemented to compensate position errors, or in attitude and heading reference systems in which the position is not required, and thus ACCs are used just for an attitude measurement done according to Earth's field 1 g sensing. ACC stochastic parameters can be estimated the same way as in the case of gyros via PSD or AVAR. More details about the calibration and estimation of deterministic errors and consecutive analyses can be found in [18, 20].

http://dx.doi.org/10.5772/63575

2.2. Inertial sensors' stochastic parameters

Various methods for stochastic error estimation and sensor modeling exist, for example PSD and auto correlation function (ACF) which are straightforward; however, these methods cannot clearly distinguish different characters of noise error sources inside the data without understanding the sensor model and its state-space representation [21]. On the contrary, AVAR is a time-domain approach to analyze time series of data from the noise terms' point of view. The AVAR was introduced by Allan in 1966 [22]. Originally, it was oriented at the study of oscillator stability; however, after its first publication, this kind of analysis was adopted for general noisy data characterization. Because of the close analogies to inertial sensors, the AVAR has been also included in the IEEE standards, e.g. [23–25], and that is why AVAR has become a standard tool for inertial sensors' noise analysis. As described in [26], the AVAR technique provides several significant advantages over the others. Traditional approaches, such as computing the sampled mean and variance from a measured data set, do not reveal the underlying error sources. Although the combined PSD/ACF approach provides a complete description of error sources, the results are difficult to interpret.

The AVAR and its results are related to five noise terms, defined in **Table 1**, whose typical performance can be seen in **Figure 5**. This kind of error sources can be identified in inertial sensor output, and whose estimation can lead to error suppression in the data [25, 27]. The five basic noise terms correspond to the following random processes: angle/velocity random walk, rate/acceleration random walk, bias instability, quantization noise, and drift rate ramp. Values of particular coefficients denoted in **Table 2** in the last column can be observed in AVAR deviation plots in a time instance corresponding to the one indicated in particular brackets. Furthermore, this basic set of random processes is extended by the sinusoidal noise and exponentially correlated (Markov) noise [27]. Generally, a total noise error can be classified as a sum of individual independent noise errors [25], and the total variance can be expressed as

$$\sigma_{total}^2 = \sigma_Q^2 + \sigma_{ARW}^2 + \sigma_{BIN}^2 + \sigma_{RRW}^2 + \sigma_{RR}^2, \tag{1}$$

where the abbreviations correspond to **Table 1**.

Type of noise	Abbreviation	Curve slope	Value of coefficients
Quantization noise	Q	−1	$Q = \sigma(\sqrt{3})$
Angular/velocity random walk	ARW	−1/2	$N = \sigma(1)$
Flicker noise/bias instability	BIN	0	$B = \sigma_{\min} / 0.664$
Rate/acceleration random walk	RRW	+1/2	$K = \sigma(3)$
Rate ramp noise	RR	+1	$R = \sigma(\sqrt{2})$

Table 1. Summary of error sources and their characterization [25].

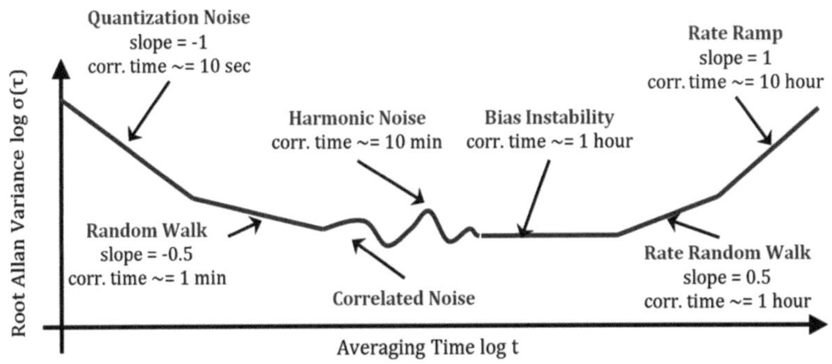

Figure 5. Allan variance/deviation plot [25].

	Parameter	DMU10 (Silicon S.)	AHRS M3 (InnaLabs)	MPU9150 (InvenSense)	DSP-3100 (KVH)	INN-204 (InnaLabs)
GYR	ARW($^\circ/\sqrt{h}$)	0.35	2.50	0.25	0.03	NA
	BIN ($^\circ/h$)	7.53	55.72	15.06	0.60	NA
ACC	VRW ($m/s/\sqrt{h}$)	0.05	0.06	0.08	NA	0.01
	BIN (mg)	0.04	0.06	0.06	NA	0.01

Table 2. Stochastic parameters of inertial sensors according to AVAR.

2.3. State-of-the-art of state estimators

Many physical systems are considered partly closed systems with no means of measuring internal signals, where only the inputs and outputs are available. However, it is often of interest to know the current value of the internal states, e.g., such that appropriate action can be taken using a control element. There might be multiple internal states and only a few measured outputs due to lack of appropriate sensors, cost, or insufficient data rate. The state estimation problem describes the need to estimate variables of interest in a model that is not otherwise directly observable [28].

The model describes the dominating dynamics of the system, while less important dynamics might be removed for simplicity. The states for navigation systems often include position, linear velocity, and attitude of the vehicle, while the inclusion of auxiliary states is possible. These auxiliary states might describe specific force of the vehicle or inertial sensor errors [29].

State estimators consist of two categories: "filters" and "observers." Filters take the stochastic approach to find the current state values and consider the measurement and state noise as well as the covariance estimate of the states. Observers use a deterministic approach based on control theory focusing on the stability of the proposed equations. In both cases, a model of the physical system is duplicated to propagate the states while comparing with the system outputs. In the literature, the terms "filter" and "observer" are used somewhat interchangeably.

The following sections will include a review of previous work on Kalman filters and nonlinear observers for navigation.

2.3.1. Kalman filter review

Modern filtering theory began around 1959–1960 with publications by Swerling [30] and Kalman [31], presenting error propagation methods using a minimum variance estimation algorithm for linear systems. The discrete method presented by Kalman has received large attention and is now a coined term in multiple fields [32].

The Kalman filter (KF) introduced a recursive algorithm for state estimation, which is optimal in the sense of minimum variance or least square error. Changing from analytical solutions to a recursive algorithm had the advantage of being easily implementable in digital computers. Another advantage was that the previous non-recursive estimation methods used the entire measurement set, whereas the recursive estimation of the KF uses current measurements as well as prior estimates to propagate the states from an initial estimate. The KF is therefore more computational efficient as it can discard previous measurements and update the state estimates with only the present measurements [29]. The KF theory was expanded upon in 1961 by Kalman and Bucy [33], introducing a continuous time variant.

The Kalman filter's stochastic approach to the state estimation problem assumes noise on the measurements as well as the state equations of the filter. It is a well-established state estimation approach [34] which excels in working with normal-distributed inputs characterized by their mean and covariance values and a linear time-varying state space model in its basic form. The KF is an estimator, which provides estimates of the state as well as its uncertainty [35]. The measurements have to be functions of the states, as the residual measurement (the difference between measured and estimated measurements) is used to update the states and keep them from diverging. The process and measurement noise is assumed to be Gaussian white noise. In some cases where the noise of the physical system cannot be confirmed to be white, the KF might be augmented, by so-called "shaping filters, with additional linear state equations to let the colored noise be driven by Gaussian white noise [28]. In addition to the recursive estimation of the model states, the Kalman filter also propagates a covariance matrix describing the uncertainties of the state estimates as well as the correlation between the various states [29].

Even though the Kalman filter was designed for linear systems, it can be applied to nonlinear systems without changing the structure or the operational principles. However, the optimality of minimal variance of the errors is lost, and the filter is no longer an optimal estimator. The kinematic equations are inherently nonlinear and thus must be addressed by nonlinear techniques or approximations to maintain the performance and stability of the modeled system. Nonlinear problems are commonly handled by the linearized KF (LKF), extended KF (EKF), or sample-based methods such as unscented KF (UKF) [36, 37]. Probably the most popular of the mentioned methods is the EKF, which has been applied in an enormous number of applications where it achieved excellent performance [38]. The EKF linearizes the model around an estimate of the current mean using multivariate Taylor expansions to adapt to the nonlinear model; however, this makes the EKF more susceptible to errors in the initial estimates and modeling errors compared to the KF.

The KF and EKF are seen as the standard theory and are therefore used as baseline for comparison when developing new methods. The KF and its variants are widely used in the navigation-related literature where a few examples are mentioned. An introduction to choice of states and sensor alignment consideration can be found in [39], while [40] considers alternative attitude error representations. For extensive details on Kalman filtering, see [28, 29, 38, 9, 41–43]. Among the extensions to nonlinear systems, other examples can be found, e.g., [44] where a method for evaluating the linearization quality is presented alongside a Kalman filter extension for nonlinear systems. The unscented Kalman filter (UKF) is an extension to nonlinear systems that does not involve an explicit Jacobian matrix, see e.g., [45]. Studies on time-correlated noise, as opposed to the white noise assumption, without state augmentation have been carried out in [46, 47]. The adaptive Kalman filter might be used in applications where tuning of the Kalman filter is uncertain at initialization, see [48–50]. If the application is not real-time critical, such as surveying, the estimate can be enhanced by use of a smoother. In [51], a forward smoother was proposed, while in [52] a backward smoother was introduced. When nonlinear systems are considered, another alternative to the EKF is the particle filter. Particle filters are based on sequential Monte Carlo estimation algorithms, which compared to the Kalman filter are more computationally demanding; however, they are noise distribution independent, see e.g. [37, 53–56]. The advantage of the particle filter is its use in nonlinear non-Gaussian systems. However, since this approach is computationally heavy in current navigation systems, it is not often used. Therefore, the particle filter is considered outside the scope of this chapter.

2.3.2. Nonlinear observer review

In comparison to the Kalman filter, the nonlinear observers have a shorter history, motivated by drawbacks of the KF when applied to nonlinear systems. These drawbacks include unclear convergence properties for nonlinear systems, difficulty of tuning, and large computational load.

Nonlinear observers are contrary to the Kalman filters based on a deterministic approach. The noise is not assumed to have specific properties, except that the difference between the measured and estimated signal is smallest when the estimate reflects the true signal. Like the Kalman filter, nonlinear observers commonly utilize an injection term consisting of the difference between measured and estimated system output to drive the observer states toward the true values.

The field of nonlinear observers has expanded within groups dealing with specific problems. Nonlinear attitude estimation has been the focus of extensive research [57–61]; see in particular [13] for an extensive survey including EKF methods. One method used has centered on the comparison of two attitude measurement vectors in the BODY frame with two corresponding vectors in an Earth-fixed or inertial frame. One such attitude observer was proposed by [62] and was later expanded upon by [63] to include a gyro bias estimate. A vector-based attitude observer was proposed by [64] which depended on inertial measurements, magnetometer readings, and GNSS velocity measurements. Expanding on this framework [65] introduced an attitude observer that utilized the derivative of the GNSS velocity as the vehicle acceleration allowing for comparison with accelerometer measurements.

Where the Kalman filter computes new gains for each iteration, some nonlinear observers have proven convergence with fixed or slowly time-varying gains, e.g. [66]. This is a computational improvement as the gain determination is the dominating computational burden, see [28; Section 5.6.1].

One of the design challenges of nonlinear observers is the requirement for proven stability. The optimality of the Kalman filter may give the user confidence in the performance and stability of the filter. However, for nonlinear observers, the stability should be explicitly stated, as the gain usually comes without any optimality guarantee. The aim is to be robust toward disturbances and poor initial estimates.

The field of nonlinear observers is recent and rapidly expanding. A few publications within navigation are mentioned here. Considerations of a nonlinear attitude estimator for use on a small aircraft was presented in [67], while a globally exponentially stable observer for long baseline navigation was presented in [68] with clock bias estimation in a tightly coupled system.

2.4. Models and preliminaries

Estimating the position, linear velocity, and attitude (PVA) of a vehicle is commonly achieved through INS/GNSS integration, where the inertial navigation system (INS) consists of an inertial measurement unit (IMU) providing inertial navigation between updates from a GNSS receiver. The GNSS receiver usually has a lower sample rate than the IMU and is used to update the PVA estimates by correcting for the drift of the inertial sensors.

2.4.1. Notation

A column vector $x \in \mathbb{R}^3$ is denoted $x = [x_1; x_2; x_3]$ with its transpose x^T and Euclidean vector norm $\|x\|_2$. The same notation is used for matrices where the induced norm is used. The skew symmetric matrix of a vector x is given as

$$S(x) = \begin{bmatrix} 0 & -x_3 & x_2 \\ x_3 & 0 & -x_1 \\ -x_2 & x_1 & 0 \end{bmatrix}.$$

A unit quaternion, $q = [r_q; s_q]$, consisting of a real part, $r_q \in \mathbb{R}$, and a vector part, $s_q \in \mathbb{R}^3$, has $\|q\|_2 = 1$. A vector $x \in \mathbb{R}^3$ can be represented as a quaternion with zero real part; $\bar{x} = [0; x]$. The product of two quaternions q_1 and q_2 is the Hamiltonian product denoted by $q_1 \otimes q_2$. The cross-product of two vectors is then represented by $\times$.

Superscripts are used to signify which coordinate frame a vector is decomposed in. Rotation between two frames may be represented by a quaternion, q_a^c, describing the rotation from coordinate frame a to c, with the corresponding rotation matrix $R_a^c = R(q_a^c) \in SO_3$, where $R(q_a^c) := I + 2s_{q_a^c} S(r_{q_a^c}) + 2S(r_{q_a^c})^2$. The attitude can also be expressed as Euler angles; $\Theta_{ca} = [\phi, \theta, \psi]^T$ with the associated rotation matrix $R_a^c = R(\Theta_{ca})$:

$$R(\Theta_{ca}) = \begin{bmatrix} c\psi c\theta & -s\psi c\phi + c\psi s\theta s\phi & s\psi s\phi + c\psi c\phi s\theta \\ s\psi c\theta & c\psi c\phi + s\phi s\theta s\psi & -c\psi s\phi + s\theta s\psi c\phi \\ -s\theta & c\theta s\phi & c\theta c\phi \end{bmatrix},$$

where the sine and cosine functions have been abbreviated, e.g. $\sin(\phi) = s\phi$. There is no difference between using $R(q_a^c)$ and $R(\Theta_{ca})$ in terms of transformation. The Euler angles are often preferred as they are more intuitively interpreted; however, they suffer from singularities (e.g., at pitch of 90°) which the quaternion representation avoids [40].

Various reference frames will be used where the Earth-centered-Earth-fixed (ECEF) frame will use notation e, while b will be used for BODY frame, n for North East down (NED), and i for Earth-centered inertial (ECI) frame.

2.4.2. Kinematic vehicle model

The vehicle model describes position, p^e, and linear velocity, v^e, as well as the attitude described either as Euler angles or quaternions. The gyroscope bias, b^b, is also included in the model and is assumed to be slowly time varying.

The kinematic equations describing the vehicle motion are [22]

$$\dot{p}^e = v^e, \tag{2}$$

$$\dot{v}^e = -2S(\omega_{ie}^e)v^e + f^e + g^e(p^e), \tag{3}$$

$$\dot{q}_b^e = \frac{1}{2} q_b^e \otimes \bar{\omega}_{ib}^b - \frac{1}{2} \bar{\omega}_{ie}^e \otimes q_b^e, \tag{4}$$

$$\dot{R}_b^n = R_b^n S(\omega_{ib}^b), \tag{5}$$

$$\dot{b}^b = 0, \tag{6}$$

where Earth rotation rate around the ECEF z-axis, ω_{ie}^{e}, is known and $g^e(p^e)$ is the plumb bob gravity vector at the vehicle position. The specific force acting on the vehicle is described by f^b. The rotational matrix from body to NED frame is denoted as R_b^n. The kinematic equations for the Euler angle and quaternion propagations have been included; however, at implementation, only one of Eqs. (4) or (5) should be used.

The vehicle is described with 6 degrees of freedom (DOF) where the BODY-frame vectors are defined as shown in **Figure 6**. An UAV is used as an example with a position vector $p^b = (x^b, y^b, z^b)$ and Euler vector $\Theta_n^b=[\phi, \theta, \psi]^T$.

Figure 6. 6 DOF UAV in the BODY frame.

The UAV can be seen as an example of a high dynamic vehicle and can be considered a challenging navigation environment, as it allows for rapid changes in attitude and heading.

2.4.3. Measurement assumptions

It is assumed that the vehicle is equipped with an IMU and a GNSS receiver, as well as a magnetometer. The following measurements are assumed available:

- Position measurement, $p^e_{\mathrm{GNSS}} = p^e$,
- Specific force measurement, $f^b_{\mathrm{IMU}} = f^b$, acting on the vehicle,
- Biased angular velocity measurement, $\omega^b_{\mathrm{ib,IMU}} = \omega^b_{\mathrm{ib}} + b^b$,
- Magnetic field measurement, $m^b_{\mathrm{IMU}} = m^b$, of the Earth's magnetic field at vehicle position.

Furthermore, knowledge of bounds on the magnitude of specific force and gyro bias, denoted as M_{f} and M_{b} respectively, is assumed. The natural magnetic field at any position is assumed known in NED and ECEF frame, as m^n and m^e, respectively.

3. Practical approaches to the state estimation problem

In the following, the two state estimators will be introduced. First, the EKF will be presented where the Allan variance is applied to tune the covariance matrices. Second, the nonlinear observer will be introduced consisting of two parts: a nonlinear attitude estimator and a translational motion observer.

3.1. Extended Kalman filter

One solution for estimating position, linear velocity, and attitude is to utilize an IMU/GPS loosely coupled integration scheme, shown in **Figure 7**, which can be done by an EKF (for details about the EKF algorithm see [69]). The 12-dimensional state vector contains position in NED frame, velocity in the BODY frame, attitude, and gyro biases. The estimation is done with respect to a control vector u consisting of measured specific forces and angular rates, and to a measurement vector y defined in Eq. (8). The measurement vector in Eq. (8) is three dimensional and includes a GNSS position in NED frame. The state and measurement vectors are given as

$$x = [p_N, p_E, p_D, v_x, v_y, v_z, \phi, \theta, \psi, b_{\omega x}, b_{\omega y}, b_{\omega z}]^{\mathrm{T}}, \tag{7}$$

$$y = [p_N, p_E, p_D]^{\mathrm{T}}, \tag{8}$$

where $p^n = (p_{\mathrm{N}}, p_{\mathrm{E}}, p_{\mathrm{D}})$ are components of position vector in NED frame; $v^b = (v_x, v_y, v_z)$ are the BODY-frame components of velocity vector; the gyroscope bias is decomposed into $b^b = (b_{\omega x}, b_{\omega y}, b_{\omega z})$; and y is the measurement vector.

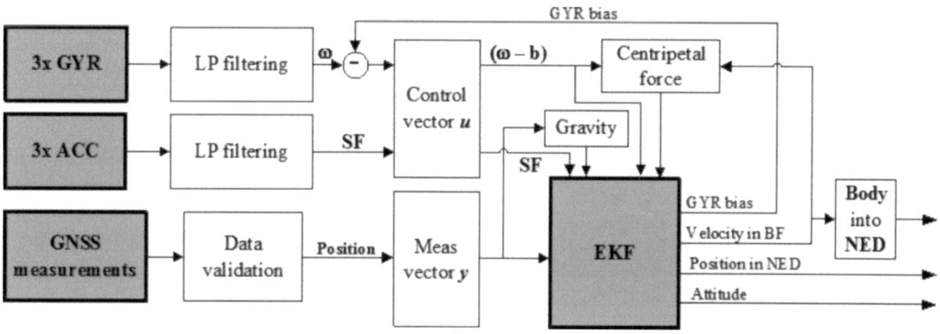

Figure 7. IMU/GNSS loosely coupled integration scheme.

(SF stands for specific force and LP is low pass)

The system function $f(x, u)$ propagates the state x and input u (i.e., accelerations and angular rates) and the measurement function $h(x)$ is used to update the EKF state with measurements (i.e., GNSS-based position). They are defined as

$$f(x,u)=\begin{bmatrix} R_b^n v^b \\ f^b + v^b \times (\omega_{ib,IMU}^b - b^b) - (R_b^n)^T g^n \\ \begin{bmatrix} 1 & \sin(\phi)\tan(\theta) & \cos(\phi)\tan(\theta) \\ 0 & \cos(\phi) & -\sin(\phi) \\ 0 & \sin(\phi)\sec(\theta) & \cos(\phi)\sec(\theta) \end{bmatrix} (\omega_{ib,IMU}^b - b^b) \\ b^b \end{bmatrix}, \tag{9}$$

$$h(x) = p^n, \tag{10}$$

where $g^n=[0;0;g]^T$ is the gravity vector and $\sec(\cdot)$ is the secant function. The process and measurement noise covariance matrices Q and R for the model used in Eqs. (9) and (10) are defined as follows:

$$Q = \mathrm{diag}(0_3, \sigma_v^2, \sigma_\omega^2, \sigma_{b_\omega}^2), \qquad R = \mathrm{diag}(\sigma_p^2) \tag{11}$$

where *diag* denotes a diagonal matrix, and σ_*^2 is a vector of element-wise squared standard deviations for velocity, angular rate, gyroscope biases, and GNSS-based position.

The system dynamic and measurement models are

$$x_k = \Phi_{k-1}x_{k-1} + \Gamma_{k-1}u_{k-1} + w_{k-1} \tag{12}$$

$$z_k = H_k x_k + v_k \tag{13}$$

where the state transition matrix, Φ_k , and control matrix, Γ_k, are linearized from Eq. (9) with respect to the state and input vector, respectively. The initial conditions are $E\langle x_0\rangle = \hat{x}_0$ and $E\langle x_0, x_0^T\rangle = P_0$. The process noise and measurement noise are assumed to satisfy $w_k \sim N(0, Q_k)$ and $v_k \sim N(0, R_k)$.

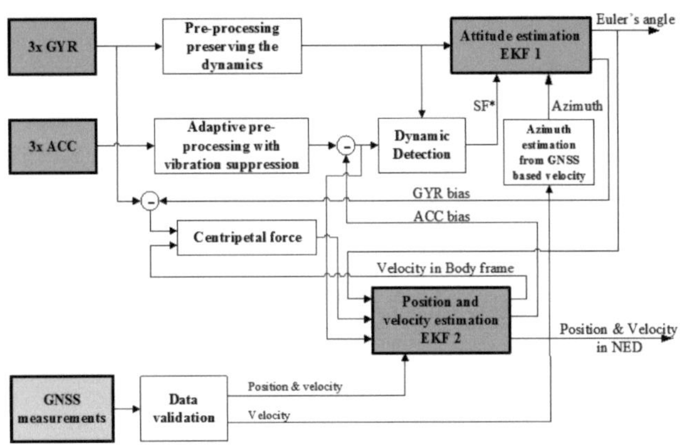

Figure 8. Enhanced IMU/GNSS integration scheme.

The state vector and covariance matrices are described by a priori and posteriori part denoted with superscripts − and +, respectively. A discrete form of the time and correction update of the state vector and covariance matrix are given as [69]:

$$\hat{x}_k^- = \Phi_{k-1}\hat{x}_{k-1}^+ + \Gamma_k u_k \tag{14}$$

$$P_k^- = \Phi_{k-1}P_{k-1}^+\Phi_{k-1}^T + Q_{k-1} \tag{15}$$

$$K_k = P_k^- H_k^T (H_k P_k^- H_k^T + R_k)^{-1} \tag{16}$$

$$\hat{x}_k^+ = \hat{x}_k^- + K_k(z_k - H_k\hat{x}_k^-) \tag{17}$$

$$P_k^+ = (I - K_k H_k)P_k^- \tag{18}$$

where the Kalman gain matrix is denoted by K_k, while the observation matrix, H_k, is the linearization of Eq. (10) with respect to the state vector.

The advantage of this approach is a straightforward implementation and satisfactory navigation performance. The motion model is corrected for the centrifugal force; therefore, it is highly preferable for applications where this force occurs frequently, e.g., during a turn. However, even when properly tuned, the estimates strongly rely on the GNSS signal availability. In the case of blocked or lost GNSS signal, the estimates begin to diverge quickly and results may become unstable as long as the filter parameters are not adjusted.

To enable enhanced positioning function of the solution within GNSS outages, it is recommended to integrate accelerometer biases into a state vector and add attitude corrections obtained from accelerometer readings. This extended solution might use the integration scheme depicted in **Figure 8**.

3.1.1. AVAR applied in Kalman filter modeling

Having a state transition matrix and observation matrix defined is one issue, but it is also very important to set driving noise in accordance to expected situation. The AVAR analysis can help to do so in terms for inertial sensors. A comparison of different grades inertial sensors from their stochastic parameters point of view is shown in **Figures 9** and **10** and further summarized in **Table 2** where two basic parameters, i.e., angular random walk (ARW) or velocity random walk (VRW), and bias instability (BIN) are picked up.

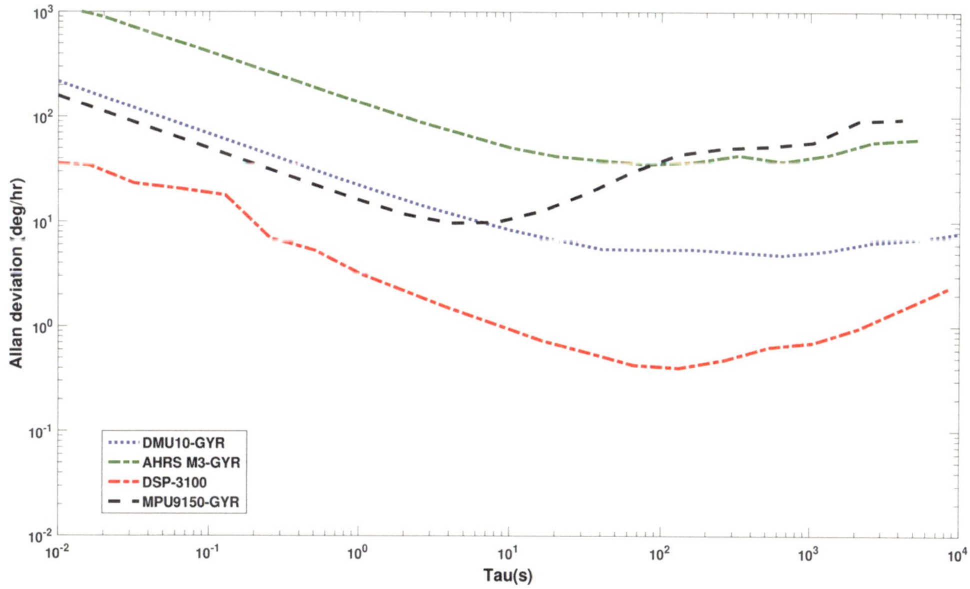

Figure 9. AVAR analysis—Allan deviation plot of several MEMS based gyros (DMU10, DSP3100—tactical grade gyros, AHRS M3, MPU9150—commercial grade gyros).

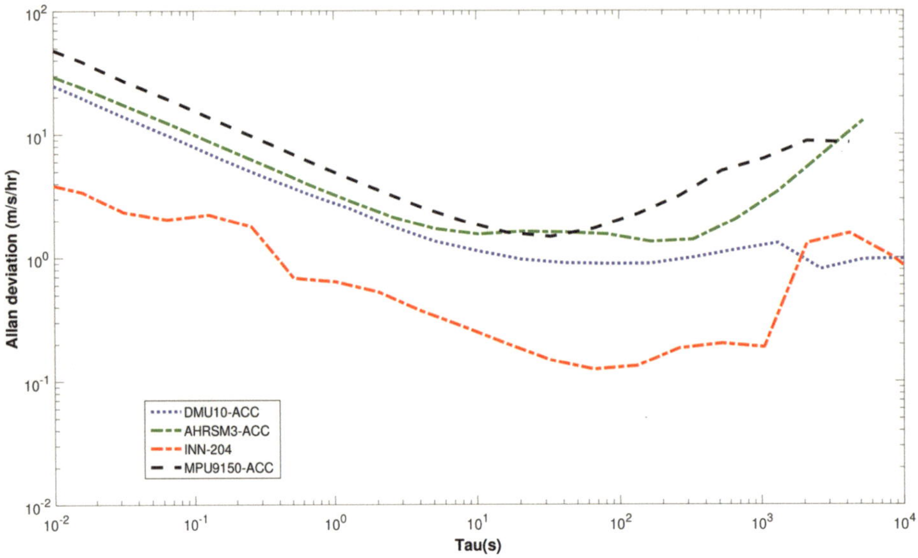

Figure 10. AVAR analysis—Allan deviation plot of several MEMS-based accelerometers (DMU10, INN204— tactical grade, AHRS M3, MPU9150—commercial grade).

According to **Table 1**, one can see that analyzed sensors differ, but the parameters still correspond to its sensor grade. However, it needs to be highlighted that the cheapest IMU MPU-9150 has parameters close to the boundary between commercial and tactical grade. So it leads to considering this unit as suitable for navigation solutions in robotics for its price and performance. From other perspectives, it is hard to say anything about the gyro sensitivity to "g" in the form of vibrations which might degrade the overall performance. Parameters ARW/VRW and BIN are generally used in covariance matrix of process noise, of course according to the particular model utilized.

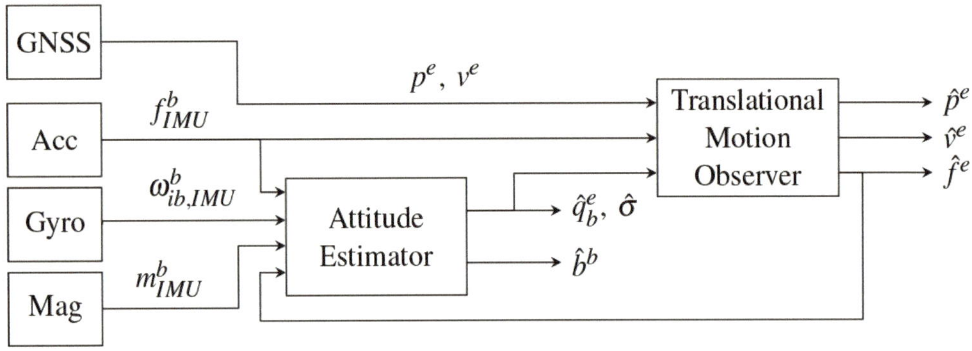

Figure 11. Block diagram of nonlinear observer.

http://dx.doi.org/10.5772/63575

3.2. Nonlinear observer

Numerous nonlinear observers have been proposed for integration of IMU and GNSS data; however, in the following, the observer proposed in [66] will be considered, estimating position and velocity in the ECEF frame and describing the attitude as a unit quaternion.

The nonlinear observer presented here has a modular structure consisting of an attitude estimator and a translational motion observer. The two subsystems are interconnected with feedback of the specific force estimate from the motion observer to the attitude estimator. An advantage of the modular design is that the stability properties of the subsystems can be investigated individually leading to the stability result of the entire observer system using nonlinear stability theory, see [66] for further details. The observer structure is depicted in **Figure 11**.

The subsystems will be explained in detail in the following sections.

3.2.1. Attitude estimation

The attitude of the vehicle is denoted by a unit quaternion, $\hat{q}_b^e$, describing the rotation between the BODY and ECEF frame. The attitude observer is a complementary filter fusing data from an accelerometer, magnetometer, and gyroscope to estimate the vehicle attitude. The nonlinear observer estimating the attitude and gyro bias, $\hat{b}^b$, is given as [58, 61]:

$$\dot{\hat{q}}_b^e = \frac{1}{2}\hat{q}_b^e \otimes \left(\bar{\omega}_{ib,IMU}^b - \bar{\hat{b}}^b + \bar{\hat{\sigma}} \right) - \frac{1}{2}\bar{\omega}_{ie}^e \otimes \hat{q}_b^e, \tag{19}$$

$$\dot{\hat{b}}^b = \mathrm{Proj}(\hat{b}^b, -k_I\hat{\sigma}), \tag{20}$$

$$\hat{\sigma} = k_1 v_1^b \times R(\hat{q}_b^e)^{\mathrm{T}} v_1^e + k_2 v_2^b \times R(\hat{q}_b^e)^{\mathrm{T}} v_2^e, \tag{21}$$

where k_1, k_2, and k_I are positive and sufficiently large tuning constants. The Proj(·,·) operator limits the gyro bias estimate to a sphere with radius $\hat{M}_b$ where $\hat{M}_b > M_b$. The injection term, $\hat{\sigma}$, consists of two vectors in BODY frame and their corresponding vectors in ECEF frame. There are various ways of choosing these vectors, but here they will be considered as

$$v_1^b = \frac{f^b}{\| f^b \|_2}, \quad v_2^b = \frac{m^b}{\| m^b \|_2} \times v_1^b, \quad v_1^e = \frac{\hat{f}^e}{\| \hat{f}^e \|_2}, \quad v_2^e = \frac{m^e}{\| m^e \|_2} \times v_1^e, \tag{22}$$

where the specific force estimate, $\hat{f}^e$, will be supplied by the translational motion observer, while the magnetic field vector, m_e, is assumed known and depends on the vehicle position.

3.2.2. Translational motion observer

The translational motion observer estimates the position and velocity of the vehicle by using injection terms based on the difference between measured and estimated position. The measurements are traditionally provided by a GNSS receiver. Additionally, the observer also estimates the specific force of the vehicle by introducing an auxiliary state, ξ. The translational motion observer is described by

$$\dot{\hat{p}}^e = \hat{v}^e + \theta K_{pp}(p^e_{GNSS} - \hat{p}^e), \tag{23}$$

$$\dot{\hat{v}}^e = -2S(\omega^e_{ie})\hat{v}^e + \hat{f}^e + g^e(\hat{p}^e_r) + \theta^2 K_{vp}(p^e_{GNSS} - \hat{p}^e), \tag{24}$$

$$\dot{\xi} = -R(\hat{q}^e_b)S(\hat{\sigma})f^b + \theta^3 K_{\xi p}(p^e_{GNSS} - \hat{p}^e), \tag{25}$$

$$\hat{f}^e = R(\hat{q}^e_b)f^b + \xi. \tag{26}$$

The observer can also be stated with additional injection terms using GNSS velocity; however, it was shown in [70] that the velocity part of the injection term is not required to achieve stability.

The constant $\theta \geq 1$ serves as a tuning parameter that should be sufficiently large to guarantee global stability of the interconnection of the translational motion observer and attitude observer. The gain matrices, K_{pp}, K_{vp}, and $K_{\xi p}$ can be chosen to satisfy $A-KC$ being Hurwitz with

$$A = \begin{bmatrix} 0 & I_3 & 0 \\ 0 & 0 & I_3 \\ 0 & 0 & 0 \end{bmatrix}, \quad C = [I_3 \quad 0 \quad 0], \quad K = \begin{bmatrix} K_{pp} \\ K_{vp} \\ K_{\xi p} \end{bmatrix}. \tag{27}$$

The translational motion observer is similar to the EKF, and the gain matrix K can therefore be determined similarly to the EKF gain, by solving a Riccati equation. However, an advantage of this nonlinear observer is that the gain matrix is not required to be determined in each iteration, but rather on a slower time scale, see [66]. This time scale can be slower than the GNSS update rate, decreasing the computational load substantially. The load can be further reduced by considering the implementation as a fixed gain observer only determining the gains at the initialization phase. It has been shown in [71] that time-varying gains aid in sensor noise suppression and gives faster convergence.

http://dx.doi.org/10.5772/63575

4. Experimental verification

Experimental measurements from flights with a fixed-wing Bellanca Super Decathlon XXL unmanned aerial vehicle (UAV) are used to verify and compare the performance of the EKF and nonlinear observer. The UAV (shown in **Figure 6**) is equipped with an ADIS 16375 IMU, supplying acceleration and angular rate measurements, a HMR2300 magnetometer, and a GARMIN 18X GPS-receiver. The inertial data are sampled at 100 Hz, while the position measurements are sampled at 5 Hz. Furthermore, the UAV is equipped with a Polar X2@e (Septentrio) GPS system consisting of three antennas, placed at the wing tips and tale, providing attitude and position estimates. The estimates of the Septentrio system are considered highly accurate and therefore used as a reference for comparison with the estimates of the EKF and nonlinear observer.

The accuracy of the reference represented by the three-antenna Septentrio GPS receiver is evaluated based on the distances among three antennas and manufacturer documentation. The resulting attitude accuracy of 1σ is 0.2° in roll angle, 0.6° in pitch angle, and 0.3° in yaw angle. Accuracy in horizontal position in standalone application is 1.1m, with SBAS corrections about 0.7m.

The goal of the following experimental verification is to compare the performance of the proposed EKF and nonlinear observer. Two datasets were used in the verification where it was desired to use the same tuning for both datasets to ensure that the state estimators were not tuned specifically for a single dataset. The performance has been evaluated by comparison with the reference position, speed, and attitude. For each of the datasets, figures showing the estimation errors are depicted comparing the state estimators.

4.1. Parameters and tuning variables

The state estimators have several parameters and tuning variables to be determined, which will be presented and explained here. In the case of coinciding, naming subscripts "EKF" and "NO" will be used.

Tuning the EKF consists of choosing reasonable Q_{EKF} and R_{EKF} matrices. While the R_{EKF} matrix relies on the accuracy of the GNSS receiver, the Q_{EKF} matrix describes the expected process noise due to accelerometer and gyro noise and instabilities and can be tuned for the application. Here they are initialized as $R_{EKF} = 14.40\, I_3$, with $Q_{EKF} = \text{blkdiag}\,(0_3, 0.0962 I_3, 0.0761 \cdot 10^{-4} I_2, 0.3047 \cdot 10^{-4}, 3.0462 \cdot 10^{-10} I_3)$. The state vector is driven by measured angular rates and specific force by inertial sensors having particular noise parameters. These parameters should be involved in the Q_{EKF} matrix.

Two versions of the fixed gain nonlinear observer are presented for comparison with the difference being the vectors used for attitude estimation: a magnetometer implementation (denoted as NLO-Mag) and a version with velocity vectors (denoted as NLO-Vel). The NLO-Vel version substitutes v_2^b and v_2^e in Eq. (22) with $v_2^b = [1;0;0]$ and $v_2^e = \hat{v}^e / \| \hat{v}^e \|_2$. This approach assumes the heading and course to be coinciding, which is mostly true for straight flight trajectories, ensuring uniform semi-global exponential stability through [72]. For flights including numerous turns, a magnetometer might be preferred as loitering, and cross-winds could affect the heading assumption.

The nonlinear observers include bound parameters which should be chosen sufficiently large $M_b = 0.0087$, while the remaining parameters are $k_1 = 0.2$, $k_2 = 0.05$, $\theta = 1$, $k_I = 0.00005$. The fixed gains are $K_{pp} = 0.38I_3$, $K_{vp} = 0.44\,I_3$ and $K_{\xi p} = 0.14\,I_3$. For the NLO-Vel, the attitude injection gain is substituted for $k_{2v} = 0.01$.

The initial values of the state vectors are chosen from the first available measurements and are similar for the three estimators (EKF, NLO-Mag, and NLO-Vel). It is important to tune the three state estimators equally and thoroughly to keep the comparison fair.

4.2. Results

Two datasets are available using the same UAV and sensor suite. The proposed state estimators are tested on both datasets to verify that they are not tuned exclusively for one dataset. The inertial measurements are preprocessed with a low-pass filter whose bandwidth is set according to vibration spectrum. Based on measured real-flight data obtained by the IMU unit and FFT analyses shown in **Figure 12**, the bandwidth of the fifth order low-pass filter was set to 5 Hz.

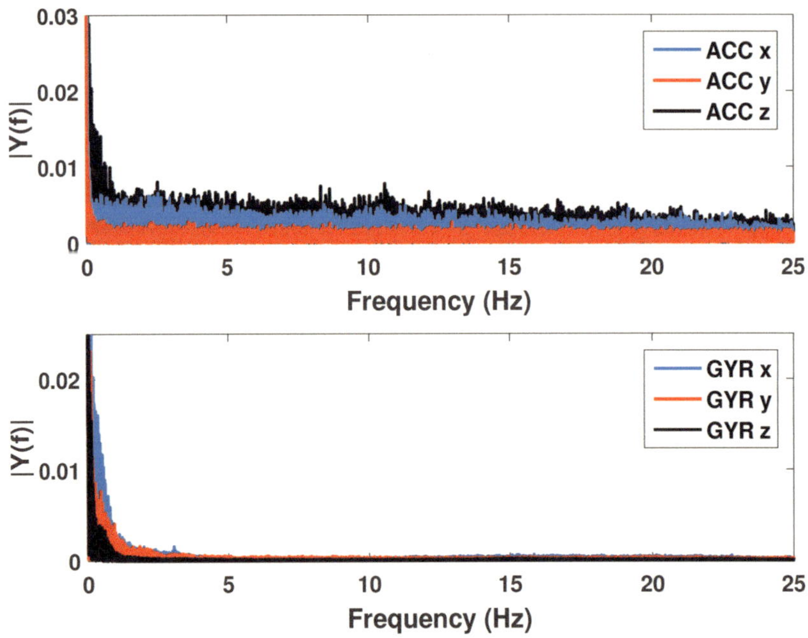

Figure 12. Amplitude spectrum of the IMU measurements from the UAV flight. Top—accelerations, bottom—angular rates.

http://dx.doi.org/10.5772/63575

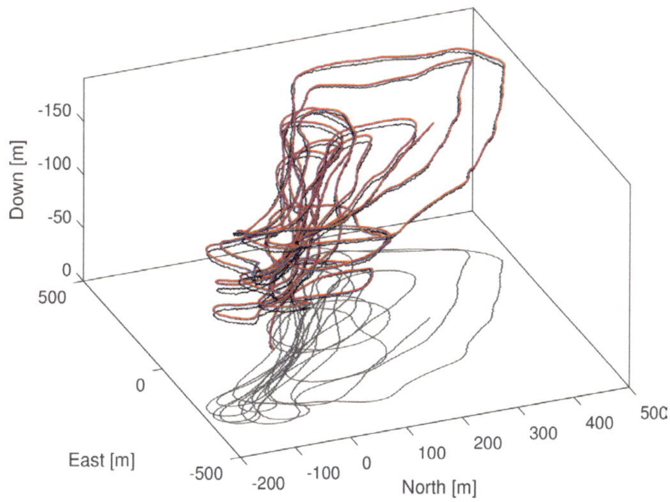

Figure 13. Vehicle trajectory (Dataset 1): Septentrio (black), EKF (red), NLO-Mag (blue dashed), NLO-Vel (green).

Results of dataset 1 can be seen in **Figures 13–16**, while the results of dataset 2 are shown in **Figures 17–20**. The occasional gap in the attitude error is due to temporary loss of reference. The findings are evaluated and summarized in **Table 2** which compares the two estimators during the two flights.

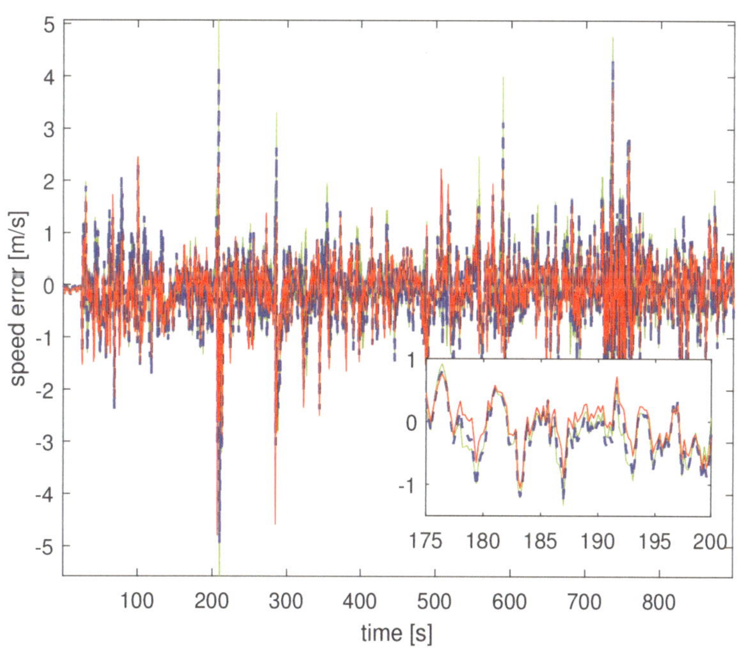

Figure 14. Speed estimation error (Dataset 1): EKF (red), NLO-Mag (blue dashed), NLO-Vel (green).

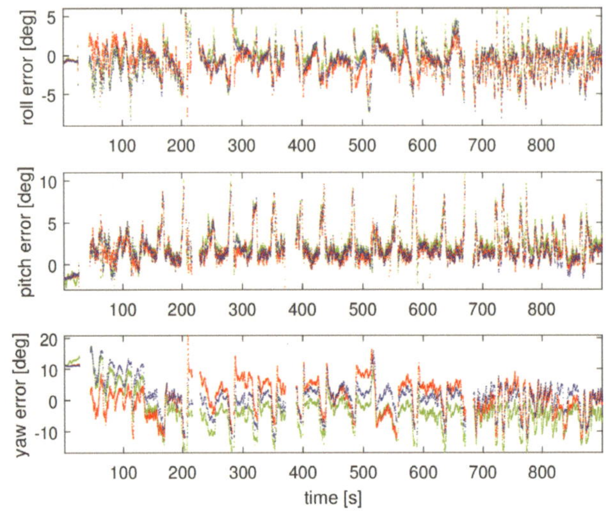

Figure 15. Attitude error (Dataset 1): EKF (red), NLO-Mag (blue dashed), NLO-Vel (green).

The trajectory of flight 1 is shown in **Figure 13**, covering an area of approximately 0.7 km^2 with a maximum altitude of 170 m. The estimation errors of speed, attitude, and position are shown in **Figures 14–16**, where the speed estimation error is centered around zero and includes a zoomed view for clarification. The attitude errors shown in **Figure 15** have similar behavior for roll and pitch for the state estimators, whereas the nonlinear yaw estimate has some systematic offset. The position errors of **Figure 16** are very similar for the state estimators attesting that the nonlinear observers have comparable results to the EKF.

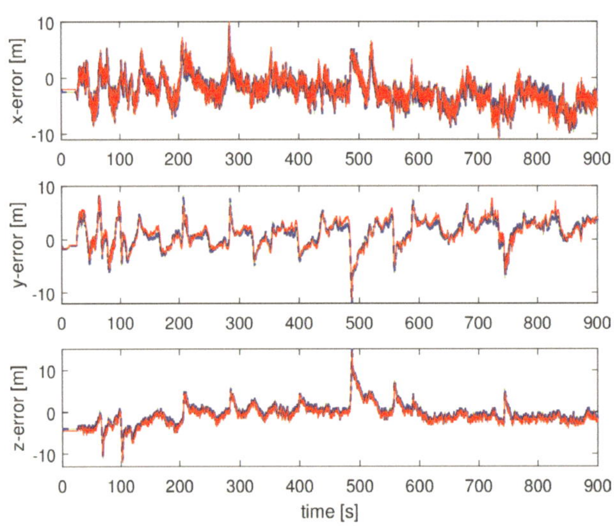

Figure 16. Position estimation error (Dataset 1): EKF (red), NLO-Mag (blue dashed), NLO-Vel (green).

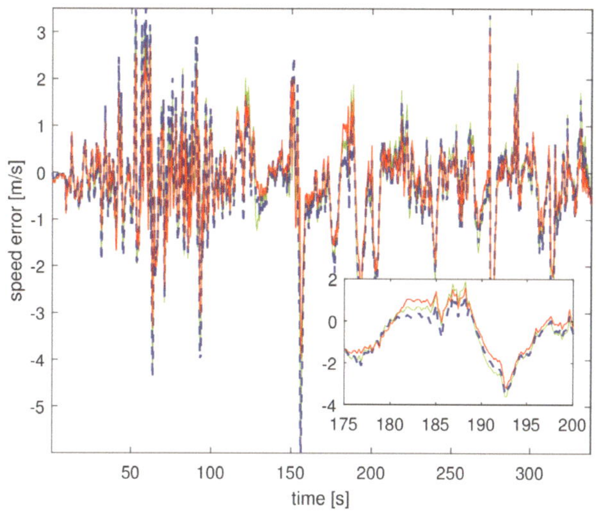

Figure 17. Speed estimation error (Dataset 2): EKF (red), NLO-Mag (blue dashed), NLO-Vel (green).

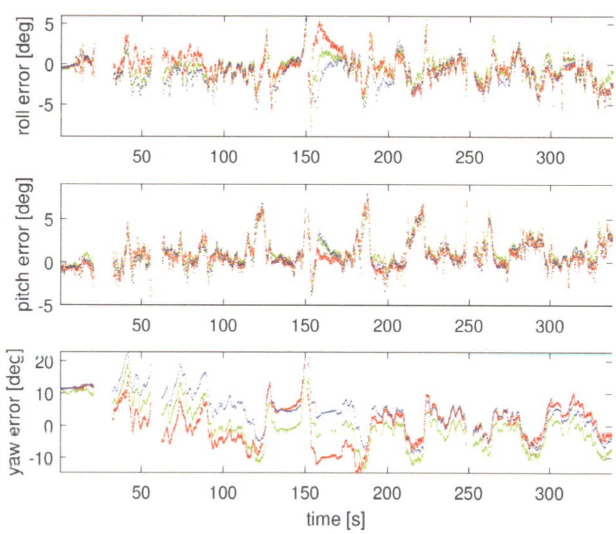

Figure 18. Attitude error (Dataset 2): EKF (red), NLO-Mag (blue dashed), NLO-Vel (green).

The second dataset consisted of approximately a third of the amount of measurements compared to Dataset 1. The speed and attitude estimation errors are shown in **Figures 17** and **18**, with comparable performance between the EKF and nonlinear observers. The position errors depicted in **Figure 19** show that an offset is present between the estimates, although the estimates follow the same pattern. Finally, the gyro bias estimates are shown in **Figure 20**. As there are no reference for the gyro biases, these are included to show the similarities across the state estimators.

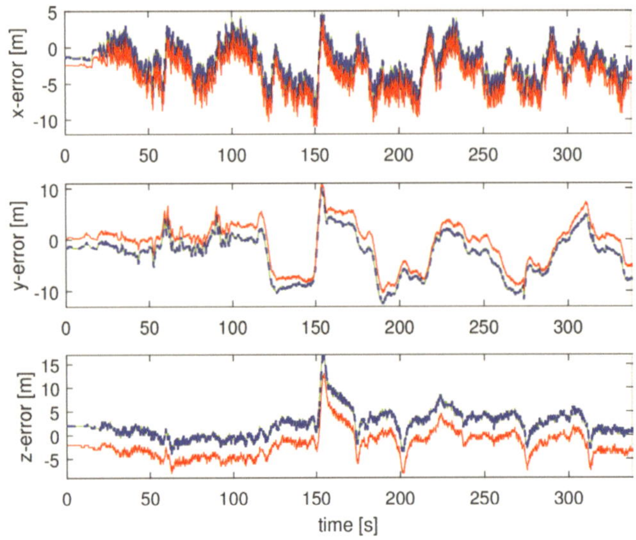

Figure 19. Position estimation error (Dataset 2): EKF (red), NLO-Mag (blue dashed), NLO-Vel (green).

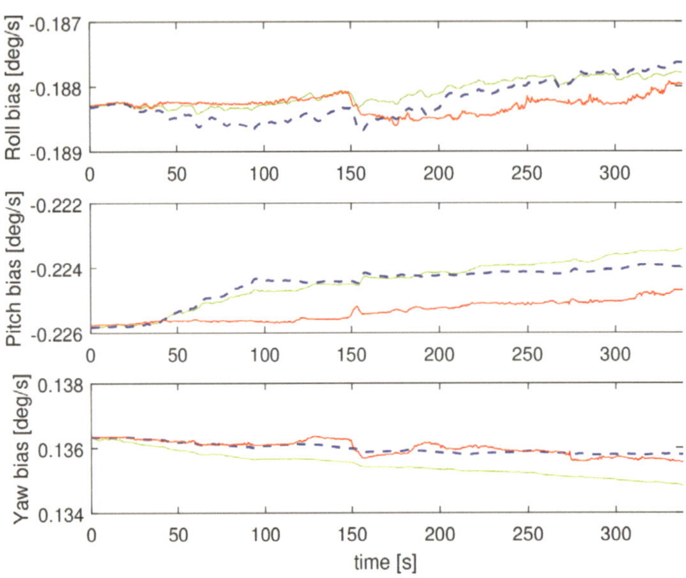

Figure 20. Gyro bias estimation (Dataset 2): EKF (red), NLO-Mag (blue dashed), NLO-Vel (green).

In summary, according to **Table 3** and previous figures, the EKF and nonlinear observers are seen to have similar performance during both compared flights. The differences can be assumed negligible, and real flight conditions are considered. The attitude estimates shown in **Figures 15** and **18** are very alike and correspond well to the reference, although the nonlinear yaw estimation is seen to have a systematic difference.

		EKF			NLO-Mag			NLO-Vel		
Dataset 1	POS RMS:	3.43	2.71	2.62	3.37	2.62	2.50	3.37	2.63	2.50
	POS STD:	2.60	2.44	2.46	2.48	2.41	2.44	2.48	2.42	2.44
	ATT RMS:	1.83	2.59	5.50	2.02	2.69	5.46	1.93	3.01	6.88
	ATT STD:	1.67	1.81	5.42	1.84	1.90	5.28	1.90	1.92	6.15
	SPE RMS:		0.60			0.67			0.69	
	SPE STD:		0.59			0.66			0.69	
Dataset 2	POS RMS:	4.43	4.40	3.43	3.38	5.00	3.63	3.37	5.00	3.63
	POS STD:	2.53	4.35	2.86	2.46	4.20	2.82	2.46	4.20	2.82
	ATT RMS:1.76	1.87	6.39	2.09	1.87	7.85	1.56	1.94	6.36	
	ATT STD:	1.70	1.66	6.34	1.73	1.67	6.28	1.46	1.59	6.36
	SPE RMS:		0.86			1.06			1.05	
	SPE STD:		0.83			1.02			1.02	

Table 3. Observer performance comparison (NED position in m, attitude in degree, and speed in m/s).

The position estimation errors depicted in **Figures 16** and **19** are within the expected bounds. From **Table 3**, it can be concluded that the three state estimators have good performances with little variation between the estimators. It can further be concluded that the tuning used gave good results for both datasets.

5. Conclusive remarks

Two methods for INS/GNSS integration have been investigated and compared: an extended Kalman filter using a 12-state vector and a nonlinear observer. The advantages and drawbacks of the methods have been presented and experimentally verified on flight data from a fixed-wing UAV. A reference system consisting of three-antenna GNSS receiver with the antennas placed at the tail and each wing tip was use for performance comparison of the presented state estimators.

The inertial sensors used in the integration schemes are considered low-cost variants with respect to the reference system utilized. As the performance of the presented methods estimates the position, linear velocity, and attitude reasonably close to the reference, it is concluded that the methods are able to overcome the vibrations, disturbances, and bias drift connected to low-cost sensors in reasonable manner and thus provide sufficiently stable and accurate navigation solution.

Acknowledgements

This work was partially supported by the EEA/Norway grant No. NF-CZ07-ICP-3-2082015 supported by the Ministry of Education, Youth and Sports of the Czech Republic and named Enhanced Navigation Algorithms in Joint Research and Education, and partially by Norwegian Research Council (projects no. 221666 and 223254) through the NTNU Centre of Autonomous Marine Operations and Systems (NTNU AMOS) at the Norwegian University of Science and Technology.

Author details

Jakob M. Hansen[1*], Jan Roháč[2], Martin Šipoš[2], Tor A. Johansen[1] and Thor I. Fossen[1]

*Address all correspondence to: jakob.mahler.hansen@itk.ntnu.no

1 Norwegian University of Science and Technology, Department of Engineering Cybernetics, Centre for Autonomous Marine Operations and Systems, Trondheim, Norway

2 Czech Technical University in Prague, Faculty of Electrical Engineering, Department of Measurement, Prague, Czech Republic

References

[1] B. Liu, Z. Chen, X. Liua, and F. Yang. An Efficient Nonlinear Filter for Space Attitude Estimation. *International Journal of Aerospace Engineering*, 1–11, 2014.

[2] M. Tarhan and E. Altug. EKF based Attitude Estimation and Stabilization of a Quadrotor UAV Using Vanishing Point in Catadioptric Images. *Journal of Intelligent & Robotic Systems*, 62(3-4):587–607, 2011.

[3] Derek B. Kingston and Randal W. Beard. Real-Time Attitude and Position Estimation for Small UAVs Using Low-Cost Sensors. *American Institute of Aeronautics and Astronautics, "Unmanned Unlimited"*, 1–9, 2004.

[4] Y. S. Suh. Attitude Estimation by Multiple-Mode Kalman Filters. *IEEE Transactions on Industrial Electronics*, 53(4):1386–1389, 2006.

[5] S. Leutenegger and R. Siegwart. A Low-Cost and Fail-Safe Inertial Navigation System for Airplanes. *IEEE Conference on Robotics and Automation*, 2012.

[6] A. Bry, A. Bachrach, and N. Roy. State Estimation for Aggressive Flight in GPS-denied Environments Using Onboard Sensing. *Proceedings of IEEE International Conference on Robotics Automation*, 2012.

[7] S. Weiss, M. Achtelik, M. Chli, and R. Siegwart. Versatile Distributed Pose Estimation and Sensor Self-calibration for an Autonomous MAV. *International Conference on Robotics and Automation (ICRA)*, 2012.

[8] J. Calusdian, X. Yun, and E. Bachmann. Adaptive-gain Complementary Filter of Inertial and Magnetic Data for Orientation Estimation. *International Conference on Robotics and Automation*, 2011.

[9] D. Zachariah and M. Jansson. Self-Motion and Wind Velocity Estimation for Small-Scale UAVs. *International Conference on Robotics and Automation*, 2011.

[10] M. Euston, P. Coote, R. Mahony, J. Kim, and T. Hamel. A Complementary Filter for Attitude Estimation of a Fixed-wing UAV. *IEEE International Conference on Intelligent Robots and Systems*, 2008.

[11] M Sotak, M. Sopata, and F. Kmec. Navigation Systems using Monte Carlo Method. *Guidance, Navigation and Control Systems*, 2006.

[12] A. Bachrach, S. Prentice, R. He, and N. Roy. RANGE - Robust Autonomous Navigation in GPS-denied Environments. *Journal of Field Robotics*, 28(5):644–666, 2011.

[13] John L. Crassidis, F. Landis Markley, and Yang Cheng. Survey of Nonlinear Attitude Estimation Methods. *Journal of Guidance, Control, and Dynamics*, 30(1):12–28, 2007.

[14] R. Munguia and A. Grau. A Practical Method for Implementing an Attitude and Heading Reference System. *International Journal of Advanced Robotic Systems*, 11(62), 2014.

[15] H. G. de Marina, F. J. Pereda, J. M. Giron-Sierre, and F. Espinosa. UAV Attitude Estimation Using Unscented Kalman Filter and TRIAD. *IEEE Transactions on Industrial Electronics*, 59(11):4465–4474, 2012.

[16] N. M. Barbour. Inertial Navigation Sensors. *NATO*. USA: Charles Stark Draper Laboratory. Cambridge, *RTO-EN-SET-116*, 2011.

[17] G. T. Schmidt. INS/GPS Technology Trends. *NATO*. USA: Massachusetts Institute of Technology. Lexington, *RTO-EN-SET-116*, 2011.

[18] J. Roháč, M. Šipoš, and J. Šimánek. Calibration of the Low-Cost Triaxial Inertial Sensors. *IEEE Instrumentation & Measurement Magazine*, (18)6:32–38, 2015.

[19] Analog Devices Inc. http://www.analog.com. Technical report, Analog Devices Inc.

[20] M. Šipoš, P. Paces, J. Roháč, and P. Novacek. Analyses of Triaxel Accelerometer Calibration Algorithms. *IEEE Sensors Journal*, 12(5):1157–1165, 2012.

[21] N. El-Sheimy, H. Hou, and X. Niu. Analysis and Modeling of Inertial Sensors Using Allan Variance. *IEEE Transactions on Instrumentation and Measurement*, 57:140–149, 2008.

[22] D. W. Allan. Statistics of Atomic Frequency Standards. *Proceedings of the IEEE*, 2(54): 221–230, 1966.

[23] IEEE Std. 1293. IEEE Standard Specification Format Guide and Test Procedure for Linear, Single-Axis, Nongyroscopic Accelerometers. Technical report, Institute of Electrical and Electronics Engineers, Available: ISBN 0-7381-1430-8 SS94679.

[24] IEEE Std. 528. IEEE Standard for Inertial Sensor Terminology. Technical report, Institute of Electrical and Electronics Engineers, Available: ISBN 0-7381-3022-2.

[25] IEEE Std. 647. IEEE Standard Specification Format Guide and Test Procedure for Single Axis Laser Gyros. Technical report, Institute of Electrical and Electronics Engineers.

[26] C. N. Lawrence. On the Application of Allan Variance Method for Ring Laser Gyro Performance Characterization. *No. UCRL-ID–115695.* Lawrence Livermore National Lab., 1993.

[27] M. Sotak. Determining Stochastic Parameters Using an Unified Method. *Acta Electrotechnica et Informatica*, 9(2):59–63, 2009.

[28] Jay A. Farrell. *Aided Navigation: GPS with High Rate Sensors*. McGraw Hill, 2008.

[29] P. D. Groves. *Principles of GNSS, Inertial, and Multisensor Integrated Navigation Systems*. Artech House, 2013.

[30] P. Swerling. First Order Error Propagation in a State-Wise Smoothing Procedure for Satellite Observations. *Journal of Astro Sciences*, (6):1–31, 1959.

[31] R. E. Kalman. A New Approach to Linear Filtering and Prediction Theory. *Transactions on American Society of Mechanical Engineers, Series D, Journal of Basic Engineering*, (82):35–45, 1960.

[32] E. Hendricks, O. Jannerup, and P. H. Sørensen. *Linear Systems Control - Deterministic and Stochastic Methods*. Springer, 2008.

[33] R. E. Kalman and R. S. Bucy. New Results in Linear and Prediction Theory. *Transactions on American Society of Mechanical Engineers, Series D, Journal of Basic Engineering*, 83:95–108, 1961.

[34] G. Dissanayake, S. Sukkarieh, E. Nebot, and H. Durrant-Whyte. The Aiding of a Lowcost Strapdown Inertial Measurement Unit Using Vehicle Model Constraints for Land Vehicle Applications. *IEEE Transactions on Robotics and Automation*, 17(5):731–747, 2001.

[35] Y. Bar-Shalom, X. R. Li, and T. Kirubarajan. *Estimation with Applications to Tracking and Navigation*. John Wiley & Sons, 2004.

[36] Z. Chen. Bayesian Filtering: From Kalman Filters to Particle Filters, and Beyond. *Statistics*, 182(1):1–69, 2003.

[37] F. Gustafsson, F. Gunnarsson, N. Bergman, U. Forssell, J. Jansson, R. Karlsson, and P. J. Nordlund. Particle Filters for Positioning, Navigation and Tracking. *IEEE Transactions on Signal Processing*, 50:425–437, 2002.

[38] M. S. Grewal, L. R. Weill, and A. P. Andrews. *Global Positioning Systems, Inertial navigation, and Integration.* John Wiley & Sons, Ltd., 2007.

[39] L. Stimac and T. A. Kennedy. Sensor Alignment Kalman Filters for Inertial Stabilization Systems. *Proceedings of IEEE PLANS*, 321–334, 1992.

[40] F. L. Markley. Attitude Error Representations for Kalman Filtering. *Journal of guidance, control, and dynamics*, 26(2):311–317, 2003.

[41] Thor I. Fossen. *Handbook of Marine Craft Hydrodynamics and Motion Control.* John Wiley & Sons, Ltd., 2011.

[42] R. G. Brown and Y. C. Hwang. *Introduction to Random Signals and Applied Kalman Filtering.* John Wiley & Sons, Inc. New York, 1998.

[43] A. Gelb, J. F. Kasper Jr., R. A. Nash Jr., C. F. Price, and A. A. Sutherland Jr. *Applied Optimal Estimation.* MIT Press. Boston, MA, 1988.

[44] A. Draganov, L. Haas, and M. Harlacher. The IMRE Kalman Filter - A New Kalman Filter Extension for Nonlinear Applications. *Proceedings of IEEE/ION PLANS*, 428–440, 2012.

[45] S. J. Julier and J. K. Uhlmann. A New Extension of the Kalman Filter to Nonlinear Systems. *Proceedings of SPIE Signal Processing, Sensor Fusion, and Target Recognition VI*, 3068:182–193, 1997.

[46] S. F. Schmidt. Application of State-Space Methods to Navigation Problems. *Advances in Control Systems: Theory and Applications*, Vol. 3, Academic Press. New York, pp. 293–340, 1966.

[47] M. G. Petovello, K.O'Keefe, G. Lachapelle, and M. E. Cannon. Consideration of Time-Correlated Errors in a Kalman Filter Application to GNSS. *Journal of Geodesy*, 83(1):51–56, 2009.

[48] R. K. Mehra. Approaches to Adaptive Filtering. *IEEE Symposium on Adaptive Processess, Decision and Control*, 1970.

[49] A. H. Mohammed and K. P. Schwarz. Adaptive Kalman Filtering for INS/GPS. *Journal of Geodesy*, 73:193–203, 1999.

[50] D. T. Magill. Optimal Adaptive Estimation of Sampled Stochastic Processes. *IEEE Transactions on Automatic Control*, AC-10:434–439, 1965.

[51] D. C. Fraser and J. E. Potter. The Optimum Linear Smoother as a Combination of Two Optimum Linear Filters. *IEEE Transactions on Automatic Control*, 7:387–390, 1969.

[52] H. E. Rauch, F. Tung, and C. T. Striebel. Maximum Likelihood Estimates of Linear Dynamic Systems. *AIAA Journal*, 3:1445–1450, 1965.

[53] N. J. Gordon, D. J. Salmond, and A. F. M. Smith. A Novel Approach to Nonlinear/Non-Gaussian Bayesian State Estimation. *Proceedings of IEE Radar Signal Process*, 140:170–113, 1993.

[54] B. Ristic, S. Arulampalam, and N. J. Gordon. *Beyond the Kalman Filter: Particle Filters for Tracking Applications*. Artech house, 2004.

[55] A. Doucet, Nando de Freitas, and N. Gordon. *Sequential Monte Carlo Methods in Practice*. New York: Springer, 2001.

[56] A. Doucet and A. M. Johansen. A Tutorial on Particle Filtering and Smoothing: Fifteen years later. *Oxford Handbook of Nonlinear Filtering (C. Crisan and B. Rozovsky, Oxford)*, pp. 656–704, 2011.

[57] J. Thienel and R. M. Sanner. A Coupled Nonlinear Space Attitude Controller and Observer with an Unknown Constant Gyro Bias and Gyro Noise. *IEEE Transactions on Automatic Control*, 48:2011–2015, 2003.

[58] R. Mahony, T. Hamel, J. Trumpf, and C. Lageman. Nonlinear Attitude Observer on SO(3) for Complementary and Compatible Measurements: A Theoretical Study. *IEEE Conference on Decision and Control*, 6407–6412, 2009.

[59] P. Batista, C. Silvestre, and P. Oliveira. Ges Attitude Observers - Part I: Single Vector Observations. *IFAC World Congress*, 2991–2996, 2011.

[60] P. Batista, C. Silvestre, and P. Oliveira. Ges Attitude Observers - Part II: Multiple General Vector Observations. *IFAC World Congress*. Milan, Italy, 2985–2990, 2011.

[61] H. F. Grip, T. I. Fossen, T. A. Johansen, and A. Saberi. Attitude Estimation Using Biased Gyro and Vector Measurements with Time-Varying Reference Vectors. *IEEE Transactions on Automatic Control*, 57:1332–1338, 2012.

[62] S. Salcudean. A Globally Convergent Angular Velocity Observer for Rigid Body Motion. *IEEE Transactions on Automatic Control*, 36:1493–1497, 1991.

[63] B. Vik and Thor I. Fossen. A Nonlinear Observer for GPS and INS Integration. *Proceedings of Conference on Decision and Control*, 3:2956–2961, 2001.

[64] T. Hamel and R. Mahony. Attitude Estimation on SO(3) Based on Direct Inertial Measurements. *Proceedings of IEEE International Conference on Robotics Automation*, 2170–2175, 2006.

[65] Minh-Duc Hua. Attitude Estimation for Accelerated Vehicles Using GPS/INS Measurements. *Control Engineering Practice*, 18:723–732, 2010.

[66] H. F. Grip, T. I. Fossen, T. A. Johansen, and A. Saberi. Nonlinear Observer for GNSS-Aided Inertial Navigation with Quaternion-Based Attitude Estimation. *American Control Conference*, 272–279, 2013.

[67] Minh-Duc Hua, G. Ducard, T. Hamel, R. Mahony, and K. Rudin. Implementation of a Nonlinear Attitude Estimator for Aerial Robotic Vehicles. *IEEE Transactions on Control Systems Technology*, 22(1):201–213, 2014.

[68] P. Batista. Long Baseline Navigation with Clock Offset Estimation and Discrete-Time Measurements. *Control Engineering Practice*, 35:43–53, 2015.

[69] M. S. Grewal and A. P. Andrews. Kalman Filtering: Theory and Practice Using MATLAB. John Wiley & Sons, 2011.

[70] H. F. Grip, T. I. Fossen, T. A. Johansen, and A. Saberi. Nonlinear Observer for Integration of GNSS and IMU Measurements with Gyro Bias Estimation. *Proceedings of the American Control Conference*, 6, 2012.

[71] T. H. Bryne, T. I. Fossen, and T. A. Johansen. Nonlinear Observer with Time-Varying Gains for Inertial Navigation Aided by Satellite Reference Systems in Dynamic Positioning. *Mediterranean Conference on Control and Automation (MED)*, 1:1353–1360, 2014.

[72] L. Fusini, T. I. Fossen, and T. A. Johansen. A Uniformly Semiglobally Exponentially Stable Nonlinear Observer for GNSS- and Camera-Aided Inertial Navigation. *Proceedings of 22nd IEEE Mediterranean Conference on Control and Automation, Italy*, 1031–1036, 2014.

Chapter 7

Time-Energy Optimal Cluster Space Motion Planning for Mobile Robot Formations

Kyle Stanhouse, Chris Kitts and Ignacio Mas

Additional information is available at the end of the chapter

http://dx.doi.org/10.5772/64615

Abstract

The motions of a formation of mobile robots along predetermined paths are optimized according to a tunable time-energy cost function using the cluster space approach to multiagent system specification and control. Upon path-parameterizing cluster state variables describing the geometry and pose of a multirobot group, an optimal control problem is formulated that incorporates formation dynamics and state constraints. The optimal trajectory is derived numerically via a gradient search, iterating over the initial value of one costate. A multirobot formation control simulation is then used to demonstrate the effectiveness of the technique. Results indicate that a substantial tradeoff is made between energy expenditure and motion time when considered as minimization criteria in varying proportions, allowing the operator to tailor mission trajectories according to desired levels of each.

Keywords: time-energy optimization, motion planning, multirobot systems, cluster control

1. Introduction

In recent years, multiagent robotic systems have been firmly established as an important topic of research owing to the continued emergence of potential applications. Cooperating teams of robots remotely operated or capable of autonomous navigation and sensing can be used to enhance or extend the functions of single-agent systems in areas such as airborne [1–3] and terrestrial-distributed mobile sensing [4–6], search and rescue [7–9], and intelligent transportation [10–12].

For mobile systems, one of the key technical considerations is the development of a technique to coordinate the motions of individual vehicles. The mutual goal of each agent is typically to establish and maintain a certain spatial configuration or to perform complicated geometrically time-varying maneuvers, a use case of particular interest to this investigation. These desired behaviors lead to a variety of formation control problems, for which a wide range of solutions have been and continue to be explored.

Notable work in this area includes the development of leader-follower strategies in which follower agents control their position relative to a designated leader to meet formation requirements [13–15]. Artificial potential fields have similarly been shown effective as a construct to establish formation-keeping forces between robots within a group [16, 17]. Cluster space, an approach that allows for intuitive specification of formation characteristics and implements control directly on these variables, has also been demonstrated successfully for a number of robotic systems [18]. Current trends in research, however, indicate a focus on the incorporation of formation requirements into the framework of optimal control, and will be discussed in this chapter.

Due to the inherent physical distribution of agents and the potential for limited information exchange, decentralized control protocols for multirobot systems are popular. However, centralized architectures, which exploit global information, are more amenable to executing specific time-varying formation trajectories. The latter pertains to a class of similar methods highly relevant to this chapter that delegate path design to an earlier step, relying on motion planning to avoid inter-agent collisions and achieve varying degrees of coordination [19–24]. Distinguished among these is [25], which frames motion coordination as a velocity-optimization problem. Formations are defined by robot-pair relative geometries as a function of distance along their respective paths, forming a constraint net. Optimal trajectories for each robot are then generated such that these formation errors are minimized. Alternatively, [26] represents the set of robot positions at a particular time as a 2D planar curve, and develops a synchronization controller that regulates robot motions to simultaneously track the prescribed formation boundaries as well as their individual paths. Although an optimal velocity signal is not derived, as in [25], the model allows for complete specification of any formation over time, provided the potentially complicated individual robot paths are attainable. For a number of conceivable formations, this is a non-trivial step and can prove prohibitive. The methods proposed in this chapter, which build from the cluster space approach to multirobot system specification, address this issue.

In contrast to the investigations discussed above, much of current research in multirobot systems is dedicated to the use of distributed optimal control techniques. These are appropriate for applications that permit simple formation specification where robots must operate with communication restrictions and local information such as relative positioning. Consensus algorithms, which draw from concepts in distributed computing and graph theory, are present in many of these approaches [27–29]. In the context of a cooperative multivehicle system, information consensus refers to the convergence of agents in a networked system to a common task or variable such as the center of a formation shape, the rendezvous time, or the direction of formation translation [30]. Also of note are distributed control-based methods to handle swarms, or multiscale dynamical systems, which are comprised of many agents [31–34]. In general, these techniques are only capable of assigning coarse-grained dynamics; specific paths and distributions are not determined a priori. Each unit is held subject to local objectives and constraints, giving rise to certain coherent macroscopic behaviors. For example, Ferrari et al. [35] model global behavior of a multiagent system using PDFs and optimizes subject to coupled local agent dynamics in such a manner that cohesion is achieved and the result requires far less computation than classical optimal control.

Energy efficiency in mobile robotics systems is of great concern given that energy sources are often carried, and practical applications require extended remote operation wherein limited resources must be conserved. Many investigations seek to address this issue through the use of motion planning techniques to reduce the energy consumption in system components. Typically, proposed energy models are incorporated into optimal control frameworks to derive trajectories that minimize energy expenditure [36–40]. For example, in a recent work, Liu and Sun [41] decompose energy consumption into three categories: kinetic energy transformation, overcoming traction resistance, and maintenance of electrical sources for the operation of sensors, on board PCs, and control circuits. This analysis is then used to generate both an energy optimal path and a velocity trajectory, which further conserves energy. While a complete and detailed energy model can be beneficial, it is well known that preventing large torque variations in motors by smoothing velocity profiles is most fundamental to energy conservation [42, 43]. This chapter concentrates on minimizing energy expenditure for a formation of mobile robots in this regard.

The investigations referenced above are geared toward individual robots and therefore do not adequately address the optimal planning of energy-efficient trajectories for multiagent systems where a holistic approach is appropriate. There are, however, a number of methods that include provisions for minimizing aggregate energy consumption. For example, Sieber et al. [44] formulate an LQR-like optimal control problem designed to move a formation of mobile robots to a goal while minimizing input energy and incorporating a provision for formation rigidity into the cost functional. Similarly, Wigstrom and Lennartson [45] generated trajectories that reduce energy consumption using pseudo-spectral optimal control, though paths are considered free. Coverage algorithms such as [40, 46] also implement energy conservation constraints while maximizing the reach of mobile robot networks and sensing. A few swarm-like methods have accounted for power consumption as well, but they minimize global energy in order to achieve stability and cohesion for the group, or to affect the size and internal coverage of the swarm [47, 48]. While these examples address the conservation of group energy, they do not deal with the generation of smooth energy-efficient velocity trajectories along predefined paths, an objective of this chapter.

The concepts of time and energy optimality in the context of trajectory generation for multi-robot systems are well represented in the literature. However, they are largely accounted for alongside formation constraints, thus the derived control signals are suboptimal with regard to time and/or energy exclusively. To our knowledge, there are no methods that achieve high precision time-varying maneuvers and successfully separate the generation of optimal trajectories for each robot in a multiagent group from formation requirements. The techniques proposed in this chapter address this shortcoming.

The contribution of this investigation is a method to generate continuous force, acceleration, and velocity profiles for a formation of mobile robots with predetermined paths, while time and energy are minimized in chosen proportions. Previously proposed methods with similar objectives have employed optimization techniques with robot level dynamic constraints and formation level cost functions. In contrast, our treatment uniquely optimizes and imposes constraints at the cluster or formation level, with favorable results. Further advantages of our technique are explained in Section 4.

We first introduce the cluster space control framework, which presents a group of mobile robots as a virtual articulating mechanism in order to facilitate characterization and to implement coordinated control of the system. A parameterization of the cluster dynamic equations is next proposed which results in a reduced second-order state space model. The structure and numerical solution of the optimization are then shown. Finally, a simulation of a three-robot cluster controller is used to verify and validate the solution trajectory, after which analysis and results are presented.

2. Cluster space

2.1. Cluster space specification of multirobot systems

The *cluster space* control framework conceptualizes a multirobot system as a single entity, a *cluster*, which is described in terms of its global position and orientation, shape, and the relative orientations of individual robots within the cluster. Based on these attributes, a set of independent state variables is defined to specify, control, and monitor the position and motion characteristics of the formation [18]. These quantities can be mathematically related to the positions and velocities of the robots in the group through a formal set of kinematic transforms, much like the end effector position and velocity of a robotic manipulator can be related to its joint angles and rotational velocities. Similarly, a set of cluster dynamic equations of motion have been defined with coefficients that are a function of dynamic properties of individual robots, allowing generalized forces in cluster space to be related to generalized forces in robot space [49]. These relationships enable the use of a feedback formation control system in which the operator can command the path(s) of a multirobot formation with respect to the cluster states, and be relieved from the task of specifying individual robot motions.

Cluster space represents a natural point of view for the operator suited for specifying well-behaved, smooth, formation state trajectories [50]. Unlike other formation control approaches that must commit to centralized or decentralized protocols, the cluster space framework accommodates a range of (de)-centralization architectures through the selection of cluster state space variables [51]. Cluster control has been experimentally demonstrated with groups of two to six robots, operating on/in land/sea/air, with both piloted and automated controls, and with a variety of compensation strategies. Ongoing work includes use of the technique for real-world applications such as gradient-based adaptive environmental sampling [52], object tracking [53], object transportation [54], and escorting in the presence of long-distance communications [55, 56].

2.2. Description of a three-robot cluster

Previous work on the cluster space state representation of a mobile multirobot system presented a generalized framework for a system of n robots each with m degrees of freedom [1]. In this study, we will consider the implementation of a group of three planar robots each with 3 degrees of freedom: two translational ($x1$, $y1$, $x2$, $y2$, $x3$, $y3$) and one rotational ($\theta 1$, $\theta 2$, $\theta 3$). These variables are referenced to a global frame and constitute the *robot space* description of the pose of the system, r. Alternatively, the pose of the system in *cluster space* can be defined by a position vector c consisting of variables that characterize the globally referenced cluster location and orientation at the formation centroid (x_c, y_c, θ_c), the geometry of the three-robot triangular formation (p, q, β), and the relative rotations of the individual robots within the cluster (φ_1, φ_2, φ_3). In general, we note that the cluster space control framework provides flexibility in the location of the cluster frame and in the selection of shape variables; for the example used in this study, the shape variables (p, q, β) represent a side-side-angle definition of the three-robot cluster. A depiction of both *robot space* and *cluster space* variables is provided in **Figure 1**.

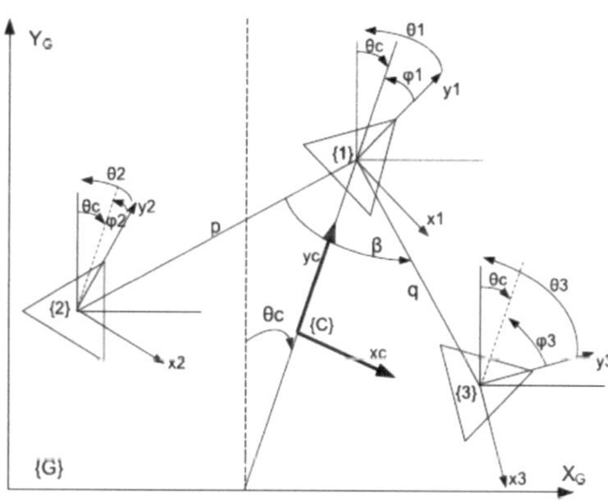

Figure 1. State space variables for a three-robot cluster.

2.3. Cluster space equations of motion

In previous work, successful cluster space control has been demonstrated through the use of a kinematic, or resolved rate, control approach in which the operator specifies desired cluster velocities, which are then converted into robot velocity commands through an inverse-Jacobian transform. This has been adequate for a number of demonstrations and applications, particularly when the robots have on-board velocity controllers, are heavily damped, and when external disturbances are minor. For more challenging cases, a dynamic cluster space controller was developed, based on the definition of cluster equations of motion. This dynamic model describes the interdependence between the positions, velocities, and accelerations of the cluster state variables and the generalized forces and torques acting on the cluster. It contains coefficients that can be mathematically related to similar coefficients in robot space dynamic equations. The details of this derivation using Lagrange formalism are presented in [54]. The resulting cluster space dynamic equation follows:

$$F = \Lambda(c)\ddot{c} + \mu(c,\dot{c}) + g(c) \tag{1}$$

where if n is the number of robots in the cluster and m the number of degrees of freedom of each robot: $F \varepsilon R^n$ is an $nm \times 1$ vector of cluster generalized forces and torques belonging to a set defined $\{F \mid F_{imin} \leq F_i \leq F_{imax}, i = 1 : nm\}$; $\Lambda(c)$ is an $nm \times nm$ symmetric, positive definite, cluster mass or inertia matrix of the quadratic form; $\vec{c}$, and its first and second derivatives $(\dot{c}, \ddot{c})$ are $nm \times 1$ vectors of cluster space state positions, velocities, and accelerations, respectively; $\mu(c, \dot{c})$ is an $nm \times 1$ vector of cluster space Centrifugal and Coriolis forces; $g(c)$ is an $nm \times 1$ vector of gravity forces in cluster space.

The cluster space equations of motion were derived using a Lagrangian formulation based on the formation's potential and kinetic energy. An alternate dynamic form for a second-order system can also be realized through a quadratic representation of the first-order cluster velocity product terms [49]:

$$F = \Lambda(c)\ddot{c} + \dot{c}^T B\dot{c}. \tag{2}$$

This version is obtained through the use of Christoffel symbols ($\Gamma^i_{jk}(c)$), which are derived from the cluster mass matrix $\Lambda(c)$, where B is an $nm \times nm \times nm$ vector of Christoffel symbols:

$$B = \left[\Gamma^1(c) \ldots \Gamma^{nm}(c)\right] \tag{3}$$

And $\Gamma^i_{jk}(c)$ is an nm x nm symmetric matrix for $i, j, k = 1$ to nm:

$$\Gamma^i_{jk}(c) = \frac{1}{2} * \left(\frac{\partial \Lambda_{ij}(q)}{\partial c_k} + \frac{\partial \Lambda_{ik}(q)}{\partial c_j} + \frac{\partial \Lambda_{kj}(q)}{\partial c_i} \right) \tag{4}$$

This form has the desired quality of being conducive to state parameterization, a necessary conversion due to the high dimensionality of the cluster state vector and equations of motion. It does not include a gravity term due to the fact that our analysis is specific to planar rovers whose gravity force is canceled out by the force normal to the surface.

2.4. Cluster space parameterization

The cluster space state vector for a formation of n robots each with m degrees of freedom is of size nm, which for the presented three-robot example corresponds to a nine-element state vector. The application of optimal control techniques to a cluster state vector of this size would result in a 2-pt boundary (shooting) problem of large order requiring an unmanageable computational effort. We therefore seek to reduce the dimensionality of the control problem and its numerical solution by defining a one-dimensional path space denoted as $s(t)$ (where $s = [0,1]$), representing the distance traveled along a specified path. The following draws from the general methodology presented in [58].

The cluster state vector, at a given time, specified as a function of the distance traveled along the path, and its first and second derivatives taken w.r.t. t are given by

$$\mathrm{c} = f(s);\ \dot{\mathrm{c}} = f_s\dot{s};\ \ddot{c} = f_{ss}\dot{s}^2 + f_s\ddot{s} \tag{5}$$

where f_s is the unit vector tangent to the path, f_{ss} is the curvature, $\dot{s}$ is the speed along the path, and $\ddot{s}$ is the acceleration. Substituting these expressions for c, $\dot{\mathrm{c}}$,, and $\ddot{\mathrm{c}}$ in Eq. (2), we obtain

$$F = \Lambda(s)\left(f_{ss}\dot{s}^2 + f_s\ddot{s}\right) + \left(f_s\dot{s}\right)^T B\left(f_s\dot{s}\right) \tag{6}$$

$$F = \Lambda(s) f_{ss}\dot{s}^2 + \Lambda(s) f_s\ddot{s} + \left(\dot{s}^T f_s^T\right) B\left(f_s\dot{s}\right). \tag{7}$$

If we let $m(s) = \Lambda(s)f_s$ and $b(s)=\Lambda(s)f_{ss} + f_s^T B f_s$:

$$F = \mathrm{m}(s)\ddot{s} + b(s)\dot{s}^2. \tag{8}$$

Using Eq. (8) we have rewritten the cluster dynamics in terms of the velocity $\dot{s}$ and acceleration $\ddot{s}$ along the cluster state paths where $m(s)$ and $b(s)$ are the coefficients of acceleration and Coriolis/Centripetal forces, respectively, expressed in cluster path coordinates. It is evident in the transformed dynamics that a linear relationship exists between F and $\ddot{s}$. This mapping is unique; therefore, we can use $\dot{s}$ as the control input, effectively reducing the $2nm$-dimensional state space to two independent states, $\dot{s}$ and $\ddot{s}$ [59]. (Note: The coefficient matrices $m(s)$ and $b(s)$ are both of size 9 × 9; each element is an expression of a size that precludes its presentation here.)

Now that we have represented the cluster dynamics in path space, inequality constraints on the magnitude of the cluster force, $\{F_{i\min} \le F_i \le F_{i\max},\ i = 1 : nm\}$ can also be converted. Substituting (8) into this expression yields nm path space inequality expressions for cluster space velocity and acceleration constraints:

$$F_{i\min} \leq \mathrm{m}_i\ddot{s} + b_i\dot{s}^2 \leq F_{i\max}, where\, i = 1:nm \tag{9}$$

The bounds on the path acceleration and velocity at a given point are

$$\dot{s} \leq \dot{s}_{\max}(s) \tag{10}$$

$$\ddot{s}_{\min}(\dot{s},s) \leq \ddot{s} \leq \ddot{s}_{\max}(\dot{s},s) \tag{11}$$

where assuming $m_i \neq 0$ and $m_jb_i - m_ib_j \neq 0$:

$$\dot{s}^2_{\max}(s) = \min_{ij}\left\{\max_{F_i,F_j}\left(\frac{m_jF_i - m_iF_j}{m_jb_i - m_ib_j}\right)\right\},\ \text{for } i,\, j = 1:nm \tag{12}$$

$$\ddot{s}_{\max}(\dot{s},s) = \min_i\left\{\max_{F_i}\left(\frac{F_i - b_i\dot{s}_2}{m_i}\right)\right\},\ \text{for i} = 1:nm \tag{13}$$

$$\ddot{s}_{\min}(\dot{s},s) = \max_i\left\{\min_{F_i}\left(\frac{F_i - b_i\dot{s}_2}{m_i}\right)\right\},\ \text{for i} = 1:nm \tag{14}$$

Again, these derivations originate in [58, 59]. By virtue of the similarity between cluster space dynamics with that of manipulators, the form of the inequality constraints is identical. They have been presented here for completeness; however, the simulations in Section 4 will not consider trajectories in which they are violated as it is outside the scope of our analysis. Furthermore, a motivating factor behind the inclusion of energy in the minimization is to depart from time-optimal trajectories and reduce the strain on robot actuators. This purpose would be defeated if robot acceleration and velocity bounds were reached.

3. Cluster space time-energy optimal control

3.1. Problem formulation

To minimize, in varying proportions, the energy and motion time of system (8) along a specified cluster path, we formulate and solve the following optimization problem. The parameterization of the cluster dynamics reduces its state space to the double integrator:

$$\dot{\vec{x}} = f(\vec{x},u) = [x_2, u]^T = [\dot{s}, \ddot{s}]^T \tag{15}$$

The objective function, denoted as J, incorporates free final time and a weighted cluster energy term:

$$\min_u (J) = \int_0^{t_f} L(x,u)dt \tag{16}$$

where $L(x, u) = 1 + \epsilon^2 F^2$, and ϵ ranges from 0 to 1. Or, in cluster path space coordinates, substituting (8) for F:

$$L(x,u) = 1 + \varepsilon^2 \left(b^T b x_2^4 + 2m^T b x_2^2 u + m^T m u^2 \right). \tag{17}$$

Subject to the boundary conditions:

$$x_1(0) = 0, x_1(t_f) = x_f, x_2(0) = 0, x_2(t_f) = 0. \tag{18}$$

The state-dependent control constraints can then be defined as

$$g_1(\vec{x},u) = u - u_{\max}(\vec{x}) \leq 0 \tag{19}$$

$$g_2(\vec{x},u) = u_{\min}(\vec{x}) - u \leq 0 \tag{20}$$

where

$u_{\max}(x) = \ddot{s}_{\max}(\dot{s}, s)$ and $u_{\min}(x) = \ddot{s}_{\min}(\dot{s}, s)$.

And the state inequality constraints:

$$h(x) = x_2 - \dot{s}_{\max}(x_1) \leq 0 \tag{21}$$

Computing $u^*(t)$, the optimal acceleration along the path gives us access to the optimal trajectory of the full dynamic system, $F^*(t), \ddot{c}^*(t), \dot{c}^*(t)$, and $c^*(t)$. The structure of the optimal control solution of this problem follows from the first-order necessary optimality conditions for the case of state-dependent control constraints and free final time [60]. Due to the refor mulation in s, the time-energy optimal control problem is convex; therefore it can be concluded that any local optimum of the problem is also globally optimal.

3.2. Structure of the optimal control solution

The optimal control (u^*) and states ($\vec{x}^*$) must satisfy the following conditions [59]:

$$H_u\left(x^*,\lambda,u^*,t\right)=0 \tag{22}$$

$$H\left(x^*,\lambda,u^*,t\right)=0 \tag{23}$$

For all t $[t_0, t_f]$, where the Hamiltonian and costates are given by

$$H\left(\vec{x},\vec{\lambda},u\right)=L\left(\vec{x},u\right)+\vec{\lambda}^T f\left(\vec{x},u\right) \tag{24}$$

$$\dot{\vec{\lambda}}=-L_x-\left(f_x\right)^T\vec{\lambda}. \tag{25}$$

Expanded using (15) and (17):

$$H=1+(\varepsilon^2\left(b^T b x_2^4+2m^T b x_2^2 u+m^T m u^2\right)+\lambda_1 x_2+\lambda_2 u+\mu_1 g_1+\mu_2 g_2\right)=0 \tag{26}$$

$$H_u=2\varepsilon^2\left(m^T b x_2^2+m^T m u\right)+\lambda_2 \tag{27}$$

$$\dot{\lambda}_1=-2\varepsilon^2(m^T m_s u^2+b^T b_s x_2^4 \tag{28}$$

$$+\left(m_s^T b+m^T b_s\right)x_2^2 u)-g_s^T\mu$$

$$\dot{\lambda}_2=-4\varepsilon^2\left(b^T b x_2^3+m^T b x_2 u\right)-\lambda_1-g_{x_2}^T\mu. \tag{29}$$

The subscript 's' denotes partial derivatives with respect to $s=x_1$. The state-dependent control constraints, $[g_1(\vec{x},u), g_2(\vec{x},u)]^T$, are incorporated into the Hamiltonian using the multipliers $\vec{\mu}=[\mu_1,\mu_2]^T$. They are positive if the associated constraint (dictated by (22)) is active and zero

http://dx.doi.org/10.5772/64615

otherwise, ensuring that the cost function can only be reduced by violating the constraints [59].

Solving $H_u = 0$ for u yields the unconstrained optimal control signal:

$$u_{\text{unc}}(t) = -\left(\frac{2\varepsilon^2 m^T b x_2^2 + \lambda_2}{2\varepsilon^2 m^T m_s}\right) \tag{30}$$

This equation is valid if the control constraints are inactive. If they are activated, the optimal control switches to the bounds $u_{\max}(t)$ or $u_{\min}(t)$:

$$u^*(t) = \begin{cases} u_{\max}(t) if\, u_{\text{unc}} > u_{\max} \\ u_{\text{int}}(t) if\, u_{\max} \geq u_{\text{unc}} \geq u_{\min} \\ u_{\max}(t) if\, u_{\text{unc}} > u_{\max} \end{cases} \tag{31}$$

3.3. Numerical solution

The control signal $u = x_2$ is found by solving a fourth-order two-point boundary value problem given by Eqs. (17), (28), and (29):

$$\dot{x}_1 = x_2 \tag{32}$$

$$\dot{x}_2 = \frac{\left(2\varepsilon^2 m^T b x_2^2 + \lambda_2\right)}{\left(2\varepsilon^2 m^T m_s\right)} \tag{33}$$

$$\lambda_1 = -2\varepsilon^2\left(m^T m_s u^2 + b^T b_s x_2^4 + \left(m_s^T b + m^T b_s\right) x_2^2 u\right) \tag{34}$$

$$\dot{\lambda}_2 = -4\varepsilon^2\left(b^T b_s x_2^3 + m^T b x_2 u\right) - \lambda_1. \tag{35}$$

If we assume initial unconstrained control, substitute initial condition $x_2(0) = 0$ into (26), and apply conditions (22) and (23), we can solve for the initial value of one of the costates:

$$\lambda_2(0) = -\frac{1 + \varepsilon^2 m^T(0) m(0) u_{\max}^2(0)}{u_{\max}(0)} \tag{36}$$

We now have initial conditions for all necessary variables with the exception of $\lambda_1(0)$. The initial condition for λ_1 must therefore be iterated, or guessed at until the correct final conditions are realized for x_1 and x_2. This numerical problem can be solved in a computationally inexpensive manner.

4. Results and analysis

The numerical methods presented in the previous sections were implemented to obtain time-energy optimal trajectories for the cluster paths presented in this section. Subsequently, a cluster space kinematic controller [18] was used to simulate the tracking of desired cluster paths and motions in the following examples for three-robot cluster rotate and translate as well as rotate and resize maneuvers.

4.1. Rotate and translate simulation

This cluster path is achieved by commanding translation from $(x_c,y_c) = (0,0)$ to $(x_c,y_c) = (10,10)$ and simultaneous formation rotation from $\theta_c = 0$ to π, with initial (c_{init}) and path-parameterized cluster state vector ($f(s)$):

$$\begin{aligned} c_{\text{init}} &= \left[x_c, y_c, \theta, \phi_1, \phi_2, \phi_3, p, q, \beta\right]^T \\ &= \left[0,0,0,0,0,0,10,10,\pi/3\right]^T \end{aligned}$$

$$f(s) = \left[10s, 10s, \pi s, 0,0,0,10,10,\pi/3\right]^T$$

The following are depictions of the cluster parameterized phase plane and path velocity vs. time for multiple values of ϵ (**Figure 2**):

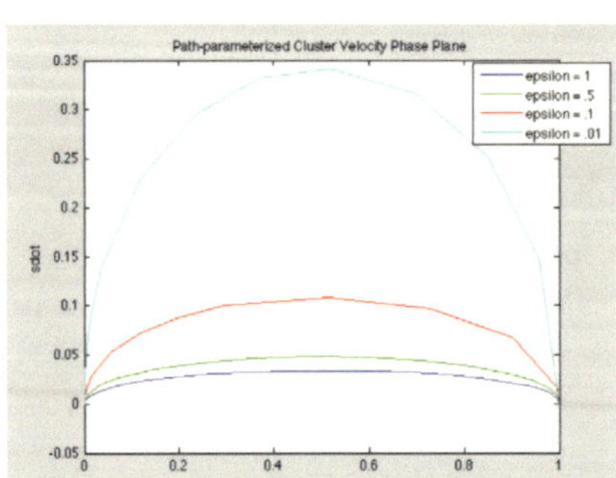

Figure 2. Cluster parameterized phase plane for multiple ϵ.

http://dx.doi.org/10.5772/64615

Upon inspection, it is apparent that the generated plots conform to intuition regarding the expected shape of an energy optimal velocity profile. Given that generally steep acceleration results in excessive energy expenditure one expects that a mobile robot would increase its speed gradually (nonlinearly) up to a peak, after which it should identically decelerate. The velocity profile should therefore be parabolic, symmetric, and continuous, unless state constraints were violated, producing a plateau (trapezoidal) effect with discontinuities at the switching point. In [61], velocity profiles that match this description were generated while minimizing total energy drawn from the batteries of a particular mobile robot. Their results, given different motor characteristics, ranged from the widely used trapezoidal profile to the parabolic profile observed in the simulation.

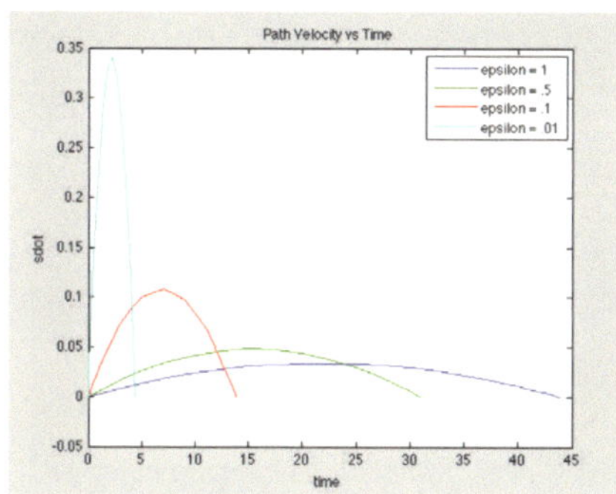

Figure 3. Velocity along the cluster parameterized path for multiple **ϵ**.

Comparison of the optimal velocity profiles for different ϵ in **Figure 3** confirms the anticipated result that increasing values of ϵ, and therefore energy contribution to the cost function, results in velocity profiles of reduced average magnitude and longer duration. In **Figure 2**, phase planes generated from successively smaller ϵ increase in magnitude as they approach the time optimal trajectory. The time-optimal boundary, though not derived as part of this study, can be determined by calculating the maximum (minimum) parameterized cluster velocity and acceleration achievable at each possible cluster configuration along the given path. However, for a small enough choice of ϵ, the generated velocity profile will sufficiently approximate this limit.

The trajectory obtained, using $\varepsilon = .5$, was executed using the simulation referenced above, to yield the snapshots in **Figure 4** of the cluster in 5-s intervals.

Notice that the paths of robots 1 and 2 (the blue and red dots, respectively) and robots 2 and 3 (the red and green dots, respectively) intersect during the course of the maneuver, indicating a need for timing constraints to avoid collision. There are methods that define multirobot coordination in this manner, referenced earlier [19, 39], that generate robot velocity profiles from given paths such that robots avoid collision. Cluster space requires no such direct provision, the timing of individual robot trajectories both to avoid inter-robot collision and maintain formation is handled implicitly. This is evident in **Figure 5**, which depicts the distinct optimal velocity profiles of each robot obtained by applying the cluster inverse kinematic transformations to the optimal parameterized velocity. They are of equal duration yet their dissimilar contours and therefore areas indicate unequal distances traveled. Additionally, due to the fact that optimizing in cluster space relieves us from imposing an explicit formation constraint in the cost function as in [25, 26], the mathematical treatment can be dedicated exclusively to optimizing time and energy.

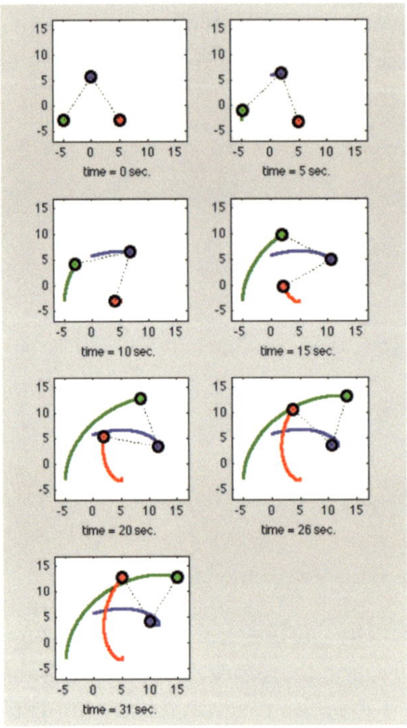

Figure 4. Snapshots of cluster rotate and translate maneuver in 5-s intervals.

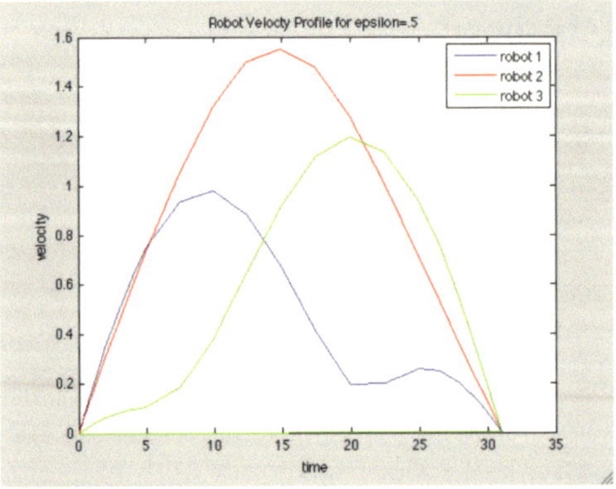

Figure 5. Individual robot velocity profiles.

4.2. Rotate and resize simulation

Another example of a cluster maneuver, rotate and resize, is obtained by commanding rotation from $\theta_c = 0$ to π and a simultaneous resize from $p, q = 10$ to $p, q = 20$, with initial (c_{init}) and path-parameterized cluster state vector ($f(s)$):

$$\begin{aligned} c_{\text{init}} &= \left[x_c, y_c, \theta, \phi_1, \phi_2, \phi_3, p, q, \beta\right]^T \\ &= \left[5, 5, 0, 0, 0, 0, 10, 10, \pi / 3\right]^T \end{aligned}$$

$$f(s) = \left[5, 5, \pi s, 0, 0, 0, 10s, 10s, \pi / 3\right]^T.$$

The following are depictions of the cluster parameterized phase plane and path velocity vs. time for multiple values of ϵ (**Figure 6**):

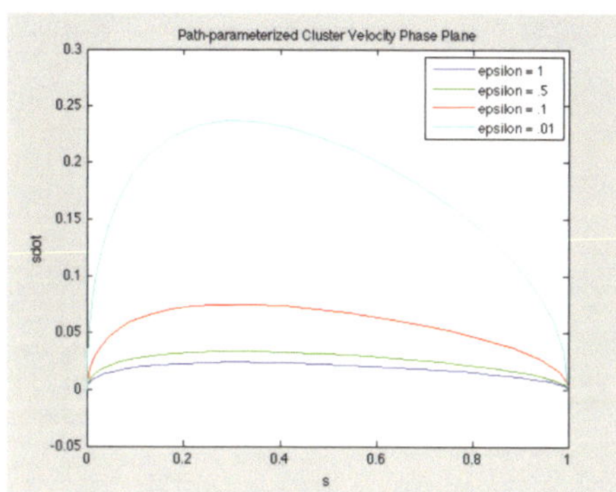

Figure 6. Cluster parameterized phase plane for multiple ϵ.

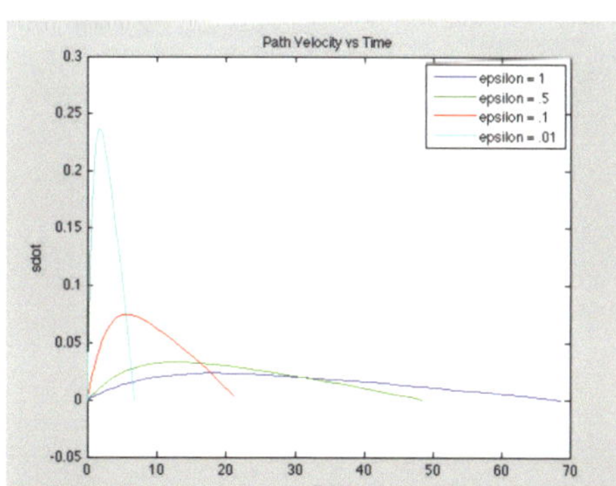

Figure 7. Velocity along the cluster parameterized path for multiple ϵ.

As with a rotate and translate case, increasing ϵ lengthens the duration of the trajectory and flattens the shape of the optimal velocity profile (**Figure 7**), suggesting reduced energy expenditure. Additionally, the phase plane of **Figure 6** appears to approach a time-optimal limit with decreasing ϵ, as anticipated. However, comparison with **Figure 2** of the general shape and relative location of the peak velocity reveals a discernable difference. The rotate and resize profile exhibits a skew toward the velocity axis resulting in a peak magnitude that occurs before the half-time trajectory point, while the rotate and translate profile is approximately symmetric about the half-time point. This effect can be explained by viewing the cluster as a virtual articulating mechanism possessing inertia. Because the cluster inertia grows as the path is followed, optimizing in cluster space produces a velocity profile that favors torque application early in the maneuver. It is also true then that the symmetric nature of the optimal velocity profile in a case such as a cluster rotate and translate often implies a constant or symmetric inertia throughout the trajectory.

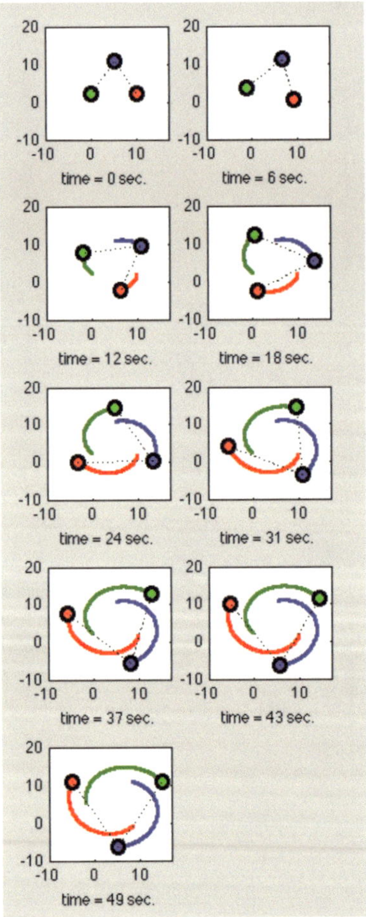

Figure 8. Snapshots of cluster rotate and resize maneuver in 5-s intervals.

The trajectory simulation for $\varepsilon = .5$ was executed to yield the following snapshots of the cluster in 5-s intervals (**Figure 8**):

The individual robot velocity profiles for this case (not pictured here) are identical due to the fact that their paths do not intersect and the chosen cluster geometry is an equilateral triangle, implying equidistant paths. Therefore, in contrast to **Figure 5**, this maneuver does not require a provision for timing to avoid inter-agent collision. A virtue of our technique is that optimizing trajectories in cluster space removes the operator from this consideration and handles both scenarios without intervention.

4.3. Time-energy tradeoff

As previously acknowledged, adjusting the weight of the energy term in the objective function results in varying duration times and energy expenditures. In order to thoroughly investigate the trade-off range, optimal solutions were obtained for a number of ϵ values in the rotate and translate case. **Figure 9** depicts the relationship between trajectory duration (T) and the cluster energy as represented in the objective function, $\int_0^{t_f} F^2 dt$, for ϵ ranging from .001 to .01. The results, normalized to $\epsilon = .001$, show that for a 10% increase in T, energy expenditure is reduced by 22.5% (point (a) in **Figure 9**). As ϵ gets larger, and T is increased further by 10%, energy is reduced by 19.8% (point (b) in **Figure 9**). The slope of this curve will of course change with cluster path and trajectory; however, we expect to always observe diminishing reductions in energy expenditure with increments of T, as we have with this case. In general, for a given cluster and trajectory, this plot can be generated and used as a tool to inform the appropriate choice of ϵ given mission requirements such as time to completion or energy limitations.

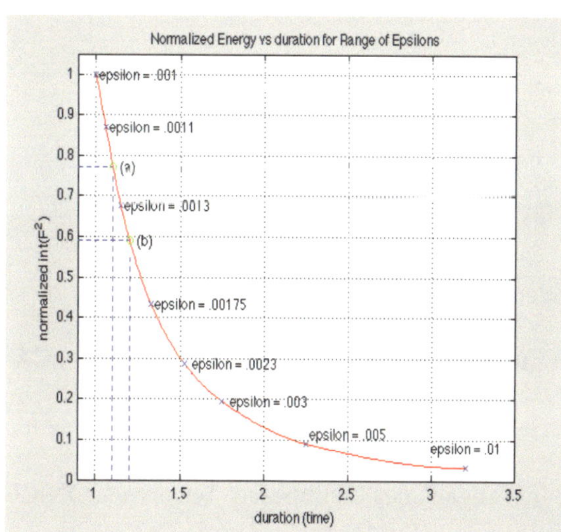

Figure 9. Energy vs. trajectory duration.

5. Conclusion

This chapter presented a technique to generate a time-energy optimal trajectory for a cluster of mobile robots with predetermined paths. The method is uniquely applied in a parameterized formation state space and incorporates the nonlinear dynamics of the cluster and actuator constraints to enable intuitive optimal trajectory planning for time varying formations. Simulation results verified the generated trajectories and demonstrated the advantages of using a time-energy tunable objective function. The ability to choose more energy-efficient trajectories reduces the strain on actuator components and energy reserves, thereby promoting longevity in both. The tool in **Figure 9** also demonstrates that the operator has access to a substantial path-specific time/energy range, within which one must choose a mission-appropriate combination.

We plan to incorporate several extensions to this work in the future. First, we will integrate cluster space obstacle avoidance [62] into the optimal trajectory-tracking controller. In addition, we will develop cluster space velocity and acceleration bounds to include the characterization of a velocity limit curve as well as a phase plane description of the admissible range. With these bounds, which are a function of the cluster states and are therefore variable in the choice of path, one could derive the time optimal cluster trajectory, approximated with small ϵ in the simulations of the previous section. We will also attempt to address practical issues such as friction, gravity, and motor dynamics into the formulation in order to make the technique more applicable to a wide range of multirobot systems. Finally, we will experimentally verify and validate the technique with several existing multirobot experimental test beds, to include terrestrial rovers, surface and underwater marine vehicles, and aerial vehicles.

Author details

Kyle Stanhouse[1], Chris Kitts[1,4] and Ignacio Mas[2,3,4*]

*Address all correspondence to: imas@itba.edu.ar

1 Robotic Systems Laboratory, Santa Clara University, Santa Clara, CA, USA

2 Instituto Tecnologico de Buenos Aires (ITBA), Buenos Aires, Argentina

3 Consejo Nacional de Investigaciones Cientificas y Tecnicas (CONICET), Buenos Aires, Argentina

4 These authors are Senior Members of IEEE

http://dx.doi.org/10.5772/64615

References

[1] D. Kingston, R.W. Beard, R.S. Holt, "Decentralized Perimeter Surveillance Using a Team of UAV's," *IEEE Transactions on Robotics*, vol. 24, no. 6, pp. 1394–1404, 2008.

[2] J. Hu, L. Xie, J. Xu, "Vision-based Multi-Agent Cooperative Target Search,", *Control Automation Robotics & Vision 12th International Conference*, pp. 895–900, 2012.

[3] J. Dong, E. Nelson, V. Indelman, N. Michael, F.Dellaert, "Distributed Rel-time Cooperative Localization and Mapping using an Uncertainty-Aware Expectation Maximization Approach," *IEEE International Conference on Robotics and Automation*, pp. 5807–5814, 2015.

[4] Y. Chen, X.C. Ding, A. Stefanescu, C. Belta, "Formal Approach to the Deployment of Distributed Robotic Teams," *IEEE Transactions on Robotics*, vol. 28, no. 1, pp. 158–171, 2012.

[5] Y. Xu, J. Choi, S. Oh, "Mobile Sensor Network Navigation Using Gaussian Processes With Truncated Observations," *IEEE Transactions on Robotics*, vol. 27, no. 6, pp. 1118–1131, 2011.

[6] Z. Yao, K. Gupta, "Distributed Roadmaps for Robot Navigation in Sensor Networks," *IEEE Transactions on Robotics*, vol. 27, no. 5, pp. 997–1004, 2011.

[7] F. Pasqualetti, A. Franchi, F. Bullo, "On Cooperative Patrolling: Optimal Trajectories, Complexity Analysis, and Approximation Algorithms," *IEEE Transactions on Robotics*, vol. 28, no. 3, pp. 592–606, 2012.

[8] S.L. Smith, M. Schwager, D. Rus, "Persistent Robotic Tasks: Monitoring and Sweeping in Changing Environments," *IEEE Transactions on Robotics*, vol. 28, no. 2, pp. 410–426, 2012.

[9] A. Macwan, G. Nejat, B. Benhabib, "Target-Motion Prediction for Robotic Search and Rescue in Wilderness Environments," *IEEE Transactions on Systems, Man, and Cybernetics-Part B: Cybernetics*, vol. 41, no. 5, pp. 1287–1298, 2011.

[10] H. La, T. Nguyen, T.D. Le, M. Jafari, "Formation Control and Obstacle Avoidance of Multiple Rectangular Agents with Limited Communication Ranges," *IEEE Transactions on Control of Network Systems*, 99, pp. 1–1, 2016.

[11] S.A. Reveliotis, E. Roszkowska, "Conflict Resolution in Free-Ranging Multi-vehicle Systems: A Resource Allocation Paradigm," *IEEE Transactions on Robotics*, vol. 27, no. 2, pp. 283–296, 2011.

[12] H. M. La, R. S. Lim, J. Du, S. Zhang, G. Yan, W. Sheng. Development of a Small-scale Research Platform for Intelligent Transportation Systems, *IEEE Transactions on Intelligent Transportation Systems*, vol. 13(4), pp.1753–1762, Dec. 2012

[13] F. Li, Y. Ding, K Hao, "A Dynamic Leader-Follower Strategy for Multi-robot Systems," *IEEE International Conference on Systems, Man, and Cybernetics*," pp. 298–303, 2015.

[14] A. Loria, J. Dasdemir, N. Alvarez Jarquin, "Leader-follower Formation and Tracking Control of Mobile Robots Along Straight Paths," *IEEE Transactions on Control Systems Technology*, vol. 24, issue: 2, pp. 727–732, 2016.

[15] S. Su, Z. Lin, A. Garcia, "Distributed Synchronization Control of Multiagent Systems with Unknown Nonlinearities," *IEEE Transactions on Cybernetics*, vol. 46, issue: 1, pp. 325–338, 2016.

[16] L.A. Valbuena Reyes, H.G. Tanner, "Flocking, Formation Control, and Path Following for a Group of Mobile Robots," *IEEE Transactions on Control Systems Technology*, vol. 23, issue: 4, pp. 1268–1282, 2015.

[17] M. M. Zavlanos, G.J. Pappas, "Potential Fields for Maintaining Connectivity of Mobile Networks," *IEEE Transactions on Robotics*, vol. 23, issue: 4, pp. 812–816, 2007.

[18] C. Kitts, I. Mas, "Cluster Space Specification and Control of Multirobot Systems," *IEEE/ASME Transactions on Mechatronics*, vol. 14, no. 2, April 2009.

[19] J. Peng, S. Akella, "Coordinating Multiple Robots with Kinodynamic Constraints along Specified Paths," *International Journal of Robotics Research*, vol. 24, pp. 295–310, Apr. 2005.

[20] S.M. LaValle, S.A. Hutchinson, "Optimal Motion Planning for Multiple Robots having Independent Goals," *IEEE International Conference on Robotics and Automation*, vol. 3, pp. 2847–2852, Apr. 1996.

[21] R. Ghabcheloo, I. Kaminer, A.P. Aguiar, A. Pascoal, "A General Framework for Multiple Vehicle Time-Coordinated Path Following Control," *American Control Conference*, pp. 3071–3076, 2009.

[22] P. Abichandani, H.Y. Benson, Moshe K., "Multiple-vehicle Path Coordination in Support of communication," *IEEE International Conference on Robotics and Automation*, pp. 3237–3244, 2009.

[23] T. Simeon, S. Leroy, J. P. Laumond, "Path Coordination for Multiple Mobile Robots: A Resolution Complete Algorithm," *IEEE Transactions on Robotics and Automation*, vol. 18, no. 1, pp. 42–49, Feb. 2002.

[24] R. Olmi, C. Secchi, C. Fantuzzi, "Coordination of multiple AGVs in an industrial application," *Proc. IEEE International Conference on Robotics and Automation*, Pasadena, CA, May 2008, pp. 1916–1921.

[25] L. Shuang, D. Son, "Coordinated Motion Planning for Multiple Mobile Robots Along Designed Paths With Formation Requirement," *IEEE/ASME Transactions on Mechatronics*, vol. 16, Issue: 6, pp. 1021–1031, December 2010.

[26] D. Sun, C. Wang, W. Shang, G. Feng, "A Synchronization Approach to Multiple Mobile Robots in Switching between Formations," *IEEE Transactions on Robotics*, vol. 25, no. 5, pp. 1074–1086.

[27] K.H.Movric, F. Lewis, "Cooperative Optimal Control for Multi-Agent Systems on Directed Graph Topologies," *IEEE Transactions on Automatic Control*, vol. 59, issue: 3, pp. 769–774, 2014.

[28] J. Wang, M. Xin, "Integrated Optimal Formation Control of Multiple Unmanned Vehicles," *IEEE Transactions on Control Systems Technology*, vol. 21, issue: 5, pp. 1731–1744, 2013.

[29] X. Sun, C.G. Cassandras, "Optimal Dynamic Formation Control of Multi-Agent Systems in Environments with Obstacles,", *IEEE Conference on Decision and Control* 15, 2015.

[30] W. Ren, R.W. Beard, E.M. Atkins, "Information consensus in multivehicle cooperative control," IEEE Control Systems, vol. 27, issue: 2, pp. 71–82, 2007.

[31] K. Elamvazhuthi, S. Berman, "Optimal Control of Stochastic Coverage Strategies for Robotic Swarms," *IEEE Conference on Robotics and Automation* 26, 2015.

[32] C. C. Cheah, S. P. Hou, J. J. E. Slotine, "Region-based Shaped Control for a Swarm of Robots," Automatica, vol. 45, no. 10, pp. 2406–2411, 2009.

[33] S. D. Bopardikar, F. Bullo, J. P. Hespanha, "A Cooperative Homicidal Chauffeur Game," Automatica, vol. 45, pp. 1771–1777, July 2009.

[34] H. Duan, Q. Luo, Y. Shi, "Hybrid Particle Swarm Optimization and Genetic algorithm for multi-UAV Formation Reconfiguration," *IEEE Computational Intelligence Magazine*, vol. 8, issue: 3, 2013.

[35] S. Ferrari, G. Foderaro, P. Zhu, T. A. Wettergren, "Distributed Optimal Control of Multiscale Dynamical Systems: A Tutorial," *IEEE Control Systems*, vol. 36, issue: 2, 2016.

[36] A. Barili, M. Ceresa, C. Paris, "Energy-saving Motion Control for an Autonomous Mobile Robot," *Proceedings of the IEEE International Symposium on Industrial Electronics*, vol. 2, pp. 674–676, 1995.

[37] A.A. El-satte, S. Washsh, A.M. Zaki, S.I. Amer, "Efficiency-optimized Speed Control System for a Separately-excited DC motor," *Proceedings of the IEEE International Conference on Industrial Electronics, Control, and Instrumentation*, vol. 1, pp. 417–422, 1995.

[38] Y. Mei, Y.H. Lu, Y.C. Hu, C.S.G. Lee, "Energy-efficient Motion Planning for Mobile Robots," *Proceedings of the IEEE International Conference on Robotics and Automation*, vol. 5, pp. 4344–4349, 2004.

[39] W. Weigui, C. Huitang, W. Peng-Yung, "Optimal motion Planning for a Wheeled Mobile Robot," *Proceedings of the IEEE International Conference on Robotics and Automation*, vol. 1, pp. 41–46, 1999.

[40] Y. Mei, Y. Lu, Y.C. Hu, G. Lee, "Deployment of Mobile Robots with Energy and Timing Constraints," *IEEE Transactions on Robotics*, vol. 22, pp. 507–522, 2006.

[41] S. Liu, D. Sun, "Minimizing Energy Consumption of Wheeled Mobile Robots via Optimal Motion Planning," *IEEE/ASME Transactions on Mechatronics*, vol. 19, issue: 2, pgs. 401–411, 2014.

[42] A. M. Hussein, A. Elnagar, "On Smooth and Safe Trajectory Planning in 2D Environments," in *Proceedings of IEEE International Conference on Robotics and Automation*, 1997, vol. 4, pp. 3118–3123.

[43] M. Yongguo, L.Y. Hsiang, Y. Hu, C.S.G. Lee, "A Case Study of Mobile Robot's Energy Consumption and Conservation Techniques," *12th International Conference on Advanced Robotics*, pp. 492–497.

[44] D. Sieber, F. Deroo, S. Hirche, "Optimal Dynamic Formation Control of Multi-Agent Systems in Environments with Obstacles," *IEEE Conference on Decision and Control*, Dec. 15, 2015.

[45] O. Wigstrom, B. Lennartson, "Towards Integrated OR/CP Energy Optimization for Robot Cells," *IEEE International Conference on Robotics and Automation*, June 7, 2014.

[46] G. Wang, M.J. Irwin, P. Berman, F. Haoying, T. La Porta, "Optimizing Sensor Movement Planning for Energy Efficiency," *Proceedings of the 2005 International Symposium on Low Power Electronics and Design*, pp. 215–220, 2005.

[47] H.J. Chang, G.C.S. Lee, Y. Lu, Y. Hu, "Energy-Time-Efficient Adaptive Dispatching Algorithms for Ant-Like Robot Systems," *Proceedings of the IEEE International Conference on Robotics and Automation*, vol. 4, pp. 3294–3299, 2004.

[48] R. Pedrami, B.W. Gordon, "Temperature Control of Energetic Swarms," *International Conference on Mechatronics and Automation*, pp. 2639–2644, 2007.

[49] I. Mas, C. Kitts, R. Lee. "Dynamic Control of Mobile Multirobot Systems: The Cluster Space Formulation," Access, IEEE, pp. 558–570, 2014.

[50] C. Kitts, P. Mahacek, T. Adamek, I. Mas. "Experiments in the Control and Application of Automated Surface Vessel Fleets." *Proceedings IEEE/MTS Oceans Conference*, pp. 1–7, 2011.

[51] I. Mas, C. Kitts, "Centralized and Decentralized Multi-Robot Control Methods using the Cluster Space Control Framework." *IEEE Conference on Advanced Intelligent Mechatronics*, pp. 115–122, 2010.

[52] P. Mahacek, T. Adamek, V. Howard, K. Rasal, C. Kitts, B. Kirkwood, G. Wheat. "Cluster Space Gradient Tracking – Control of Multi-Robot Systems." *Proceedings ASME Information Storage and Processing Systems Conference, Santa Clara CA*, June 2011.

[53] S. Dayanidhi, R. Beetem, C. Kitts, "Initial Experiments in Multirobot-Based Tracking Networks," *ASME-ISPS/JSME-IIP Joint International Conference on Micromechatronics for Information and Precision Equipment*, 2012.

[54] I. Mas, C. Kitts, "Multi-Robot Object Manipulation Using Cluster Space Control," *2010 ASME Information Storage and Processing Systems Conference*, 2010.

[55] P. Mahacek, C. Kitts, I. Mas, "Dynamic Guarding of Marine Assets Through Cluster Control of Automated Surface Vessel Fleets." *IEEE/ASME Transactions on Mechatronics*, vol. 17, no. 1, pp. 65–75, 2012.

[56] I. Mas, S. Li, J. Acain, C. Kitts, "Entrapment/Escorting and Patrolling Missions in Multi-Robot Cluster Space Control." *Proc IEEE International Conference on Intelligent Robots and Systems*, pp. 5855–61, 2009.

[57] H. Choset, K. Lynch, S. Hutchinson, G. Kantor, W. Burgard, L. Kavraki, S. Thrun, "Principles of Robot Motion – Theory, Algorithms, and Implementations," Cambridge, MA, The MIT Press, 2005, pp. 353–356.

[58] Z. Shiller, H. Lu, "Computation of Path Constrained Time Optimal Motions With Dynamic Singularities," *Transactions of the ASME Journal of Dynamic Systems, Measurement, and Control*, vol. 114, no. 1, pp. 34–40, 1992.

[59] Z. Shiller, "Time-Energy Optimal Control of Articulated Systems with Geometric Path Constraints," *Transactions of the ASME. Journal of Dynamic Systems, Measurement and Control*, vol. 118, no. 1, pp. 139–143, March 1996.

[60] A.E. Bryson, Y.C. Ho, "Applied Optimal Control, Optimization, Estimation and Control," Hemisphere Publishing, New York, 1975.

[61] C. Kim, B. Kim, "Minimum-Energy Translational Trajectory Generation for Differential-Driven Wheeled Mobile Robots," *Journal of Intelligent Robotic Systems*, vol. 49, pp. 367–383, 2007.

[62] I. Mas, C. Kitts. "Obstacle Avoidance Policies for Cluster Space Control of Nonholonomic Multi-Robot Systems," *IEEE/ASME Transactions on Mechatronics*, vol. PP, no. 99, pp. 1–12, 2011.

Chapter 8

Laser Graphics in Augmented Reality Applications for Real-World Robot Deployment

Gerald Seet, Viatcheslav Iastrebov,
Dinh Quang Huy and Pang Wee-Ching

Additional information is available at the end of the chapter

http://dx.doi.org/10.5772/63635

Abstract

Lasers are powerful light source. With their thin shafts of bright light and colours, laser beams can provide a dazzling display matching that of outdoor fireworks. With computer assistance, animated laser graphics can generate eye-catching images against a dark sky. Due to technology constraints, laser images are outlines without any interior fill or detail. On a more functional note, lasers assist in the alignment of components, during installation.

We propose the use of lasers in the generation of graphics for augmented reality applications. Whilst the unfilled line drawings may be considered as a disadvantage, the ability to project images in a bright outdoor environment is an advantage, particularly in a natural environment. This chapter describes the use of laser outline graphics to augmented reality applications for a proposed 'industrial' application in a shared-environment of humans and robots. This implementation demonstrates a novel application and reaffirms the efficacy of laser graphics in providing notification to third party humans in the environment.

When a mobile robot is able to indicate its intentions, humans in its vicinity can better accommodate its actions to avoid possible conflicts. A framework for implementing human-robot interface is proposed. Wearable transparent LCD displays offer a high definition graphical interface to the human supervisor to allow for robot control, whilst laser generated notifications allows both the supervisor and other humans in the shared environment to be informed of the robots intentions, without the need of wearable devices. This leads to a more inclusive interaction for all humans in the shared environment.

Keywords: laser graphics, human-robotic interaction (HRI), wearable technology, mobile robot control, augmented reality

1. Introduction

In the past, robots were deployed primarily in industrial scenarios where tasks were repetitive and in a fixed sequence under a structured and well-constrained condition. Robots were programmed and debugged 'off-line', before their programs were ported to the shop floor. The tasks may be expected to be repeated many hundreds of thousands, or even millions of times, 24 hours a day, and continuously for a number of years. A typical example of such a scenario would be in the manufacturing of automobiles. In this 'high volume and low-mix' manufacturing application, the robot actions are explicitly defined and programmed for the robot to execute. The cost and time spent to program and commission each robot would be negligible considering the number of automobiles produced and the relatively long product cycles.

In 'low volume and high-mix' applications, the use of robots may be unattractive due to the relative complexity of robot programming and the related setup costs. As we expand on the scope of robotic applications, a different mode of interaction is evolving. Recently, more robots are deployed in semi-structured manufacturing environment [1] or in domestic environment like in homes [2]. Industrial robots are becoming more collaborative in their interactions with humans and are designed to work with humans in the same environment, without the provision of safety enclosures. In a home setting, it is best that the robot takes on the role of a compliant friend who is able and willing to do our bidding. Tasks required of the robot in such an environment are expected to be different and non-repetitive, at least within a time scale of hours or minutes.

Hence, it is important that the task of programming and interaction between human and robot become more natural and intuitive, evolving to an interaction that is typically associated to that between human. The different operating scenarios require new paradigms in the implementation of a Human-Robot Interface (HRI) such that data are clearly presented and easily accessible to all relevant parties. Industrial robots typically present their data on a computer display and often use a keyboard, mouse, or touch pendants as its input device. This setup is not favourable because it causes the human operator to have a divided attention between following the task procedure, visually, and simultaneously monitoring the important parameters on the computer display [3]. Furthermore, in an industrial setting, the need for protective wear would make the usage of the mouse and keyboard or touch pendants untenable.

In this chapter, we propose a framework that uses laser graphics to develop an augmented reality application for HRI. This envisages the evolution of HRI to be more natural with the robot providing a greater contribution in formulating the desired outcome. This invariably moves the robot and human towards a relationship exhibiting greater interaction in terms of frequency and quantity of information exchanged during their interactions. In an environment shared between humans and robots, the human needs to be able to recognise the intention of robots and vice versa. This is to avoid the possibility of conflicts and accidents.

For human and robot to interact meaningfully, through mutual understanding, a mechanism for dialog between robots and humans must exist [4, 5]. The human should be able to define and refine his needs. In addition, the robot is required to deliberate, formulate solutions and present appropriate options, in a form suitable for human understanding. On the other hand, a robot should be able to understand its human partner through common conversational gestures frequently used by humans [6, 7], such as by pointing and gazing. There must also be a common frame of spatial referencing [8, 9] to avoid ambiguities.

Augmented reality (AR) technologies [10] are used in our proposed framework to help the human and robot to define their communication and intentions [11]. Two types of AR technologies – the 'see-through' AR [12, 13] and spatial AR [14, 15] – are applied to enable the human to manipulate the robot in a way that the robot can understand. Similarly, these AR technologies help the robot to convey information so that the human has a better understanding of what the robot is doing and in its intentions. In research efforts involving the programming and controlling of mobile robots, there are some that make use of 'see through' AR technologies [11, 16], and others that utilize the spatial AR technologies [8, 17].

In the Mercedes-Benz autonomous concept car programme, laser generated graphics was proposed to indicate when it was safe for a human to cross its path by projecting a moving 'pedestrian crossing'. In this application, the advantages of laser in being able to project images in a natural environment were exploited effectively. In addition, the need for a mobile robot, or vehicle, to communicate its intention to humans was also elaborated.

The desire for a more natural and intuitive interface for human-robot interaction was studied by a number of researchers including [18]. These incorporated laser pointing to assist in defining targets and projected imagery to enhance interaction. In addition, others explored human mimicking interactions [19], while others focused on projectors devices [20].

The design of HRI had previously focused only on the interaction between the robot and the human that is controlling the robot. This is a natural omission, as humans and robots had previous been kept separate as a feature of design. As humans and robots intrude into the other's space, the needs of both humans and robots outside the intended interaction need to be considered. In the sharing of resources, it is important that all parties are aware of the other's intentions. This chapter identifies the need and describes the use of laser-based line graphics in the provision of such a function.

2. Proposed human-robot interaction framework

Figure 1 presents our proposed framework for Human-Robot Interaction. The components of the framework and their inter-linkages are presented in the schematic. The central element is the interaction kernel, which directs the control flow of the application and updates the model data such as the motion commands, motion trajectories and robot status. In each update loop, the human user could use the *user interface*, which consists of a multi-modal handheld device, to provide control commands or task information for the robot to execute.

The user interface uses two different types of AR technologies to display the robot information for the human user to visualize. The human user would be wearing a see-through AR glasses so that he or she can view more information regarding the robot, as well as the task in the real world. In addition, the robot has an on-board laser projector to provide for spatial AR. The robot projects its status and intentions as words or symbols onto the physical floor or wall, depending on the nature of the desired notification.

In the *task support* module within the framework, the human interacts with the robot through a dialogue Graphical-User-Interface (GUI), moving towards a defined task, which is acceptable to the human and executed by the robot. The role of the robot is enhanced from the traditional role of a dumb servant to that of a competent 'partner'. To highlight the need to maintain the higher status of the human in the decision-making hierarchy, we refer to this partnership as a master-partner relationship. It is inferred that the human makes the decisions whilst the robot assists the human user by considering the information on intended task, as well as the task constraints to provide appropriate task support to the human. The robot should be capable of providing suggestions to its human master and be able to learn and recognise the human's intentions [21]. The robot must also be imbued with a knowledge base to allow it to better define the problem. Only with these capabilities will the robot be able to elevate its role towards that of a collaborating partner.

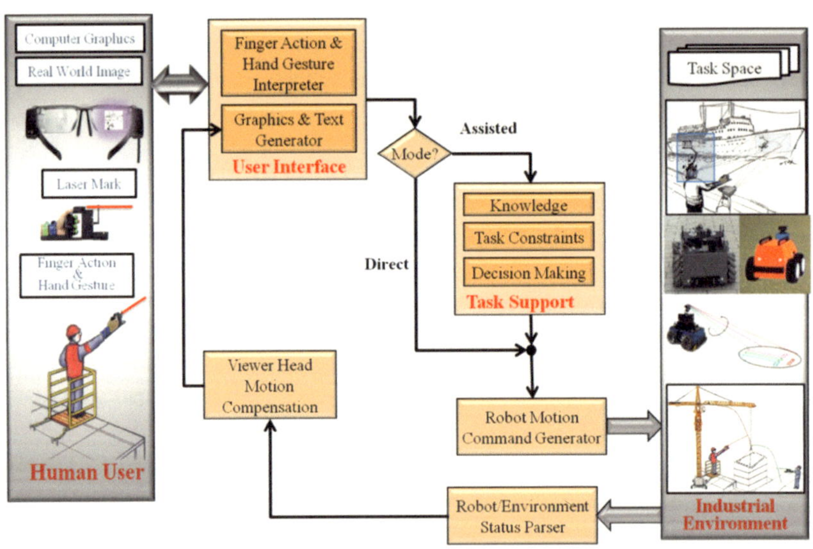

Figure 1. A framework for dynamic human-robot interactions.

Under the assisted mode, the human's cognitive load may be expected to be lower [22, 23] than if he were be responsible for all aspects of the task. In the performance of a task where the robot is unable to assist the human, in an appropriate manner, the human may elect to proceed without robot assistance. This would be the direct mode where the human operator determines the path, trajectory and operation parameters of the task. The model data, such as generated robot motion commands, planned trajectories, as well as the robot status are updated accordingly, and visualized consistently through the laser projection, the augmented reality display or a 2D graphical user interface.

This chapter presents, as concept verification, the development and evaluation of a user interface module, which uses laser graphics to implement the spatial AR technology to display the robot provided information for the human user to visualize.

3. Hardware configuration of the user interface module

The implementation of the user interface module is illustrated in **Figure 2**. It has been used to enable the robot to perform a waypoint navigation task, in a known environment, where local features may change and obstacles moved. Intrinsic to the implementation are the hardware devices for the human to interact with the robotic system.

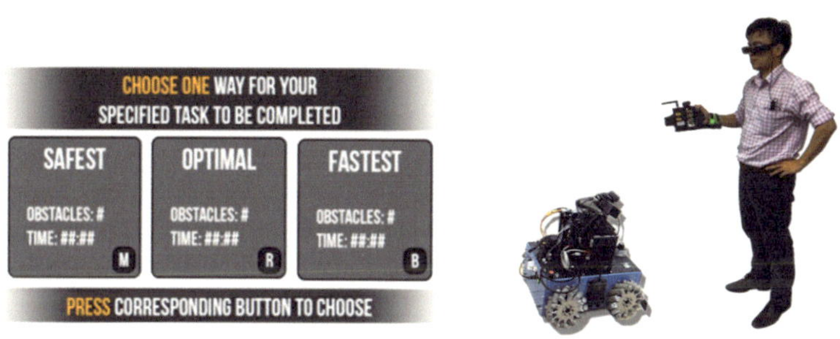

Figure 2. The GUI menu (left): an operator controlling a robot with a handheld device and a wearable display device (right).

3.1. Wearable transparent display

The human is provided with an Epson Moverio BT-200. It is a wearable device with a binocular ultra-high resolution full colour display. It incorporates a front facing camera and motion sensors. The motion sensors capture the user's head motion and the camera supports target tracking in the observer's field of view. This device is depicted on the human operator in **Figure 2**. Wearing the device allows the user to view his environment as well as any augmented data that is generated, overlaid on the real-world scene. Through the display, the robot system can provide status information and selection menus for the operator to select. The advantage of a wearable transparent LCD display is that it provides the human with an unimpeded view of his environment.

3.2. Multi-modal handheld device

A novel multimodal wireless single-handheld device is as depicted in **Figure 3**. It comprises five units of spring loaded finger paddles, a nine-axis inertial measurement unit (IMU), a laser pointer and a near-infrared LIDAR sensor. With the spring-loaded linear potentiometers, a position-to-motion mapping is programmed to map individual finger displacement to a certain motion command for the robot. The hand motions of a user are sensed by the IMU, and gesture recognition is applied to interpret the human gestures. This would allow the user to interact with the robot through gestures.

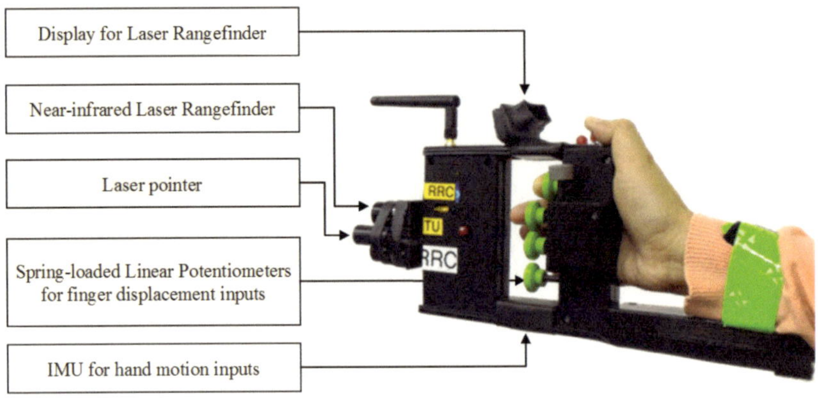

Figure 3. Multimodal single-handed human input device.

A laser pointer is included on the device so that a user can point and define a particular waypoint location or a final destination. The LIDAR sensor is included to assist in user localization by the robot.

This device supports a one-handed gloved operation and was designed for use in an industrial scenario. In an industrial setting, the hands of a human operator are, frequently, gloved, and the use of double-handed devices is viewed as being undesirable from safety considerations.

3.3. Robotic platform

The human-robot partnership framework is implemented on MAVEN [24, 25]. The robot, shown in **Figure 4**, is a holonomic robot with four mecanum wheels. It hosts an on-board computer for controlling the drive motors and its other robotic services. The Linux OS has been installed as the robot's operating system for the embedded computer. The robot operating system (ROS) is installed along with the Linux OS.

The maximum forward and lateral speeds of the robot have been limited to 0.6 m/s while the on-the-spot rotational speed has been limited to 0.9 rad/s. The various sensor and behaviour modules that have been installed on the robot include a Hokuyo laser rangefinder module, a USB camera module, an MJPEG server, a localization system, map server, as well as a path planning and a navigation module.

The robot is provided with a laser projection-based spatial AR system, which enables the projection of line graphics and text onto a suitable surface. The projected images can be used to augment reality in the traditional manner or to provide indications of the robot's intention or status. In the context of a moving robotic platform, it can project the robot's intention to

http://dx.doi.org/10.5772/63635

move, turn, or stop. The intended path that is planned by the robot can also be projected on to the floor or road surface. During interactions, the laser graphics are used to project markers to confirm destinations or to place virtual objects for the human to confirm its desired position and orientation.

The ability to recognise the robotic platform's intentions allows humans (and other robots) to adjust their motion to avoid conflicts. This would enhance the safety of humans in the vicinity of the robot.

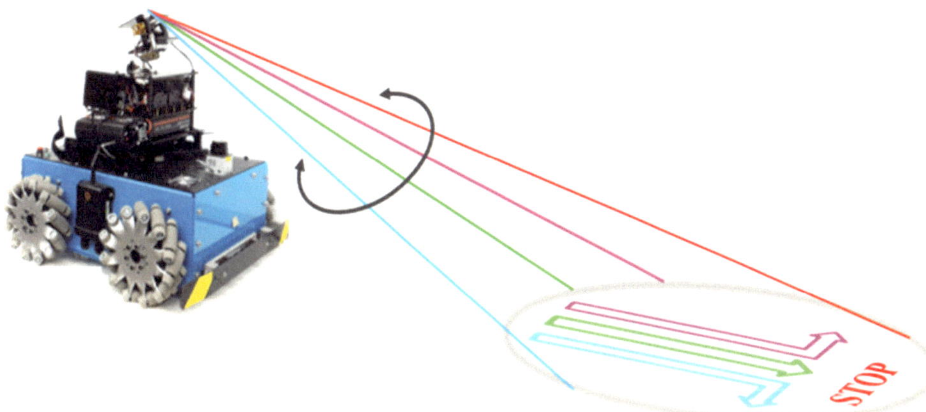

Figure 4. Robot with a laser projection-based spatial AR system.

4. Laser projection-based AR system

A laser projection system has been installed on the robot to provide for a spatial augmented reality. It facilitates the presentation of projected digital AR information such as the robot motion behaviours and its motion trajectories onto the floor surface of the work environment.

Figure 5 depicts a schematic of the implemented laser projection system. To create laser graphics, two tiny computer-controlled mirrors are used to direct the laser beam onto a suitable surface. The first mirror rotates about the horizontal axis while the second mirror rotates about the vertical axis. A pair of galvanometers is used to produce the rotating motions, which subsequently aims the laser beam to any point on a square or rectangular raster. The position of the laser point is controlled by changing the electric current through its coil in a magnetic field. The shaft, of which the mirror is attached to, will rotate to an angle proportional to the coil current. In this manner, by combining the motions of the two galvanometers in orthogonal planes, the x-y position of the projected laser spot can be changed.

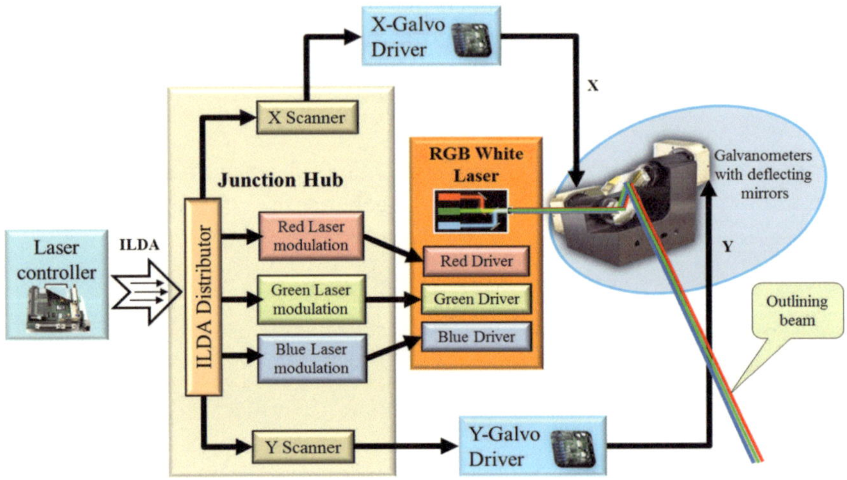

Figure 5. Schematic diagram of the laser writer display system.

The computer 'connects the dots' by rotating the mirrors at a very high speed. This causes the laser spot to move sufficiently fast from one position to another, resulting in a viewer seeing a single outline drawing. This process is called 'scanning' and computer-controlled mirrors are galvanometer 'scanners'. The scanners move from point to point at a rate of approximately 30–40 kpps. To add more detail to a scene, additional sets of scanners can be used to overcome the limitations in scanning speeds.

A multicolour laser projection system would consist of red, green and blue lasers, each with its individual driver and optics. The drivers also control the intensity of each laser source independently. Red, green and blue laser beams are mixed in the transparent mirror system and the combined beam is subsequently projected onto the mirrors of the galvanometers. Together, these three laser diodes combine their output to produce a white or an 'infinitely' varied coloured beam.

4.1. Image generation

An International Laser Display Association (ILDA) interface can be used to import custom graphics, text and effects into laser animation format. Files containing scalable vector pictures or videos are loaded to the graphic controller in a special format. The control software converts these files into a list of sequential points, each of which is characterised by the angular deflections of the galvanometers in the vertical and horizontal planes. The intensity of laser radiation is also controlled via the interface.

The ILDA laser control standard produces a sequence of digital-to-analog converter (DAC) outputs on differential wire pairs with average amplitude of ± 24V for the galvanometer control and ±5V for the laser diode drivers. When the galvanometers receive a new value for mirror deflections, it drives the mirrors to the next desired angular position.

4.2. Transformations for inclined surface projection

As the projector frame is not necessarily perpendicular to the projection surface, there will be visible distortions in the source image (Projector Frame) projected onto the surface (Projected Image Frame) as shown in **Figure 6**.

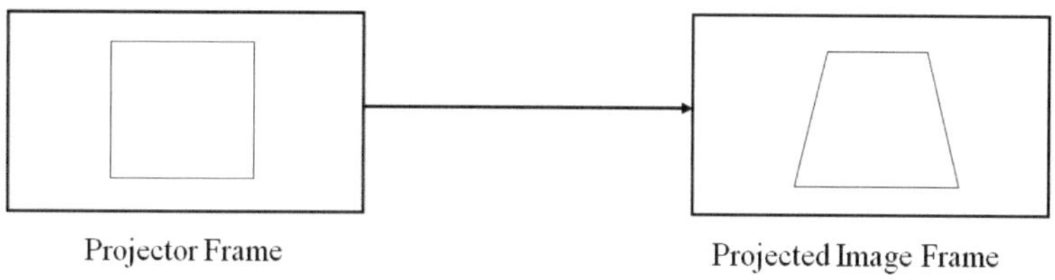

Figure 6. Projected image is distorted when projected onto an oblique surface.

This distortion needs to be corrected through a pre-warping process that is applied to the projection image, before being projected to the particular surface.

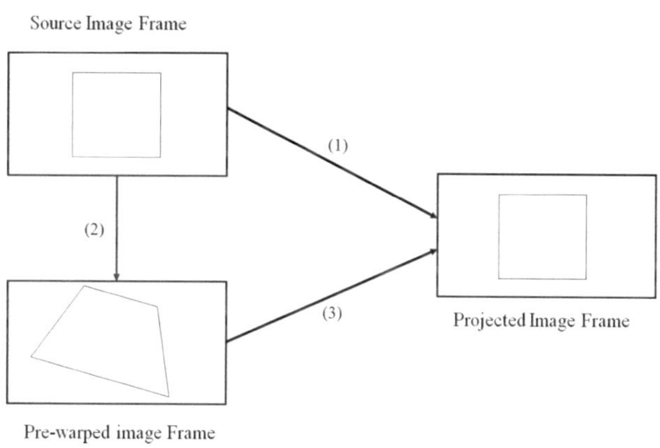

Figure 7. Projected image distortion correction.

The pre-warping process can be accomplished using the concept of homography in computer vision [26], as shown in **Figure 7**. The idea is to find the transformation matrix between the source image frame and the projected image frame, as illustrated in **Figure 7** step (1). Typically, a square outline will be projected onto the desired surface and its corner's positions will be estimated. The corner positions of this square, in the source projector frame, are also calculated. The appropriate transformation matrix will be determined using the corresponding points. The details of this method are described in the paper by Rahul, Robert and Matthew [27]. The next step, as seen in **Figure 7** step (2), is to multiply this matrix with our original image to obtain the pre-warped image frame. Finally, as depicted in **Figure 7** step (3), this image is projected onto the inclined surface to produce the undistorted square image (Projected Image Frame).

4.3. Camera-projector calibration and software

Our multimodal handheld device is equipped with a laser pointer, which is used to project a marker point for indicating a reference or indicating a chosen item. The marker and its projected position are identifiable by the camera, located close to the laser writer. To determine the position of the marker, a calibration process is performed to obtain the necessary transformation parameters.

The calibration process is performed by using the laser writer to draw a square with known parameters onto the floor within a region, in the camera field of view. The camera captures the image of the square that was projected on the floor. The corners of the square in the camera frame are subsequently extracted. These values, together with the known projector frame, are used to obtain the projector-camera homography. **Figure 8** illustrates the procedures.

Figure 8. Projection of calibration target and corner extraction.

Figure 9 shows the camera-projector system and demonstrates the use of the laser marker to indicate a position. The system confirms the position of the marker by responding with the projection of an arrow head that points to the marker location. In this scenario, the robot will project an arrow that follows the laser marker indicated by the user.

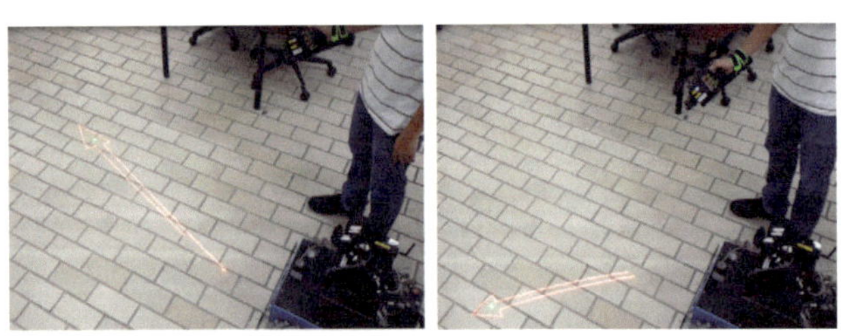

Figure 9. Laser input detection and arrow projection.

5. An implementation

Within the proposed human-robot framework, we identify three different groups of people that may interact with the robot. These people, who are known as interactants, are grouped

http://dx.doi.org/10.5772/63635

according to their roles, relating to the robot actions, and on the nature of the information they may require.

Group 1 Interactants – *Operator* who is responsible for the control and supervision of the robot. The operator is equipped with an LCD display and the Multi-Modal Human Input Device (HID). A wearable see-through transparent LCD display is provided to allow the operator an unobstructed visual awareness of the environment. The LCD display functionality allows for the projecting of high-resolution dialog actions proposed by the partner-robot. In addition, relevant information required by the operator to allow for timely intervention would also be displayed. This information includes communication strength and battery health. With the augmented view, the operator is able to provide the necessary support and commands to the robot. Navigation through the menu options and selection of specific items are executed using the Multi-Modal Human Input Device.

Group 2 Interactants – *Observers* who are interested in monitoring the tasks being executed. This group of humans may be equipped with the transparent LCD displays where they can share the actions of the human in controlling the tasks. Without the wearable displays, this group of observers would only be able to share in the notifications by the robot through the 'Laser Writer'. They would, however, not be permitted or able to control the actions of the robot, as control is only permitted through the Multi-Modal Human Input Device. This restriction provides a clear differentiation between the two groups.

Group 3 Interactants – *Passerby* who are in the vicinity of interaction, but who are not directly related to, or interested in the task being executed by the robot. The interest of this group of passerby arises from the sharing of common space and the need to accommodate the robot's motion. Predominantly, the interest may be restricted to one of avoiding the robot and its workspace. Their interest is in the near-term actions of the robot as in the robot's current actions or in its next action. They need to be provided with the ability to identify the robot's actions or its intention. The robot can support this need for situation awareness by this group through the visual prompts generated by the laser writer (**Table 1**).

	Operator (Group 1 interactant)	**Observer (Group 2 interactant)**	**Passerby (Group 3 interactant)**
Wearable transparent LCD display	Provided	May be provided	Not provided
Multimodal HID	Provided	Not provided	Not provided
Remark	Fully involved in controlling the robot. View laser indications for status feedback	Involved in collaborating with either the operator or robot. View laser indications for status feedback	Not involved in the operation. Views laser indications to avoid collisions

Table 1. Framework for multimodal AR GUI.

Figure 10 shows the dialog generated by the wearable display. Each robot is equipped with a unique augmented reality marker for facilitating the process of overlaying the computer-generated information over the real scene. The operator, with the multimodal hand controller, will have the capability of visually selecting the options suggested by the robot to complete the task. This offers a more intuitive and convenient way to interact with the robot.

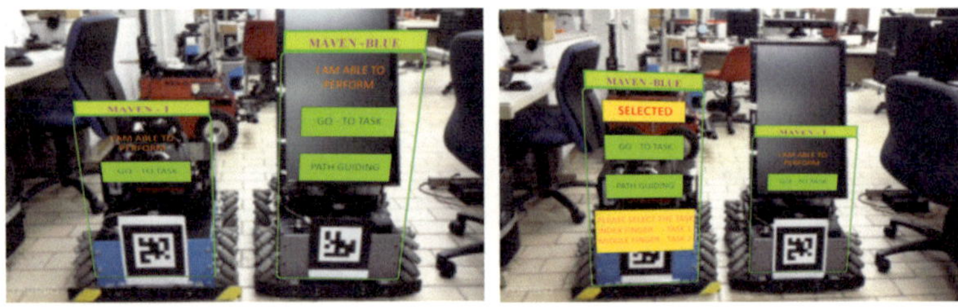

Figure 10. Dialog menu for Group 1 interactants.

While the GUI on the wearable display enables the operator to intuitively control and monitor the robot, other humans who share the same working environment (Passerby) might encounter difficulties in accommodating this robot's motion and inadvertently cross into its intended path. The laser writer is provided to overcome this problem by providing passersby with visual indications of the robot's intentions. **Figure 11** shows one possible implementation for this system. In this scenario, this laser system allows the robot to project the direction of its trajectory, before moving from one point to another. Particularly, the robot is able to indicate: 'stop', 'forward', 'backward', 'turn left' and 'turn right' as text or line-graphic symbols.

Figure 11. Laser notification to all interactants and LCD menus.

6. Discussion

The proposed framework for the dual-modality AR system was implemented for deployment in a laboratory setting where humans and mobile robots are expected to coexist and operate in close proximity. The mobile platform was 'let loose' within the laboratory without any prior briefing to the other laboratory users. Third party human-robot interaction was observed, and a 'qualitative' feedback was solicited from the 'unsuspecting' human. Two scenarios were

http://dx.doi.org/10.5772/63635

evaluated. The first scenario was without the laser writer augmentation, and the second with the laser writer function activated. In this second scenario, the robot projected its action and indicated its motion trajectory.

Invariably the response was positive, especially when the robot was stationary and the human was unsure of the response required of them. In instances when the robot's route was identified, humans would find an alternative route in an attempt to avoid the robot. The human response of avoiding the robot's workspace enabled the robot the option of increasing its speed resulting in better operational performance.

As a natural extension of the unmanned deployment of mobile robots, the framework of providing notification of the robot's actions is recognised to be supportive to the application of autonomous transportation in an urban scenario. In such a scenario, vehicles are larger and environments shared with humans who are less familiar with robot interaction. The ability to improve on the passerby's awareness of the robot's actions could be achieved using the laser writer information system described in this chapter. When the vehicle is avoiding or stopping for the human, appropriate indications may be provided to the human by use of the laser notification system.

7. Conclusion

In this chapter, we highlight the application of laser-generated outline-graphics as a viable addition to 'augmented-reality'. Its images are bright and of high contrast. This lends itself to applications in natural environments, both indoors and outdoors, where the ambient lighting is expected to be relatively bright. Whilst unfilled graphics may be considered as a deficiency, when attempting to generate visually appealing GUI menu with filled graphical images, it can be used effectively to project images and outline text onto the surrounding surfaces. These images can be viewed without the aid of any wearable devices in a natural environment.

In addition, we proposed a design framework for GUI implementation in human-robot shared environments. In this framework, we identify specific requirements of the first party, human in direct control of the robot and the requirements of the third party, humans in the operating vicinity of the robot. The needs of each group are different and can be optimally addressed using different AR modalities. The use of laser-generated line graphics was deployed as a means of projecting messages and notifications to humans in the vicinity. This is perceived as supporting a safer working environment that humans and robots can share. In addition, humans are also enabled with a better awareness of the robot's actions and this reduces the possibility of accidents. The behaviour of humans avoiding the robot's workspace, where such options exist, produces opportunities for faster platform speeds and improved task efficiencies.

Mobile robots should indicate its intension as it executes its task to humans in its vicinity. This is a necessary requirement when the general public and mobile robots share and intrude into the workspace of the other. In recent times, we witness the deployment of automated transportation of food items in restaurants and of humans in autonomous vehicles. These are deployments in shared environments. The issues highlighted are relevant and worthy of consideration in their design and implementation. The laser writer provides a simple and effective way to improve on passerby situation awareness in a naturally bright environment.

Acknowledgements

This research was carried out at the School of Mechanical & Aerospace Engineering and BeingThere Centre, Nanyang Technological University. The BeingThere Centre is supported by the Singapore National Research Foundation under its International Research Centre @ Singapore Funding Initiative and administered by the IDM Programme Office. In addition, the A*STAR Industrial Robotics Program is gratefully acknowledged.

Author details

Gerald Seet[1*], Viatcheslav Iastrebov[1], Dinh Quang Huy[2] and Pang Wee-Ching[2]

*Address all correspondence to: mglseet@ntu.edu.sg

1 Robotics Research Centre, School of Mechanical & Aerospace Engineering, Nanyang Technological University, Singapore

2 BeingThere Centre, Institute of Media Innovation, Nanyang Technological University, Singapore

References

[1] E. Guizzo and E. Ackerman, "How rethink robotics built its new baxter robot worker," in *IEEE Spectrum*, 2012.

[2] J. Forlizzi and C. DiSalvo, "Service robots in the domestic environment: a study of the roomba vacuum in the home," in *Proceedings of the 1st ACM SIGCHI/SIGART Conference on Human-Robot Interaction*, 2006, pp. 258–265.

[3] A. Olwal, J. Gustafsson, and C. Lindfors, "Spatial augmented reality on industrial CNC-machines," in *Electronic Imaging 2008*, 2008, pp. 680409–680409.

[4] T. Fong, C. Thorpe, and C. Baur, "Collaboration, dialogue, human-robot interaction," in *Robotics Research*, Springer, 2003, pp. 255–266.

http://dx.doi.org/10.5772/63635

[5] M. E. Foster, M. Giuliani, A. Isard, C. Matheson, J. Oberlander, and A. Knoll, "Evaluating description and reference strategies in a cooperative human-robot dialogue system," in *IJCAI*, 2009, pp. 1818–1823.

[6] G. Hoffman and C. Breazeal, "Robots that work in collaboration with people," in *AAAI Symposium on the Intersection of Cognitive Science and Robotics*, 2004.

[7] C. L. Sidner, C. D. Kidd, C. Lee, and N. Lesh, "Where to look: a study of human-robot engagement," in *Proceedings of the 9th International Conference on Intelligent User Interfaces*, 2004, pp. 78–84.

[8] T. Tenbrink, K. Fischer, and R. Moratz, "Spatial strategies in human-robot communication," *KI*, vol. 16, no. 4, pp. 19–23, 2002.

[9] M. Skubic, D. Perzanowski, A. Schultz, and W. Adams, "Using spatial language in a human-robot dialog," in *Robotics and Automation, 2002. Proceedings. ICRA'02. IEEE International Conference on*, 2002, vol. 4, pp. 4143–4148.

[10] R. T. Azuma, et al., "A survey of augmented reality," *Presence*, vol. 6, no. 4, pp. 355–385, 1997.

[11] P. Milgram, S. Zhai, D. Drascic, and J. J. Grodski, "Applications of augmented reality for human-robot communication," in *Intelligent Robots and Systems '93, IROS '93. Proceedings of the 1993 IEEE/RSJ International Conference on*, 1993, vol. 3, pp. 1467–1472.

[12] M. Billinghurst, R. Grasset, and J. Looser, "Designing augmented reality interfaces," *ACM Siggraph Computer Graphics*, vol. 39, no. 1, pp. 17–22, 2005.

[13] T. Collett and B. A. MacDonald, "Developer oriented visualisation of a robot program," in *Proceedings of the 1st ACM SIGCHI/SIGART Conference on Human-Robot Interaction*, 2006, pp. 49–56.

[14] O. Bimber and R. Raskar, *Spatial Augmented Reality: Merging Real and Virtual Worlds*. CRC Press, 2005.

[15] O. Bimber and R. Raskar, "Modern approaches to augmented reality," in *ACM SIGGRAPH 2006 Courses*, 2006.

[16] T. Pettersen, J. Pretlove, C. Skourup, T. Engedal, and T. Lokstad, "Augmented reality for programming industrial robots," in *Mixed and Augmented Reality, 2003. Proceedings. The Second IEEE and ACM International Symposium on*, 2003, pp. 319–320.

[17] M. Zaeh and W. Vogl, "Interactive laser-projection for programming industrial robots," in *Proceedings of the 5th IEEE and ACM International Symposium on Mixed and Augmented Reality*, 2006, pp. 125–128.

[18] J. Park and G. J. Kim, "Robots with projectors: an alternative to anthropomorphic HRI," in *Proceedings of the 4th ACM/IEEE international conference on Human robot interaction*, 2009, pp. 221–222.

[19] A. Strauss, A. Zaman, and K. J. O'Hara, "The IMP: an intelligent mobile projector," in *Enabling Intelligence through Middleware*, 2010.

[20] G. Reinhart, W. Vogl, and I. Kresse, "A projection-based user interface for industrial robots," in *Virtual Environments, Human-Computer Interfaces and Measurement Systems, 2007. VECIMS 2007. IEEE Symposium on*, 2007, pp. 67–71.

[21] S. K. Sim, K. W. Ong, and G. Seet, "A foundation for robot learning," in *Control and Automation, 2003. ICCA'03. Proceedings. 4th International Conference on*, 2003, pp. 649–653.

[22] C. Y. Wong, G. Seet, S. K. Sim, and W. C. Pang, "A framework for area coverage and the visual search for victims in usar with a mobile robot," in *Sustainable Utilization and Development in Engineering and Technology (STUDENT), 2010 IEEE Conference on*, 2010, pp. 112–118.

[23] W. C. Pang, G. Seet, and X. Yao, "A study on high-level autonomous navigational behaviors for telepresence applications," *Presence: Teleoperators and Virtual Environments*, vol. 23, no. 2, pp. 155–171, 2014.

[24] W. C. Pang, G. Seet, and X. Yao, "A multimodal person-following system for telepresence applications," in *Proceedings of the 19th ACM Symposium on Virtual Reality Software and Technology*, 2013, pp. 157–164.

[25] G. Seet, P. W. Ching, B. Burhan, C. I-Ming, V. Iastrebov, W. Gu, and W. C. Yue, "A design for a mobile robotic avatar – modular framework," in *3DTV-Conference: The True Vision – Capture, Transmission and Display of 3D Video (3DTV-CON), 2012*, 2012, pp. 1–4.

[26] R. Hartley and A. Zisserman, *Multiple View Geometry in Computer Vision*. Cambridge University Press, 2003.

[27] R. Sukthankar, R. G. Stockton, and M. D. Mullin, "Smarter presentations: Exploiting homography in camera-projector systems," in *Computer Vision, 2001. ICCV 2001. Proceedings. Eighth IEEE International Conference on*, 2001, vol. 1, pp. 247–253.

Chapter 9

Dynamic Optimized Bandwidth Management for Teleoperation of Collaborative Robots

Chadi Mansour, Mohamad El Hariri, Imad H. Elhajj,
Elie Shammas and Daniel Asmar

Additional information is available at the end of the chapter

http://dx.doi.org/10.5772/63725

Abstract

A real-time dynamic and optimized bandwidth management algorithm is proposed and used in teleoperated collaborative swarms of robots. This method is effective in complex teleoperation tasks, where several robots rather than one are utilized and where an extensive amount of exchanged information between operators and robots is inevitable. The importance of the proposed algorithm is that it accounts for Interesting Events (IEs) occurring in the system's environment and for the change in the Quality of Collaboration (QoC) of the swarm of robots in order to allocate communication bandwidth in an optimized manner. A general dynamic optimized bandwidth management system for teleoperation of collaborative robots is formulated in this paper. The suggested algorithm is evaluated against two static algorithms applied to a swarm of two humanoid robots. The results demonstrate the advantages of dynamic optimization algorithm in terms of task and network performance. The developed algorithm outperforms two static bandwidth management algorithms, against which it was tested, for all performance parameters in 80% of the performed trials. Accordingly, it was demonstrated that the proposed dynamic bandwidth optimization and allocation algorithm forms the basis of a framework for algorithms applied to real-time highly complex systems.

Keywords: bandwidth allocation, collaborative robots, teleoperation, optimization

1. Introduction

Teleoperation, or remote tele-manipulation of robots in inaccessible environments such as deep sea and outer space [1, 2], includes a wide variety of applications such as micro- and nano-teleoperation [3], tele-surgery [4], etc. A network-based teleoperation system involves distant interactions between human operators and remote robotic systems [5, 6]. In addition, swarms of robots are widely employed in complex tasks that cannot be performed by a single robot or in tasks that are better achieved by cooperation of robots such as localization in formations [7], target tracking [8], mapping and localization [9], object pushing [10], area exploration for search and rescue [11], etc. When swarms of mobile robots are teleoperated, specific network requirements should be satisfied in order to guarantee a minimum quality of control, which results in efficient task execution. Research done on teleoperated systems showed that constraints such as bandwidth and CPU processing cause the Quality of Service (QoS) to degrade to an extent that may severely affect performance [12, 13]. To address this problem, various bandwidth management algorithms have been presented for distributed multimedia systems in order to maintain a performance that guarantees an adequate QoS [6]. However, the literature rarely tackled the problem of managing bandwidth based on sensory feedback and the quality of collaboration among robots. Accordingly, a real-time dynamic optimized bandwidth management for teleoperation of collaborative robots is introduced in this paper. The proposed method accounts for interesting events (IEs) and the change in the quality of collaboration (QoC) between robots in order to optimize the allocation of bandwidth between acting agents, where necessary, in a given environment. The developed optimization technique showed outstanding performances when implemented on a system of two collaborating humanoid robots, and thus could be considered a basis for a framework for highly complex algorithms implemented in systems involving real-time bandwidth optimization, where multiple users control multiple collaborating robots.

Different types of resource management algorithms are used to solve the bandwidth allocation problem in robotics systems. Such applications in networked control systems fall into two main categories: static [14] and dynamic [15] bandwidth allocation. Static methods cannot adapt to changes in the system state (surrounding environment, collaboration quality, etc.). Alternatively, dynamic bandwidth allocation algorithms increase performance at the cost of increased computation. Mourikis et al. [7] address the problem of resource allocation in formations of mobile robots localizing as a group. The goal is to determine the frequency at which each individual sensor should be used in order to attain the highest possible localization accuracy. The set of frequencies mentioned is obtained by solving an optimization problem that maximizes the accuracy matrix expressed in terms of the sensors' frequencies. However, the problem is solved offline and the algorithm does not account for any dynamic events that might occur. Sugiyama et al. [8] propose a bandwidth reservation algorithm for multi-robot systems in a target tracking mission. The interesting information, corresponding to a survivor's detection, is sent to the base station with wideband signals such as dynamic picture images. The final call is left to operators to decide whether the received images indicate a real victim, by allowing/preventing the corresponding robot to reserve the bandwidth affecting the flow of various signals from other robots to the base station. In this approach, the operator's intervention is crucial in allocating bandwidth and thus the allocation process is not fully automated. Xi et al. [5] developed a bandwidth allocation mechanism based on online measured task dexterity index of dynamic tasks so that operators can control remote manipulators efficiently and smoothly even under poor network quality. However, the executed task is simple and does not require the collaboration of multiple robots to be performed. Thus, the quality of collaboration factor is not considered in the bandwidth allocation. Finally, in [10], a bandwidth management algorithm is introduced and the rate of feedback is regulated based on the amount of activities occurring in the environment. The work shows that during complex tasks, the operator's performance is affected by the rate of feedback of information. It is also confirmed that a higher sampling rate is required to maintain the same level of performance obtained when the environment is less dynamic. Yet, the implemented algorithm does not impose any constraint on the total bandwidth of the system. In addition, the notion of monitoring changes in QoC to allocate bandwidth is not mentioned since the task execution only requires the use of a single robot. To the best of the author's knowledge, there was limited research addressing bandwidth management for the specific application of collaborative robots teleoperation. In 2015, Ricardo and Guilherme designed a Dynamic Bandwidth Management Library to control the frequency of individual sensors present in a robotic environment performing a certain task [16]. This work is seeking a universal Dynamic Bandwidth Management Library designed to be used on a system with a variable number of heterogeneous robots performing any collaborative task that requires communication transactions such as the exchange of sensor data between involved agents.

Accordingly, the main contribution of the work presented in this paper is in accounting for (a) IEs occurring in the robotic swarm's environment and (b) changes in QoC among the swarm of robots in real-time optimized bandwidth management of teleoperated collaborative robots. Consequently, assessing the multi-robot swarm dynamics, the stability of the robotic swarm and the effects of packet loss and transmission delay under the proposed algorithm falls out of the scope of this paper. Factoring the latter into the proposed algorithm is possible but will alter the emphasis from the main contribution of dynamic optimized bandwidth management.

A literature review of the most relevant work in bandwidth management was presented in Section 1. The general problem is formulated in Section 2. The formulation is then implemented on an application in which an operator drives two collaborating robots. Section 3 describes the experimental set-up and the corresponding results. Finally, conclusions are presented in Section 4.

2. Problem formulation

The focus of the work presented is on real-time dynamic managing of the 'User' to 'Robot' and 'Robot' to 'User' (U2R/R2U) and 'Robot' to 'Robot' (R2R) communication channels, where actuation commands, system state and sensory data including video frames are exchanged. A general formulation of the problem that can be applied to swarms with a variable number of collaborating robots is presented first and is then implemented on a formation of two humanoid robots performing a collaborative task. The formulation is concluded by presenting the solution of the optimization problem.

2.1. The general formulation

The goal of the optimized dynamic bandwidth allocation algorithm is to optimize communication at each time event based on information related to the occurrence of *IEs* in the robot's surroundings and to changes in *QoC* among robots. The idea comes from the fact that when sudden changes occur in the robot's environment, the teleoperator needs to be updated more frequently in order to retain the same level of performance. In other words, the rates of information exchanged between operator(s) and robots are updated based on changes in task conditions that may affect the performance severely. Similarly, the increase in *R2R* communication compensates for any decay in the performance of the collaborative task of the robotic swarm. Hence, if the environment is less dynamic and the robots are collaborating well, the corresponding communication rates will be decreased. On the other hand, if the environment is more dynamic and the robots are collaborating poorly, the corresponding communication rates will be increased.

Let the collaborative task be executed by n robots and let x_i, where $i \in \{1, 2, \ldots, r\}$, be the communication rates to be optimized. Then, we define X as an r-dimensional vector of communication rates such that

$$X = \begin{bmatrix} x_1 & x_2 & \cdots & x_r \end{bmatrix}^T$$

Where the elements of X are classified into 3 sets of rates:

- Feedback Rates: R2U communication rates
- Collaboration Rates: R2R communication rates
- Command Rates: U2R communication rates

The rates x_i are subject to practical constraints that bound each of them with a minimum and maximum value x_{imin} and x_{imax}, respectively. In addition, the sum of bandwidth consumed by all channels is bounded by the total bandwidth of the system B_{max}. Hence, for all the rates, we have the following constraints:

$$x_{imin} \leq x_i \leq x_{imax}, \text{where } i \in \{1,2,\ldots,r\} \tag{1}$$

and

$$\sum_{i=1}^{r} w_i . x_i \leq B_{max} \tag{2}$$

where w_i's are weights associated with each rate corresponding to the rate of information (bps) sent at each time event on each channel. We define the vector $W = \begin{bmatrix} w_1 & w_2 & \ldots & w_r \end{bmatrix}^T$ as an r-dimensional vector of weights corresponding to each channel. We also define P as an m-dimensional observation vector that is composed of two main sets of components such that

$$P = \begin{bmatrix} p_1 & p_2 & \cdots & p_t & p_{t+1} & \cdots & p_m \end{bmatrix}^T = \begin{bmatrix} IE_{tx1} \\ QoC_{(m-t)x1} \end{bmatrix}, \text{ where } t < m$$

each element in P is an observation related to interesting events occurring in the robot environment or to the quality of the collaborative task executed by the robotic swarm. The elements of *IE*, $\{p_1, p_2, \ldots, p_t\}$, and of *QoC*, $\{p_{t+1}, p_{t+2}, \ldots, p_m\}$, are variables that dictate the choice of the communication rates. All the elements of P are normalized in the interval [0: 1]. If the ith observation is reflecting a slightly changing environment or a high collaborative performance, then p_i would be close to 0. On the contrary, if the ith observation reflects a highly changing environment or significant degradation in the collaborative performance, then p_i would be closer to 1.

Distance to obstacles, speed and displacement of robot are potential examples of *IEs* that could be monitored in order to allocate bandwidth. Moreover, any observation that tracks error in a collaborative task could also contribute to the bandwidth management algorithm.

In the formulation presented in this work, a mapping equivalent to the one presented in our previous work [17] is applied, however, with additional constraints that transform the problem from a simple matrix multiplication to a linear optimization problem. The new constraints bound the set of feasible communication rates x_i depending on the choice of the minimum and maximum rate for each channel x_{imin} and x_{imax} and the maximum bandwidth of the system B_{max}.

In this mathematical formulation, s_i is defined in the interval [0: 1] for all $i \in \{1, 2, \ldots, r\}$ to be as follows:

$$x_i = a_i s_i + b_i \tag{3}$$

where

$$a_i = x_{imax} - x_{imin} \tag{4}$$

and

$$b_i = x_{imin} \tag{5}$$

Then, an r-dimensional vector S is defined as follows:

$$S_{r*1} = \begin{bmatrix} s_1 & s_2 & \dots & s_r \end{bmatrix}^T.$$

Hence, at each time instant, the algorithm solves for the s_i's and then uses the mapping in (3) to get the rates x_i's.

$$X_{r*1} = A_{r*r}.S_{r*1} + B_{r*1} \tag{6}$$

S is related to P using the mapping matrix M as shown in (7):

$$S_{r*1} = M_{r*m}.P_{m*1} \tag{7}$$

where

$$A_{r*r} = \begin{bmatrix} a_1 & \dots & 0 \\ \vdots & \ddots & \vdots \\ 0 & \dots & a_r \end{bmatrix}; \; B_{r*1} = \begin{bmatrix} b_1 \\ \vdots \\ b_r \end{bmatrix}; \; S = \begin{bmatrix} s_1 \\ s_2 \\ \vdots \\ s_r \end{bmatrix};$$

$$M = \begin{bmatrix} m_{11} & m_{12} & \cdots & m_{1m} \\ m_{21} & m_{22} & \cdots & m_{2m} \\ \vdots & \vdots & \ddots & \vdots \\ m_{r1} & m_{r2} & \cdots & m_{rm} \end{bmatrix}; \; P = \begin{bmatrix} p_1 \\ p_2 \\ \vdots \\ p_m \end{bmatrix}$$

In (7), the elements of the matrix M are selected based on the relation between the observations and the rates. Each row of the matrix M can be interpreted as the weights of the observations in P affecting the corresponding rate in X. Since s_i's are selected in the range [0: 1], then choosing the sum of the coefficients in each row of M to be equal to 1 ensures that the result of multiplying any row of M by the vector P represents a weighted average of the observations that results in a value in the [0: 1] range. Thus, for a specific environment, M can be initialized once at the beginning of the set of trials. However, in order to improve the performance, M could also be updated dynamically based on the quality of the task previously executed. Thus, an improvement in the overall performance is achieved while maintaining an equivalent level of bandwidth consumption.

Since the sum of bandwidth consumed on all channels is bounded by the maximum bandwidth of the system, B_{max}, the allocation of the rates on different channels will be formulated as a linear optimization problem. Thus, the problem formulation becomes:

Minimize:

http://dx.doi.org/10.5772/63725

$$\|S - M.P\|_1 \tag{8}$$

subject to

$$\sum_{i=1}^{r} w_i.x_i \leq B_{max} \tag{9}$$

and

$$0 \leq s_i \leq 1 \forall i \in \{1,2,\ldots,r\} \tag{10}$$

Since $x_i = a_i s_i + b_i$, then Eq. (9) can be written as follows:

$$\sum_{i=1}^{r} (w_i.a_i.s_i + w_i.b_i) \leq B_{max} \tag{11}$$

which is equivalent to

$$\sum_{i=1}^{r} w_i.a_i.s_i \leq B_{max} - \sum_{i=1}^{r} w_i.b_i \tag{12}$$

The constraint in (12) can be expressed in terms of matrix multiplication as follows:

$$W^T.A.S \leq B_{max} - W^T.B \tag{13}$$

Therefore, for any collaborative task, it is sufficient to set the parameters in Eq. (13) in order to define the linear optimization problem. Thus, optimization techniques can be applied to solve the defined problem.

2.2. Mathematical formulation

In order to solve the aforementioned problem, a simple change of variable is first performed. We let:

$$Z = S - M.P \tag{14}$$

Therefore, the problem formulation becomes minimize:

$$\|Z\|_1 \tag{15}$$

subject to

$$W^T.A.Z \leq B_{max} - W^T.B - W^T.A.M.P \tag{16}$$

and

$$-M.P(i) \leq z_i \leq 1 - M.P(i) \forall i \in [1:r] \tag{17}$$

Since the problem is an L1 norm problem, it needs to be slightly modified in order to get rid of the absolute value that complicates the solution of the problem. Thus, the problem can be translated to minimize:

$$\sum_{i=1}^{r} t_i \tag{18}$$

subject to

$$W^T.A.Z \leq B_{max} - W^T.B - W^T.A.M.P \tag{19}$$

and

$$-M.P(i) \leq z_i \leq 1 - M.P(i) \forall i \in [1:r] \tag{20}$$

and

$$-T \leq Z \leq T \tag{21}$$

where T is an r-dimensional vector containing all the t_i's, which are dummy variables that are introduced to avoid the use of absolute value in the formulation and replace it by a simple minimization of a summation. Hence, at each time instant, the constraint matrix is formed, and then the optimization problem is solved. Since the problem is linear, it is solved efficiently using interior point method. In the experiments performed in this work, the optimization problem was solved in 10–15 iterations for an average time period nearly equal to 100 ms. Thus, by scheduling the rates' update at every 5 s, each communication rate is computed by averaging the values calculated in the last 50 iterations. On the other hand, it is worth noting

http://dx.doi.org/10.5772/63725

here that the internal model matrix M grows quickly with the increase in the number of robots used, which would affect the complexity of the problem and the speed of convergence. Therefore, in the following section, the developed optimization technique was tested on a system of two collaborating humanoid robots. The proposed method could be considered a basis for a framework for developing highly complex algorithms for systems involving real-time bandwidth optimization, where multiple users control multiple collaborating robots in various scenarios.

2.3. Dynamic optimized bandwidth algorithm experimental verification

In order to illustrate the use of the formulation, we apply it for the case, where an operator drives two collaborating robots. The mission consists of navigating a delimited path to reach a predetermined destination while avoiding obstacles and preserving a formation. The formation is characterized by a distance of 60 cm that separates the two robots, while keeping alignment nearly zero as in **Figure 1**.

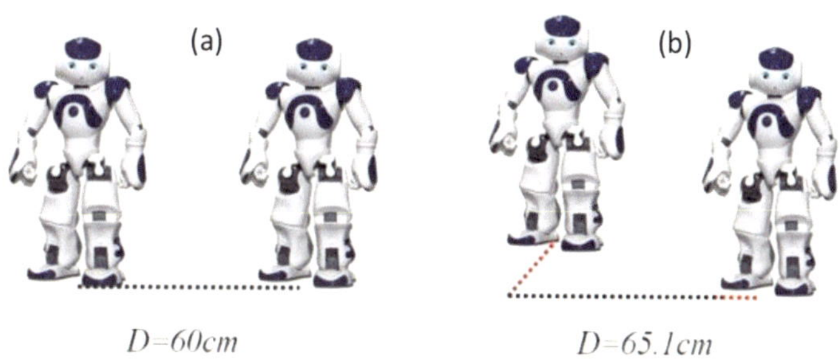

Figure 1. Two robots in a formation requiring a fixed distance D = 60 cm. (a) Without error (b) with vertical and horizontal error.

The robots' feedback includes visual and haptic data reflecting the environmental conditions. Ultrasonic sensors mounted on each robot allow the detection of obstacles in the navigation path. Each returns an integer value, indicating the distance to the nearest detected obstacle, which is fed back to the operator in the form of haptic feedback. In addition, the camera mounted on top of each robot provides visual feedback of the area in front of the formation. In order to allocate bandwidth based on changes in task conditions, *IEs* such as the distance to obstacles with respect to each robot and the speed of the swarm are monitored. Also, to maintain a high performance at high speeds, the speed of the swarm is one of the dynamic events that are monitored. Moreover, *QoC* factors are measures of how well the robots are collaborating together to efficiently accomplish the predetermined task. Specifically, the errors in position (Δx and Δy) between the robots indicating the deviation of the robots from the required formation are the quantities reflecting the change in the quality of the collaborative task, which need to be monitored. During the task execution, the visual and haptic feedback rates, the rates of commands and the rates of *R2R* communication are allocated based on real-time observations related to occurrence of *IEs* in the environment and to changes in *QoC* between robots. Hence, the elements of *X* are defined as follows:

- x_1: Rate of visual feedback from R1
- x_2: Rate of haptic feedback from R1
- x_3: Rate of visual feedback from R2
- x_4: Rate of haptic feedback from R2
- x_5: Rate of R2R collaboration information exchange
- x_6: Rate of commands generation

The observations in p_i's are also defined below:

- p_1: Distance to obstacle from R1
- p_2: Distance to obstacle from R2
- p_3: Speed of formation (Combined from R1 and R2)
- p_4: Positioning error in the horizontal direction Δx
- p_5: Positioning error in the vertical direction Δy.

Each element of *P* is normalized with respect to the maximum value that the sensors could measure (in case of p_1 and p_2), to the maximum value that the system can reach (in case of p_3) or to the maximum allowed error beyond which the task built upon the formation would start being affected (in case of p_4 and p_5). It is worth mentioning that p_3 is introduced in order to detect any sudden changes in the system dynamics that could be captured by monitoring the speed of the formation, which would also require significant changes in video feedback.

Knowing that the maximum distance detected by the robots' sensors is 2.5 m, p_1 and p_2 are defined as follows:

$$p_1 = \frac{2.5 - Distance\ from\ R1}{2.5} \qquad p_2 = \frac{2.5 - Distance\ from\ R2}{2.5}$$

Given that v_1 and v_2 are the average forward/backward and sideway speed of the formation, respectively, and knowing the maximum forward and sideway speed that the robots could reach is 6 and 4 cm/s, respectively, p_3 is defined as follows:

$$p_3 = \frac{\sqrt{v_1^2 + v_2^2}}{\sqrt{v_{1max}^2 + v_{2max}^2}}$$

where the exact values used in this case are v_{1max}=6 cm / s and v_{2max}=4 cm/s.

Finally, p_4 and p_5 are defined with respect to positioning error in the horizontal (Δx) and vertical (Δy) directions, respectively, as follows:

$$p_4 = min\left(\frac{|e_x|}{e_{x_{max}}},1\right); \quad p_5 = min\left(\frac{|e_y|}{e_{y_{max}}},1\right).$$

Additionally, the maximum allowed positioning error in the horizontal and vertical directions $e_{x_{max}}$ and $e_{y_{max}}$ for the application are estimated to be 2.5 and 5 cm, respectively.

Furthermore, the experiment is performed for a maximum bandwidth B_{max} equal to 1.4 Mbps. This value is deemed to be convenient for such small swarm with very few sensory data and video frame to exchange. Moreover, the imaging resolution of the robot's cameras is 160 × 120 × 3, which implies that each frame is formed of 3 matrices and thus, $w_1 = w_3 = 160 \times 120 \times 3 \times 8$ *bits* = 460,800 *bits*. As for the four remaining rates, the size of the data packets sent to/by each agent is considered to be 1500 *Bytes*. Hence, $w_2 = w_4 = 1500 \times 8 = 12000$ *bits*. But since the collaboration information exchanged requires both robots to send one packet, and since the operator generates a packet for each robot as command, then $w_5 = w_6 = 2 \times 12000$ *bits* = 24000 *bits*. Consequently, *W* is represented as follows:

$$W = \begin{bmatrix} 460800 & 12000 & 460800 & 12000 & 24000 & 24000 \end{bmatrix}$$

In addition, the matrices *A* and *B* are set based on the allowed minimum and maximum of these rates. During experimentation, a frequency of 1 Hz is allocated as minimum for all rates. As for the maximum rates of cameras, it is equal to the total bandwidth from which the minimum consumption of all other sensors is removed. For this application, x_{1max} and x_{3max} are chosen to be equal to 2 Hz. The force feedback (x_{2max} and x_{4max}), collaboration information exchange (x_{5max}), and commands rates (x_{6max}) are accorded a maximum rate of 5 Hz. Thus, using Eqs. (4) and (5), matrices *A* and *B* are calculated to be as follows:

$$A = \text{diag}\left(\begin{bmatrix} 1 & 4 & 1 & 4 & 4 & 4 \end{bmatrix}\right); \quad B = \begin{bmatrix} 1 & 1 & 1 & 1 & 1 & 1 \end{bmatrix}$$

Finally, a fixed matrix *M* is adopted in the implementation as described earlier. Each row of *M* contains the weights of the observations that affect the corresponding rate. The weights for all set of observations corresponding to each rate are allocated in a way to have the sum of each row remain equal to 1. It is worth mentioning that for the speed of the formation, weights of 25% are allocated in the rows of *M* corresponding to all the *R2U* and *U2R* rates (x_1 to x_4), since it was found to be the least dynamic observation. However, the rate of commands was deemed to be mostly dependent on the speed of the formation p_3. Thus, *M* would be as follows:

$$M = \begin{bmatrix} .75 & 0 & .25 & 0 & 0 \\ .75 & 0 & .25 & 0 & 0 \\ 0 & .75 & .25 & 0 & 0 \\ 0 & .75 & .25 & 0 & 0 \\ 0 & 0 & 0 & .5 & .5 \\ .25 & .25 & .5 & 0 & 0 \end{bmatrix}$$

In the experiments, the aforementioned values of A, B, M, W and B_{max} are adopted while solving the optimization problem and allocating rates.

3. Experimental set-up and results

In this section, the suggested algorithm is evaluated against two static allocation algorithms: 'Equal Bandwidth' and 'Equal Rates'. The equal bandwidth method allocates bandwidth equally among the different streams, while the equal rates method divides the available bandwidth such that the different streams would have equal rates of transmission.

The experimental set-up is composed of two NAO humanoid robots driven in a certain formation. An operator drives the two robots using a force feedback joystick (Microsoft SideWinder Force Feedback 2) and communicates with the robot through a router connected to a PC as shown in **Figure 2**. The operator sends real-time commands to the agents that return back haptic feedback as proximity measures and visual feedback from cameras mounted on each robot. The signal flows within the different components of the system as illustrated in **Figure 3**. The collaborative task that is to be accomplished by the robots is to maintain a certain formation while traversing an environment and avoiding collision with potential obstacles.

Figure 2. Experimental set-up.

The formation is characterized by a fixed distance (D = 60 cm) separating the two robots, while maintaining the error in the vertical direction nearly zero. Accurate position of the humanoids is calculated using the aid of the Inertial Unit built in the robots, which is made of 2-axis gyrometers with 5% precision (*angular speed* ~500°/s) and a 3-axis accelerometer with 1% precision (*acceleration* ~2G).

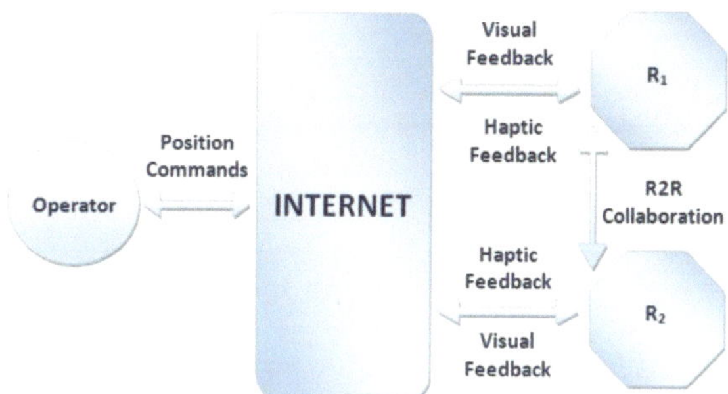

Figure 3. Teleoperation experimental set-up of humanoid robots.

3.1. Testing scenarios

3.1.1. Scenario 1: path with no obstacles

The first scenario consists of driving the swarm in the space delimited by dashed lines, shown in **Figure 4**, and reaching the final destination (dashed rectangular region on the right side) with no obstacles in the path. Obviously, with the absence of obstacles, the shortest path in this case is moving from 'Start' to 'End' in a straight line. The operators tend to adopt the shortest path, the straight line in this case, to reach the final destination.

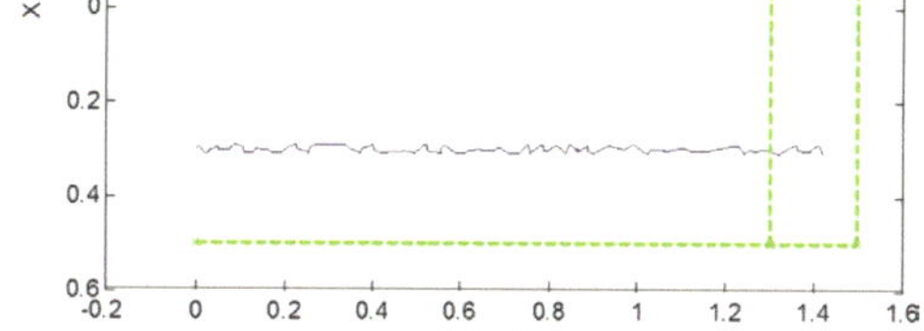

Figure 4. Scenario 1—path with no obstacles.

3.1.2. Scenario 2: obstacle in front of robot 1

In the second scenario, an obstacle is detected in front of Robot 1 (dashed square) when driving the formation in the delimited area as shown in **Figure 5**. The operator should steer to the left in order to avoid the collision with the obstacle. The swarm is forced to steer toward the left since by steering in the opposite direction Robot 2 would then exit the delimited path.

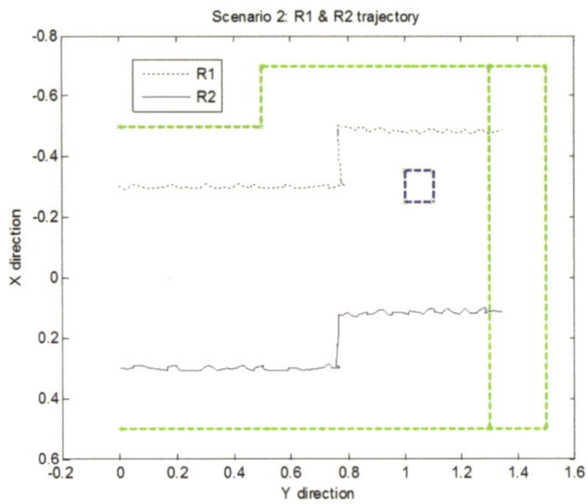

Figure 5. Scenario 2—obstacle in front of R1.

3.1.3. Scenario 3: obstacle in front of robot 2

The third scenario features an obstacle in front of Robot 2 (dashed square) when driving the formation in the delimited area. In this case, the operator steers to the left in order to avoid the collision of Robot 2 with the obstacle as shown in **Figure 6**.

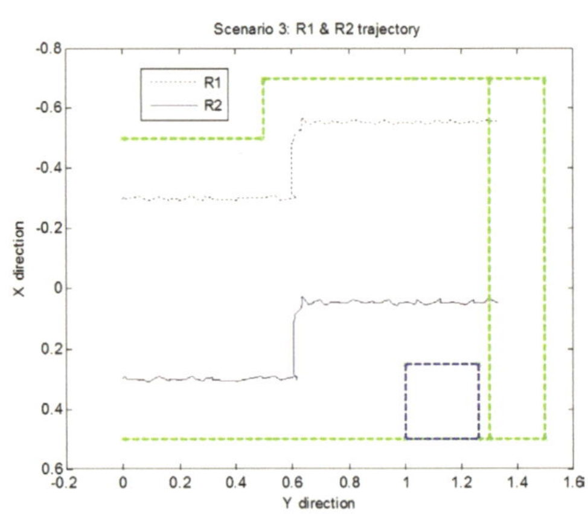

Figure 6. Scenario 3—obstacle in front of R2.

http://dx.doi.org/10.5772/63725

3.2. Testing and results

In order to evaluate the suggested algorithms, for each scenario, teleoperators drove the swarm under the Equal Bandwidth, Equal Rates and Optimized Bandwidth method. In each trial, the following performance parameters are collected: the completion time in *seconds*, the average speed of each robot in cm/s, the average deviation of each robot from the shortest path (Esp1 and Esp2) in , the average error in the formation in the horizontal and vertical directions (Δx and Δy) in *cm*, the maximum errors in both directions as well as the average bandwidth in Mbps consumed. The first static algorithm divides the available bandwidth equally among the 6 communication channels, whereas the second applied algorithm allocates equal rates to all communication channels. The computed rates for each channel for the static algorithms are reported in **Table 1**. It is worth noting that since the image frame size is much greater than the other data exchanged, allocating that equal bandwidth to all communication channels reduces the cameras' frame rate to around 1 Hz, while allocating equal rates to all communication channels increases frame rates to 1.5 fps; however, it drops the rates of all other data exchanged by a factor of five.

	x_1	x_2	x_3	x_4	x_5	x_6	Total bandwidth (Mbps)
Equal bandwidth	1	8	1	8	8	8	1.43
Equal rates	1.5	1.5	1.5	1.5	1.5	1.5	1.42

Table 1. Computed rates (in Hz) of both static algorithms.

The path in front of the formation can be visualized by the cameras located on the forehead of each robot. Robots R_1 and R_2 navigate inside a delimited path while avoiding obstacles to reach the final destination. Moreover, a force feedback that corresponds to the distance to obstacles in front of the formation is calculated based on values measured by the ultrasonic sensors mounted on each robot. In order to evaluate the suggested algorithm, four teleoperators drove the formation under the three allocation methods in the three defined scenarios for a total number of runs equal to 36. Under each scenario, the bandwidth methods were *randomly* selected for each driver. Additionally, two training runs were performed by each user in order to get familiar with the task performed and experiment the haptic and visual feedback before executing the official runs. Results for the three mentioned scenarios are recorded in **Tables 2–4**. Runs, which included visible slippage by the robots, were repeated in order not to bias the results.

Referring to **Table 2**, at an average bandwidth consumption less than that of both static algorithms, dynamic optimized bandwidth allocation method results in a better performance in the first scenario. With a reduction in bandwidth consumption of around 70 Kbps, the operator performs better when using the proposed dynamic algorithm than when applying the static ones. With the dynamic bandwidth algorithm, the average trial duration is 1.9 s less than the best static allocation method. Moreover, the driving performance improved significantly, since the parameters measuring the average error with respect to the shortest path improved in addition to the average speed of both robots. The instantaneous error to the shortest path has decreased by around 0.15 cm for both robots, while the average speed of both robots is around 0.35 cm/s higher. As for the parameters reflecting the quality of the executed collaborative task, we remark that the dynamic algorithm performs better than both static methods. The average errors in the horizontal and vertical directions are smaller as well as the maximum error is in both directions. For instance, the average horizontal error decreased by 10% for around 0.05 cm, while the average vertical error decreased by 40% (0.19 cm). It is worth noting that for most parameters the proposed method has a lower standard deviation indicating a more consistent performance.

		Duration (s)	Speed R1 (cm/s)	Speed R2 (cm/s)	Esp1 (cm)	Esp2 (cm)	Avg horiz error (cm)	Avg vert error (cm)	Max horiz error (cm)	Max vert error (cm)	Average bandwidth (Mbps)
Equal bandwidth	Average	35.2	4.53	4.73	0.52	0.53	0.75	0.48	2.63	3.95	1.40
	Std Dev	1.13	0.17	0.23	0.15	0.19	0.56	0.23	1.28	1.30	0.00
Equal rates	Average	34.2	4.71	4.91	0.43	0.47	0.54	0.59	1.99	2.92	1.40
	Std Dev	5.26	0.77	0.77	0.09	0.10	0.37	0.18	1.25	1.21	0.00
Optimized rates	Average	32.3	5.04	5.27	0.37	0.41	0.49	0.29	1.32	1.33	1.33
	Std Dev	4.39	0.45	0.45	0.05	0.05	0.34	0.18	0.79	0.69	0.0015

Table 2. Results of Scenario 1 at 1.4 Mbps.

In Scenario 2, the advantage of dynamic bandwidth allocation is also demonstrated by the collected results in **Table 3**. With a reduction in bandwidth consumption of around 70 Kbps, the operators perform better when using the proposed dynamic algorithm than when applying the static ones. With the dynamic bandwidth algorithm, the average trial duration is 3.7 s less than the best static allocation. Moreover, the driving performance improves significantly, since the parameters measuring the average error with respect to the shortest path improve in addition to the average speed of both robots. The instantaneous error to the shortest path decreased by around 0.4 cm, while the average speed of both robots is around 0.35 cm/s higher. Additionally, the maximum errors in the horizontal and vertical directions decreased by around 0.35 cm. It is worth noting here that the runs performed with 'Equal Rates' method lead to a better average horizontal and vertical error; however, with this method, high peaks of errors are reached in both directions that almost reach the tolerated bounds of 2.5 and 5 cm.

The rates of visual feedback of robots R_1 and R_2 during Scenario 3 for the different adopted bandwidth algorithms are presented in **Figure 7**. Additionally, the rates of haptic feedback of robots R_1 and R_2 and the collaboration and commands rate during Scenario 3 for the different bandwidth algorithms are presented in **Figures 8** and **9**, respectively.

Furthermore, we examine the percentage of runs in which the suggested algorithm outperforms the two static algorithms for each performance parameter in all scenarios. In other words, we count the number of times a user performed better according to a parameter when adopting the dynamic algorithm versus when driving with each of the static algorithms. Percentage of best performance for task duration, average speed of each robot (Speed R1, Speed R2), error of each robot with respect to the shortest path (Esp1 and Esp2) and maximum alignment and separation errors in the formation are recorded in **Table 5**. From the collected results, it can be seen that operators perform better with the dynamic allocation algorithm than the static algorithms at a minimum of 67% of the runs (Esp1 and Esp2 with Equal Bandwidth and Max horizontal/vertical error with Equal Rates). The suggested algorithm reaches a success rate of 92% for the speed R1 with respect to 'Equal Rates' static allocation.

		Duration (s)	Speed R1 (cm/s)	Speed R2 (cm/s)	Esp1 (cm)	Esp2 (cm)	Avg horiz error (cm)	Avg vert error (cm)	Max horiz error (cm)	Max vert error (cm)	Average bandwidth (Mbps)
Equal bandwidth	Average	45.3	4.01	4.17	7.49	7.51	0.54	0.51	2.60	3.39	1.40
	Std Dev	1.49	0.34	0.32	2.10	2.02	0.24	0.14	0.72	0.61	0.00
Equal rates	Average	46.1	4.15	4.34	5.92	5.98	0.37	0.42	2.25	4.66	1.40
	Std Dev	3.83	0.03	0.07	0.31	0.33	0.20	0.37	1.64	4.88	0.00
Optimized rates	Average	41.6	4.49	4.66	5.62	5.46	0.65	0.49	2.23	2.61	1.33
	Std Dev	2.21	0.21	0.24	0.82	0.89	0.15	0.23	0.53	1.96	0.0002

Table 3. Once more, the results collected during the experiments performed in Scenario 3 have shown the advantages of the suggested dynamic allocation method as depicted in **Table 4**. The parameters measuring the quality of the collaborative task and the driving performance show real improvements. It is worth noting that in Scenario 3, the average task duration with the dynamic bandwidth method is equal to the average duration with equal bandwidth method. However, three of the four drivers have performed better with the suggested algorithm than with static bandwidth allocation. Only one user performed the trial with a total duration of 53 s. This trial biased the calculated average, which was reflected by the standard deviation value.Results of Scenario 2 at 1.4 Mbps.

		Duration (s)	Speed R1 (cm/s)	Speed R2 (cm/s)	Esp1 (cm)	Esp2 (cm)	Avg horiz error (cm)	Avg vert error (cm)	Max horiz error (cm)	Max vert error (cm)	Average bandwidth (Mbps)
Equal bandwidth	Average	47.5	4.09	4.23	7.55	7.52	0.66	0.63	7.66	4.93	1.4
	Std Dev	2.31	0.38	0.45	2.31	2.39	0.64	0.27	7.62	1.95	0.00
Equal rates	Average	49.3	3.99	4.00	9.03	9.09	0.66	0.88	2.35	6.32	1.4
	Std Dev	2.47	0.16	0.21	1.27	1.16	0.25	0.54	0.98	6.04	0.00
Optimized rates	Average	47.5	4.15	4.27	6.78	6.81	0.36	0.23	1.43	2.90	1.33
	Std Dev	4.40	0.23	0.28	1.16	1.15	0.24	0.15	0.76	2.52	0.0010

Table 4. Results of Scenario 3 at 1.4 Mbps.

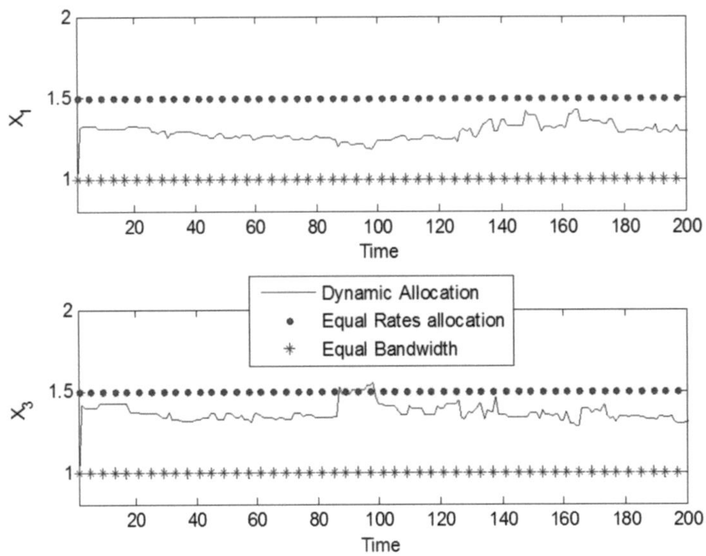

Figure 7. Visual rates of R1 & R2 in Scenario 3.

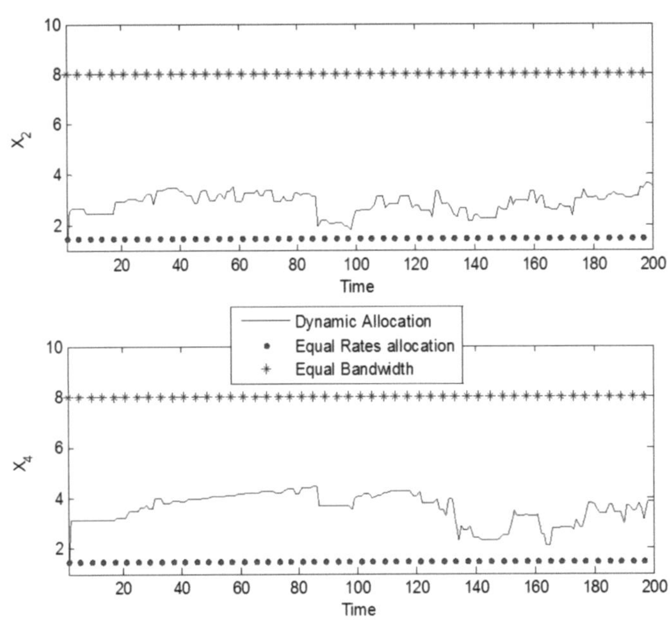

Figure 8. Haptic rates of R1 & R2 in Scenario 3.

Finally, the advantages of the proposed dynamic bandwidth optimization and management scheme over legacy bandwidth management schemes are clearly expressed in the results in terms of performance improvement and conserving network resources. Since the proposed

http://dx.doi.org/10.5772/63725

algorithm is scalable and not limited to a single task, the improvement in performance is greatly realized in critical situations, where the collaborative task requires high levels of accuracy especially in cases involving human safety.

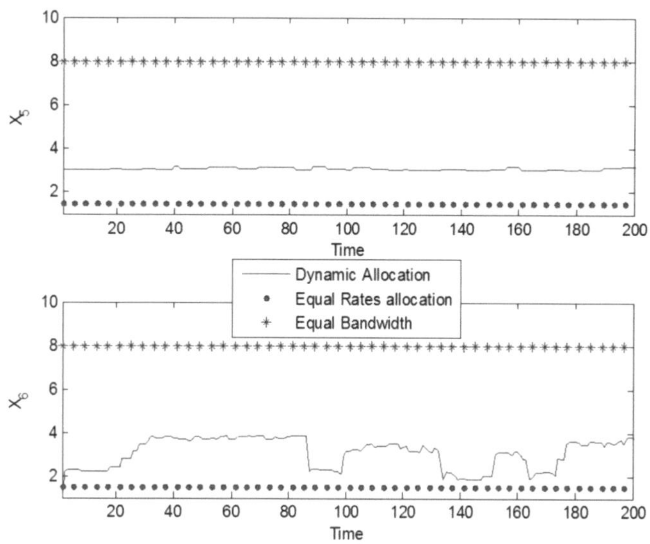

Figure 9. Collaboration & commands rates in Scenario 3.

	Duration (%)	Speed R1 (%)	Speed R2 (%)	Esp1 (%)	Esp2 (%)	Max horizontal error (%)	Max vertical error (%)
Dynamic vs. equal bandwidth	83	75	83	67	67	83	83
Dynamic vs. equal rates	75	92	83	83	83	67	67

Table 5. Percentage of best performance.

4. Conclusions

In this work, dynamic optimized bandwidth management in teleoperated collaborative robotics is tackled. The focus was on managing all communication channels, where actuation commands, system state and sensory data are exchanged. This was achieved by monitoring the interesting events occurring in the robots' environment and the changes in quality of collaboration among them. Effective completion of the collaborative task with lower bandwidth consumption and better performance was accomplished proving that the proposed method could be the basis of a framework for developing more complex algorithms applied to highly complex systems.

Acknowledgements

This research was funded and supported by AUB University Research Board and the Lebanese National Council for Scientific Research.

Author details

Chadi Mansour[1], Mohamad El Hariri[1*], Imad H. Elhajj[1], Elie Shammas[2] and Daniel Asmar[2]

*Address all correspondence to: me129@aub.edu.lb

1 Electrical and Computer Engineering, Faculty of Engineering and Architecture, American University of Beirut, Beirut, Lebanon

2 Mechanical Engineering, American University of Beirut, Beirut, Lebanon

References

[1] Lee D., Spong M.W. Semi autonomous teleoperation of multiple cooperative robots for human robot exploration. In: Proceedings of the AAAI 2006 Spring Symposium; Palo alto, California, USA, 2006.

[2] Schenker P.S., Huntsberger T.L., Pirjanian P., Baumgartner E.T., Tunstel E. Planetary rover developments supporting mars exploration, sample return and future human-robotic colonization. Autonomous Robots. 2003; 14(2–3):103–126. doi:10.1023/A:1022271301244

[3] Hwanga G., Szemesb P.T., Andoc N., Hashimotod H. Development of a single-master multi-slave tele-micromanipulation system. Advanced Robotics. 2007; 21(3–4):329–349. doi:10.1163/156855307780132054

[4] Thomson J.M., Ottensmeyer M.P., Sheridan T.B. Human factors in telesurgery: effects of time delay and asynchrony in video and control feedback with local manipulative assistance. Telemedicine Journal. 1999;5(2):129–137. doi:10.1089/107830299312096.

[5] Wai-keung Fung et al., "Task driven dynamic QoS based bandwidth allocation for real-time teleoperation via the Internet," Intelligent Robots and Systems, 2003. (IROS 2003). Proceedings. 2003 IEEE/RSJ International Conference on, Las Vegas, Nevada, USA 2003, pp. 1094–1099 vol.2. doi: 10.1109/IROS.2003.1248790

[6] Ehajj I., Xi N., Fung W.K., Liu Y.H., Hasegawa Y., Fukuda T. Supermedia-enhanced internet-based telerobotics. Proceedings of the IEEE. 2003; 91(3):396–421. doi:10.1109/JPROC.2003.809203

[7] Mourikis A.I., Roumeliotis S.I. Optimal sensor scheduling for resource-constrained localization of mobile robot formations. IEEE Transactions on Robotics. 2006; 22(5):917–931. doi:10.1109/TRO.2006.878947

[8] Sugiyama H., Tsujioka T., Murata M. Integrated operations of multi-robot rescue system with Ad Hoc networking. In: 1st International conference on wireless communication, vehicular technology, information theory and aerospace & electronic systems technology, 2009. Wireless VITAE 2009; 17–20 May 2009; Aalborg. IEEE; 2009. p. 535–539. doi:10.1109/WIRELESSVITAE.2009.5172502

[9] Bhuvanagiri S., Krishna M.K., Achar S. Coordination in ambiguity: coordinated active ocalization for multiple robots. In: AAMAS (Demo Paper); Estoril, Portugal, 2008; 1707–1708. doi: 10.1016/j.robot.2009.09.006

[10] Andre T., Bettstetter T. Collaboration in multi-robot exploration: to meet or not to meet? Journal of Intelligent & Robotic Systems. 2015; 1–13. doi:10.1007/s10846-015-0277-0

[11] Hu Y., Wang L., Liang J., Wang T. Cooperative box-pushing with multiple autonomous robotic fish in underwater environment. Control Theory & Applications. 2011;5(17): 2015–2022. doi:10.1049/iet-cta.2011.0018 (IET. Nov. 17)

[12] Wang X., Chen M., Huang H.M., Subramonian V., Lu C., Gill C.D. Control-based adaptive middleware for real-time image transmission over bandwidth-constrained networks. IEEE Transactions on Parallel and Distributed Systems. 2008;19(6):779–793. doi:10.1109/TPDS.2008.41

[13] Phillips-Grafflin C., Suay H.B., Mainprice J., Alunni N., Lofaro D., Berenson D., Chernova S., Lindeman R.W., Oh P. From autonomy to cooperative traded control of humanoid manipulation tasks with unreliable communication. Journal of Intelligent & Robotic Systems. 2015; 1–21. doi:10.1007/s10846-015-0256-5

[14] Chaskar H.M., Madhow U. Fair scheduling with tunable latency: a round-robin approach. IEEE/ACM Transactions on Networking. 2003; 11(4):592–601. doi:10.1109/TNET.2003.815290

[15] Tipsuwan Y., Kamonsantiroj S., Srisabye J., Chongstitvattana P. An auction-based dynamic bandwidth allocation with sensitivity in a wireless networked control system. Computers & Industrial Engineering. 2008; 47:114–124. doi:10.1016/j.cie.2008.08.018

[16] Julio R.E., Bastos G.S., Dynamic bandwidth management library for multi-robot systems. In: 2015 IEEE/RSJ International Conference on Intelligent Robots and Systems (IROS), Hamburg. 2015; 2585–2590. doi:10.1109/IROS.2015.7353729

[17] Mansour C., Elhajj I.H., Shammas E., Asmar D. Event-based dynamic bandwidth management for teleoperation. In: 2011 IEEE international conference on robotics and biomimetics (ROBIO), 7–11 Dec. 2011; Karon Beach, Phuket. IEEE; 2011. p. 229–233. doi: 10.1109/ROBIO.2011.6181290

Chapter 10

Series Elastic Actuator: Design, Analysis and Comparison

Arnaldo Gomes Leal Junior,
Rafhael Milanezi de Andrade and
Antônio Bento Filho

Additional information is available at the end of the chapter

http://dx.doi.org/10.5772/63573

Abstract

In general, actuators are built to be as stiff as possible to increase the bandwidth. When a robot works in a structured environment, its automation is easier than in a non-structured environment in which case its modeling is quite difficult and presents a high computational effort. To overcome this difficulty, series elastic actuator (SEA) has been applied in compliant robotic grasping. Unlike rigid actuators, a SEA contains an elastic element in series with the mechanical energy source. Such an elastic element gives SEAs tolerance to impact loads, low mechanical output impedance, passive mechanical energy storage, and increased peak power output. The spring has to be able to support the loads, but it cannot be too stiff; otherwise, system impedance will be high. This chapter describes a comparison between two types of SEA, an electric series elastic actuator (ESEA) and a hydraulic series elastic actuator (HSEA), for four-legged dynamic robot application. The parameters employed in the comparison are bandwidth, output impedance, time response, power density, and dynamic range. The results indicate that HSEA is a better actuator than ESEA for a weight carrying four-legged dynamic robot because of its higher power density and dynamic ratio with desirable output impedance, time response, and bandwidth.

Keywords: SEA, actuator, force control, dynamic robot, legged robot

1. Introduction

1.1. Force control

Over the years, the premise for constructing machines and actuators is "Stiffer is better" since a stiffer actuator has larger bandwidth force control and accurate position control. On the other hand, the force control is difficult since it develops high forces from small displacement and a little position error can result in a large force error. For this reason, the sensor has to be very precise with good resolution. Moreover, a stiffer actuator can lead to accidents with humans in a non-structured environment.

The majority of standard robot actuation systems cannot create precise force on the robotic joints due to friction, stick-slip, backlash and reflected inertia through the transmission, cogging in motors, and pressure drop in hydraulic circuits, among others. However, even with these nonlinearities and noises on force measurements, some robots are well aligned for a position or trajectory control since the mass of the robot and actuators acts as a low-pass filter for the force on position output (Robinson, 2000).

In some tasks, such as precise load positioning and human joints actuation, a good force control is required. In this case, the force noise on the actuators will hinder the application of the standard robot actuation. To achieve a good force-controlled system, some parameters have to be measured which are force bandwidth, mechanical output impedance, dynamic range and force density. All these parameters are related with each other and are further discussed in Section 4.

Although standard robots and stiff actuators are not good in force control, the humans are. Moreover, humans have no problems in contact hard surfaces and perform precise operations. Based on this bioinspiration, Williamson (1995) began to sacrifice the bandwidth and the stiffness for more stable force control. The method applied is to place an elastic element in series with the power source which are, commonly, a DC motor with gear reduction or a hydraulic cylinder (**Figure 1**).

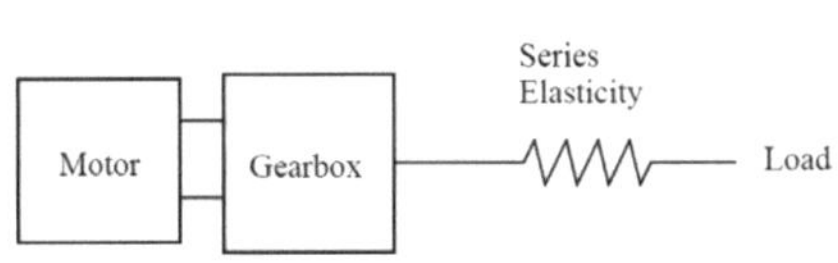

Figure 1. Series elastic actuator schematic (Williamson, 1995).

1.2. Series elastic actuators

Series elastic actuator (SEA) has been successfully applied in a number of applications for almost 20 years (Pratt and Williamson, 1995). As widely reported by a number of researchers (see Pratt and Williamson, 1995; Arumugom et al., 2009; Paluska and Herr, 2006; Paine et al., 2013), SEA provides many benefits in force control of robots. Series elastic actuators have an elastic element attached with the mechanical energy source output. This configuration presents advantages over rigid actuators such as low mechanical output impedance, tolerance to impact loads, increased peak power output, and passive mechanical energy storage. These properties enable the application of SEAs in legged actuation systems and human orthotics (Pestana et al., 2010; Parietti et al., 2011).

http://dx.doi.org/10.5772/63573

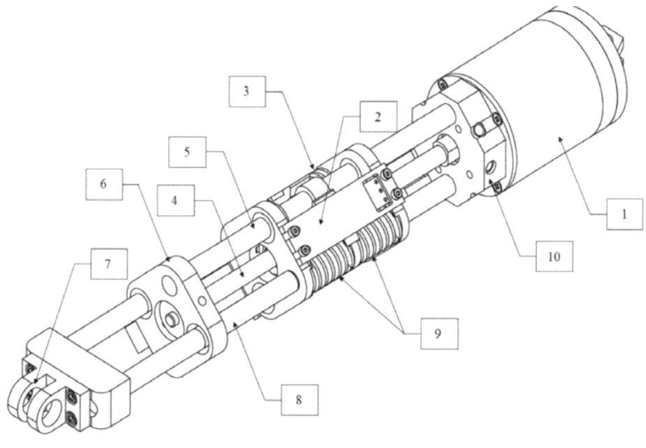

Figure 2. Electric SEA (1) DC motor with gearbox; (2) moving sub assembly, (3) ball nut plate, (4) ball screw, (5) guide rod, (6) support plate, (7) load connection terminal, (8) moving rod, (9) and springs (10) motor angle and rods plate (Pratt et al., 2002).

Figure 2 shows a linear electric driven SEA (Pratt et al., 2002). The DC motor with gearbox (1) drives the ball screw (4) which moves the ball nut plate (3). Ball nut plate (3) then moves the springs (9) against the moving sub-assembly (2) which is fixed to the moving rods (8), thus causing the displacement of the load connection terminal (7) and then moves the load. The guide rods (5) are fixed to the motor flange (10) and to the support plate (6) through plain bearings for the ball screw and for the moving rods (8). They move together with the moving sub-assembly (2) in which they are fixed. It follows that all the force exerted by the spindle motor assembly on the load is directly supported by the springs, which are the elastic element purposely inserted to lower the stiffness of the actuator load interface.

The SEA's designs are made under various trade-offs, and the decision parameters often are power output, volumetric size, weight, efficiency, Back-drivability, impact resistance, passive energy storage, backlash, and torque ripple. Furthermore, iterations may be necessary on components design of a SEA. Curran and Orin (2008), Hutter et al. (2009), Kong et al. (2009), and Ragonesi et al. (2011) propose rotary designs based primarily on commercially available off-the-shelf components. Lagoda et al. (2010) and Diftler et al. (2011) design a compact rotary SEA using a harmonic drive and a high-stiffness planar spring. A compact actuator can be achieved by placing linear springs coupled to rotary shafts between the motor and the chassis ground as presented in Hutter et al. (2011) and Torres-Jara and Banks (2004). Furthermore, this compact actuator employs springs with low stiffness. In order to reduce the torque requirement on the spring, Kong et al. (2012) and Taylor (2011) placed the spring within the reduction phase. A promising design for SEA that is widely researched throughout the years is to apply ball screws as the primary reduction mechanism followed by a cable drive which remotely drives a revolute joint. This design takes advantage of the ball screw high efficiency even for large speed reductions (85–90%), backdrivability, impact tolerance, and do not introduce torque ripple (Edsinger-Gonzales and Weber, 2004; Gregorio et al., 1997; Pratt and Pratt, 1998). Paine et al. (2013) presented a good review of some recent progress in series elastic actuators. In this chapter, we presented a different design of SEA. In order to avoid accidents risks and components contamination, the internal components were encapsulated in a unique hollow tubular structure. This allows the substitute of the heavy structure of solid steel tubes of the reference model (Pratt et al., 2002), permitting the applications in wearable robots, exoskeletons and prostheses.

Regarding the control of series elastic actuators, as can be expected, the design of the controller is closely related to the hardware employed. Therefore, a controller parameter such as the output force can be estimated by measuring spring deflection (Kong et al., 2010), or by applying strain gauges (Pratt and Williamson, 1995). Nevertheless, a PID approach for force control suffers from stability issues if friction and backlash are too large. For this reason, an impedance control is developed, which basically is an innermost control structure with position feedback to treat the force control as a simple position or velocity control (Pratt et al., 2004; Lagoda et al., 2010). In this chapter, such idea was adopted to develop the proposed SEA controller.

1.3. Series elastic actuator background

The first robot designed with SEA was the Spring Flamingo (Pratt and Pratt, 1998) which is a planar bipedal walking robot. The SEAs drive six degrees of freedom mechanism which has three degree of freedom in each leg representing hip, knee and ankle. **Figure 3** shows the Spring Flamingo, it can walk at 1.25m/s on flat terrain. Moreover, it can walk over uphill and downhill terrains up to 15 degrees slope.

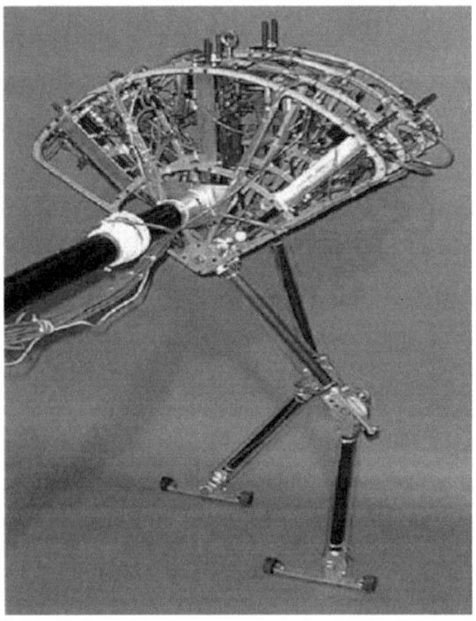

Figure 3. Spring Flamingo (Pratt and Pratt, 1998).

The extension of the work started with the Spring Flamingo is the M2 (Robinson et al., 1999) which is a three-dimensional bipedal walking robot. For this reason, it has twelve degrees of freedom, six on each leg. Regarding one leg, the M2 has three degrees of freedom on the hip, two degrees of freedom on the ankle, and one degree of freedom on the knee. This configuration provides three-dimensional movements with the same characteristics of the Spring

Flamingo. The M2 project is far more complex than the Spring Flamingo, as can be seen in **Figure 4**.

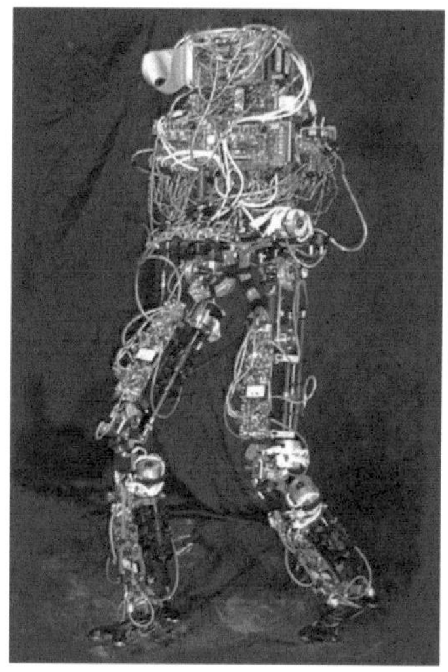

Figure 4. M2 (Robinson et al., 1999).

Another lower limb robot developed with SEA is the Corndog (Krupp, 2000) which is a planar running robot with one fore and one aft leg. Therefore, it represents only the half of a dog and employs electro-mechanical series elastic actuator (**Figure 5**).

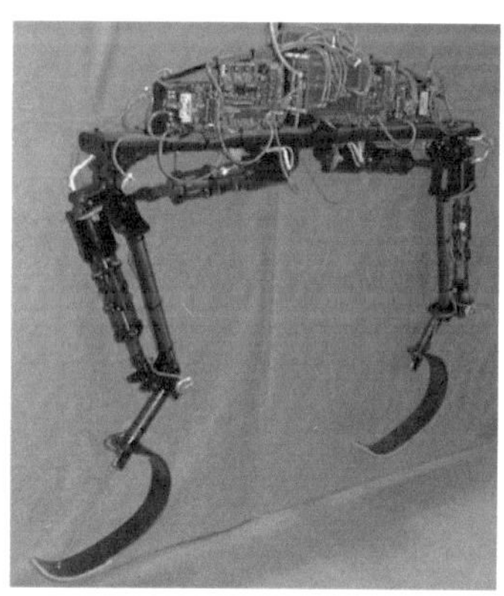

Figure 5. The Corndog (Krupp, 2000).

Series elastic technology is also applied on upper limb robots. The COG robot is a humanoid with upper torso and head (Brooks et al., 1999). It has 6 degrees of freedom on its arms which are operated with SEAs. Moreover, COG has the ability to hammering and turning a crank (**Figure 6**).

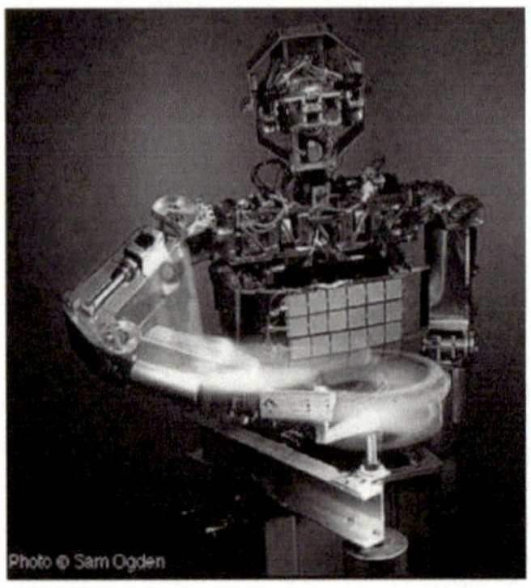

Figure 6. COG humanoid robot (Brooks et al., 1999).

In addition to the application on dynamic robots, SEAs have been widely studied and applied on rehabilitation and assistive technologies. Regarding assistive technologies for gait rehabilitation, a compact rotary SEA for knee joint assistive exoskeleton is developed (Kong et al., 2010). Unlike the majority of SEAs with motor and gear reduction as the motor part, a worm gear is used for motor reduction due to it compactness. However, it increases the friction of the system, which makes it more difficult to control (**Figure 7**). A robust control algorithm inspired by disturbance observer is implemented for the system control.

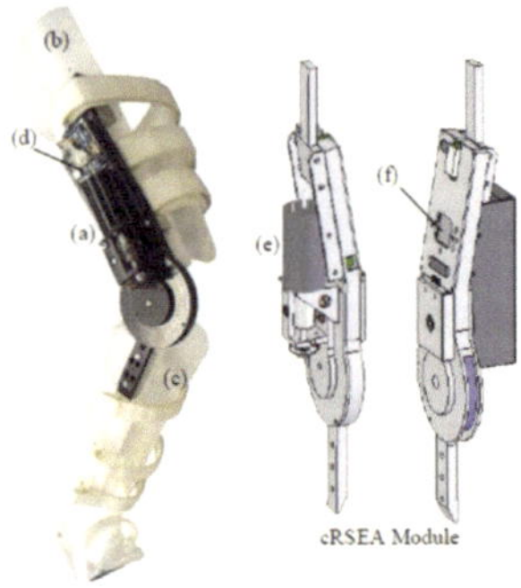

Figure 7. Proposed exoskeleton for knee assistance. (a) Rotary SEA module, (b) a thigh brace, (c) a calf brace, (d) a motor driver, (e) a DC motor, and (f) an controller (Kong et al., 2010).

Lagoda et al. (2010) designed a Series elastic actuated joint for robotic gait rehabilitation training to be applied at LOPES (Lower Extremity Powered Exoskeleton) which helps an impaired patient to training his/her gait with an assisted-as-need (AAN) approach which only provides actuated torque as needed to help the patient to complete the gait cycle.

Series elastic actuator is also employed on assistive technologies, a prosthetic knee is developed and the stiffness adjustments are described to achieve the optimum values for peak power and energy consumption for walking and running (Grimmer and Seyfarth, 2011). Results have shown large reductions in peak power and energy requirements with the proper spring selection. Moreover, SEA can mimic the human knee behavior better than the passive approaches.

Au et al. (2007) proposed a prosthetic ankle comprised of SEA that matches size and weight of the human ankle, comparing with the passive approaches, SEA ankle prosthesis enhances the capability of mimicking the human gait (**Figure 8**).

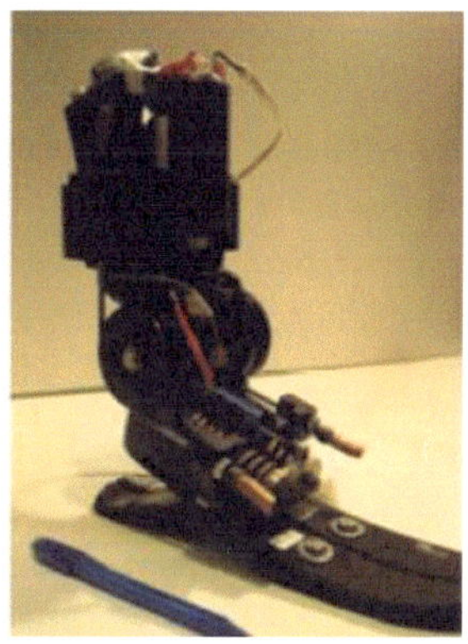

Figure 8. Ankle-foot prosthesis (Au et al., 2007).

Another approach for human-friendly actuators for assistive robotic application is to add a variable damping in the SEA design as shown in **Figure 9**. It can be achieved with magnetorheological fluids (MR fluids).

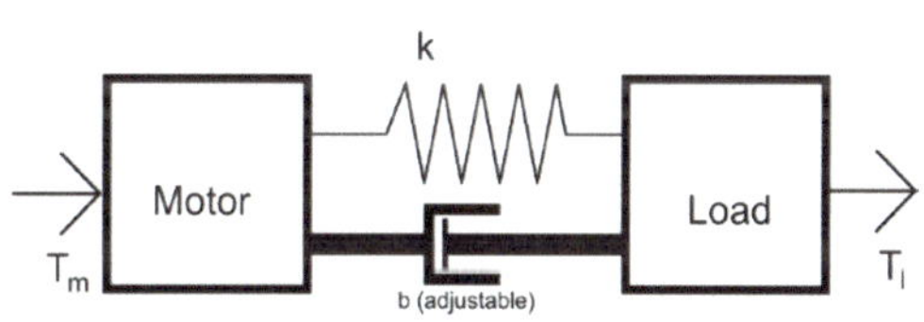

Figure 9. Series elastic actuator with controlled damping (Iyer, 2012).

Although this configuration allows a human-friendly device that can be applied to design legs with agile locomotion (Garcia et al., 2011), the investigation of such devices are beyond this chapter analysis, but will be investigated in further works.

2. Electric series elastic actuator

2.1. Electric series elastic actuator design

Over the years, digital prototyping has been successfully applied in engineering and manufacturing allowing virtually explore a complete product before it's built (Thurfjell et al., 2002;

Soyguder and Alli, 2007; Wang, 2011). As reported by Johansson et al. (2004), the digital prototyping has been important to the development of industrial robots, reducing the time of robot programming phase. According to the authors, the robot task can be simulated in a virtual model of the workcell when still only a digital prototype of the workpiece exists and the risk of technical failure for a transition can be reduced. In this way, before manufacture of the proposed serial elastic actuator, its digital prototyping is presented in this chapter.

2.1.1. First design

The characteristics desired for the actuator are enclosed internal components, to avoid component contamination and accidents risks; size adequate for wearable robots, exoskeletons and prostheses applications, emulating natural muscles; aesthetically suited to applications in wearable robots, exoskeletons and prostheses; and smart enough for applications in unstructured environment.

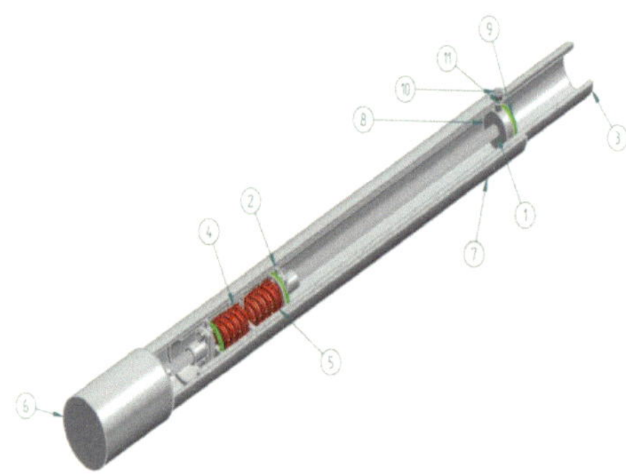

Figure 10. First prototype showing internal actuator components: ball screw (1); ball nuts (2); movable arm (3); springs (4, 5); motor & gearbox (6); base tube (7); bearing screw (8); O-ring (9); Allen screws (10); guide ring (11).

The first prototype design is shown in **Figure 10**. There are two ball nuts (2) bolted to plates with a tab and grooved flanges for O-rings (9) installation. The flanges have a central hole and are connected by means of an extender, which is a tube with external male thread at both ends. The springs (4 and 5) slide around the extender and press on both sides a plate fixed to the tube base (7) of the retractable arm (3).

2.1.2. The moving sub-assembly

Figure 11 shows the exploded view of the moving sub assembly, a set of pieces that behaved as a single device after assembled. Two bolted flanges (4, 3) enclose the ball nut (1). Tubes (2) are used to grip a flange on the nut (1). The flange (4) has threaded holes for Allen screws (5) and flange (3), countersunk holes for these screws and thread. In the central bore of each flange is a threaded guide tube (6), with the pre stressed springs. The circuit board (7) is fixed over the flanges with the sensor for acquiring the deformation of the spring against the linear encoder attached to the flanges (3) and (4).

http://dx.doi.org/10.5772/63573

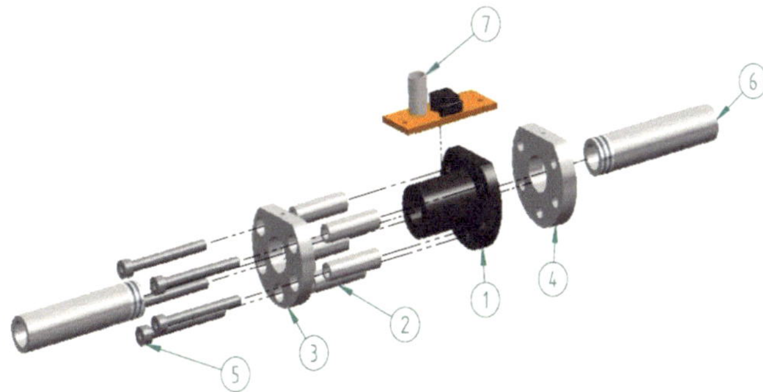

Figure 11. Exploded view of the final design action sub assembly components.

2.1.3. DC motor and gearbox

It was chosen based on the forces for actuator operation and for the spring assembly, according to the planetary gearbox, and the ball screw torque force conversion. The DC motor is a 60 mm, brushless, 400 W; the gearbox is a planetary gearhead 62 mm diameter, 8–50 N.m torque and the ball screw is a 10 mm diameter and 3 mm pitch rolled steel.

2.1.4. Final design

Figure 12 shows the actuator final design. The iterative digital prototyping design process done resulted in a set of desirable characteristics for the actuator very similar to those originally intended, which are 425 mm long when retracted, with 130 mm range with maximum diameter of 60 mm. For the sake of wider range of operation, the model has interchangeability of springs. The ball screw has 10 mm diameter and 3 mm pitch. Main model structure of carbon steel and aluminum alloy tubes allows a final weight of 3.7 kg. The operating voltage of the ESEA is 48 V. These model's properties allows a 450 N maximum load capacity and 13.4 cm/s continuous speed.

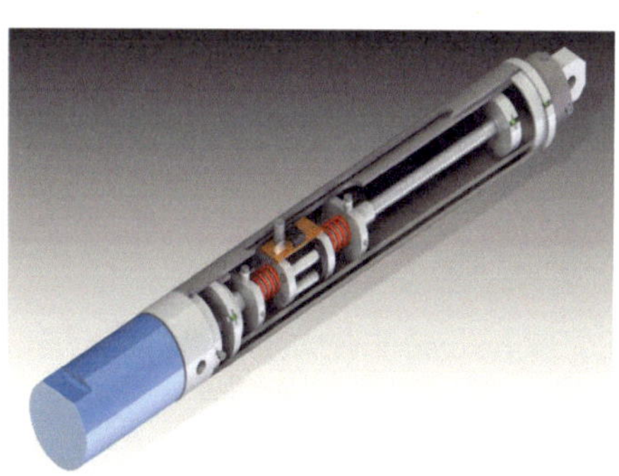

Figure 12. Actuator final design with internal components.

2.2. Dynamic model

The dynamic model of the actuator was constructed from the digital prototype data to run a simulation in order to verify its dynamic behavior, bandwidth and validate the project settings, as presented in **Figure 13**.

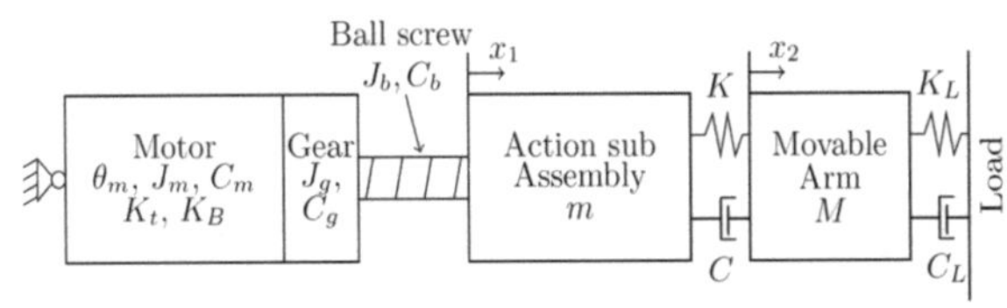

Figure 13. Actuator's mechanical model.

Referring to **Figure 13**, the actuator dynamic model equations are described as follows:

$$\mathrm{L_a}\frac{di_a}{dt} + \mathrm{R_a i_a} + \mathrm{K_b}\omega = \mathrm{V_a} \tag{1}$$

In Eq. (1), L_a is the armature inductance, i_a is the armature current, R_a is the armature resistance, K_b is the back Electromotive Force (back-EMF) constant, ω is the angular velocity, and V_a is the armature voltage.

$$J_T\ddot{\theta}_m + C_T\dot{\theta}_m = k_t i_a \tag{2}$$

Referring to Eq. (2), J_T is the total inertia of the actuator, C_T is the total viscous friction coefficient, θ_m is the angular displacement velocity of the motor axis, $\ddot{\theta}$ is the motor angular acceleration, and k_t is the motor torque proportional constant.

The SEA's total inertia referred in Eq. (2) is the summation of all inertia effects presented by each component as depicted in Eq. (3).

$$J_T = \left[J_m + J_g + n^2 J_b + \left(\frac{np}{2\pi}\right)^2 m \right] \tag{3}$$

where J_m is the rotor inertia, J_g is the gear inertia, n is the gear ratio, J_b is the ball screw inertia, p is the ball screw pitch, m is the action device sub assembly mass.

A similar approach is applied to evaluate the total viscous friction coefficient presented in Eq. (2). The summation of all the friction effects results in the total viscous friction coefficient (Eq. (4))

$$C_T = \left[C_m + C_g + n^2 C_b + C\left(\frac{np}{2\pi}\right)^2 \right] \tag{4}$$

Referring to Eq. (4), C_m is the rotor viscous friction coefficient, C_g is the gear reduction viscous friction coefficient, C_b is the ball screw viscous friction coefficient, C is the action sub assembly viscous friction coefficient.

In Eqs. (5) and (6), Newton's second law is applied on the movable arm and action sub assembly systems to evaluate the action sub assembly and movable arm displacement in order to estimate the spring deflection.

$$m\ddot{x}_1 + C_b x_1 = m\left(\frac{np}{2\pi}\right)\ddot{\theta}_m + C_b\left(\frac{np}{2\pi}\right)\dot{\theta}_m = f \tag{5}$$

$$M\ddot{x}_2 + C_L x_2 + K_L x_2 = f \tag{6}$$

where f is the force between action sub assembly and the movable arm, M is the movable arm sub assembly mass, C_L is the load interface viscous friction coefficient, and K_L is the load interface elastic constant. The term $\left(\frac{np}{2\pi}\right)$ is the linear displacement of the movable arm reflected to motor axis rotation, and $\left(\frac{np}{2\pi}\right)^2$ is the inertia and viscous friction components of the action device reflected to motor axis.

Once the actuator dynamic model equations are described, the closed loop PID transfer function block diagram in the frequency domain can be made, as shown in **Figure 14**.

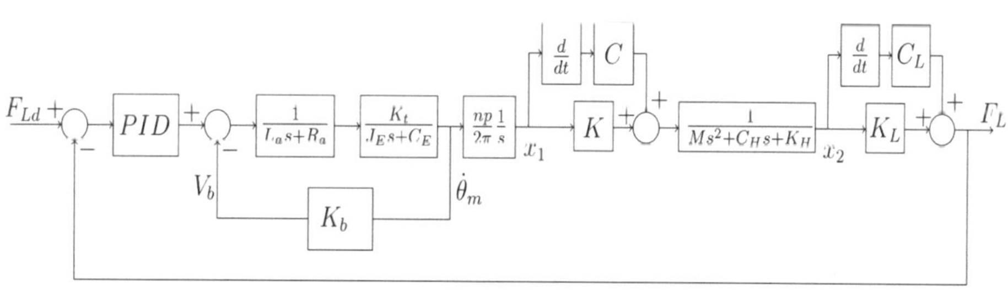

Figure 14. Actuator's closed loop transfer function block diagram.

2.3. Results and discussion

Once the actuator dynamic model has been described, the parameter values had to be established. The parameters were obtained by the digital prototype design and data sheets of some components, i.e. motor and ball screw (**Table 1**).

Spring constant is the most important parameter in the system. To define the spring constant, we set the minimum impedance level and a maximum bandwidth with respect to the previous selected components of the system. Iteration was made to choose the best value with upper and lower bounds.

After knowing all the actuator's parameters, the PID controller could be designed. The controller was tuned by applying the pole placement and phase margin method (Tang et al., 2010). The same PID control tuning strategy is applied in the controller of the linear serial elastic hydraulic actuator (Leal Junior et al., 2015). For this reason, applying the same tuning method in the electric SEA helps on further responses comparison between the hydraulic and the electric SEA.

Parameter	Value	Units
Armature inductance (La)	8.2E–04	(H)
Armature resistance (Ra)	1.03	(Ω)
Total inertia (JT)	9.51E–05	(kg m^2)
Mass (M)	1.261	(kg)
Total viscous friction (CT)	4.594E–05	(Ns/m)
Arm and load stiffness (KH)	3.336E+06	(N/m)
back-EMF constant (Kb)	1.469E–01	(Vs/rad)
Load stiffness (KL)	3.303 +06	(N/m)
Load damping (CL)	8.08	(Ns/m)
Spring stiffness (K)	3.27E+04	(N/m)
Spring viscous friction (C)	8.000E–01	(Ns/m)
Torque proportional constant (KT)	1.470E–01	(Nm/A)
Gear ratio (n)	0.001	(1)
Ball screw pitch (p)	0.003	(m)

Table 1. Parameters value of the entire ESEA system.

Desired pair of poles is estimated by representing time domain specification, e.g. percent overshoot, settling time. However, in some cases the dominant poles lose its dominant position in the result of the closed loop system. To overcome this problem, a phase margin is employed to guarantee the closed loop system robustness.

Assuming a pair of poles as

$$p_{1,2} = -\alpha \pm j\beta \tag{7}$$

The process G(s) is given and the controller has its conventional configuration with all positive gains as follows.

$$C(s) = K_p + \frac{K_i}{s} + K_d s \tag{8}$$

Substituting one of the poles into the characteristic equation of the closed loop equation which is the transfer function denominator of the process G(s).

$$1 + G(p_1)C(p_1) = 0 \tag{9}$$

Moreover, the definition of phase margin gives:

$$G(j\omega_g)C(j\omega_g) = -e^{j\varphi_m} \tag{10}$$

Substituting and splitting the complex values and the real ones of Eqs. (9) and (10) gives two real functions of (Eqs. (11) and (12)).

$$f_1(\omega) = \mathrm{Re}\left[\frac{-p_1}{G(p_1)}\right] + \frac{\alpha^2 - \beta^2}{2\alpha\beta}\mathrm{Im}\left[\frac{-p_1}{G(p_1)}\right] + \frac{\alpha^2 + \beta^2}{2\alpha}\mathrm{Re}\left[\frac{-e^{j\varphi_m}}{G(j\omega)}\right] \tag{11}$$

$$f_2(\omega) = \frac{\omega^2}{2\alpha}\mathrm{Re}\left[\frac{-e^{j\varphi_m}}{G(j\omega)}\right] - \frac{\omega^2}{2\alpha\beta}\mathrm{Im}\left[\frac{-p_1}{G(p_1)}\right] - \omega\,\mathrm{Im}\left[\frac{-e^{j\varphi_m}}{G(j\omega)}\right] \tag{12}$$

The positive values of the intersection between $f_1(\omega)$ and $f_2(\omega)$ are the first approximation for integral gain (K_i). This first value was applied on Eqs. (13) and (14) to evaluate the proportional and derivative gains. Proportional and derivative gains are obtained by substituting the well-known PID controller equation C(s), presented in Eq. (8), in Eqs. (9) and (10).

$$K_p = \frac{2\alpha}{\alpha^2 + \beta^2}\left(K_i - \mathrm{Re}\left[\frac{-p_1}{G(p_1)}\right] - \frac{\alpha^2 - \beta^2}{2\alpha\beta}\mathrm{Im}\left[\frac{-p_1}{G(p_1)}\right]\right) \tag{13}$$

$$K_d = \frac{1}{\alpha^2 + \beta^2}\left(K_i - \mathrm{Re}\left[\frac{-p_1}{G(p_1)}\right] - \frac{\alpha}{\beta}\mathrm{Im}\left[\frac{-p_1}{G(p_1)}\right]\right) \tag{14}$$

The final integral gain is the maximum value between Eqs. (15) and (16).

$$K_{i,1} = \left(\mathrm{Re}\left[\frac{-p_1}{G(p_1)}\right] - \frac{\alpha}{\beta}\mathrm{Im}\left[\frac{-p_1}{G(p_1)}\right] \right) \quad (15)$$

$$K_{i,2} = \left(\mathrm{Re}\left[\frac{-p_1}{G(p_1)}\right] - \frac{\alpha^2 - \beta^2}{2\alpha\beta}\mathrm{Im}\left[\frac{-p_1}{G(p_1)}\right] \right) \quad (16)$$

To achieve a robust control with large bandwidth it was set an overshoot about 10% and a settling time of 0.02 s. Phase margin was set 60°. **Table 2** shows the integral, proportional and derivative gains of the ESEA's controller.

Parameter	Value
Proportional gain (Kp)	2.915
Integral gain (Ki)	4.129
Derivative gain (Kd)	0.007

Table 2. Controller parameters.

Figure 15 shows the Bode diagram of the system, there is a good frequency band of 40 Hz, compatible with the Paine's results (Paine et al., 2013). On the other hand, the phase angle is quite high at 40 Hz, greater than 90°. Yet it is possible to apply the actuator in several systems where force control is necessary, as in some robot arms in industrial plants. Resonance frequency is greater than 200 Hz. This implies that the effects due to vibrations are efficiently filtered into the frequency band of the actuator.

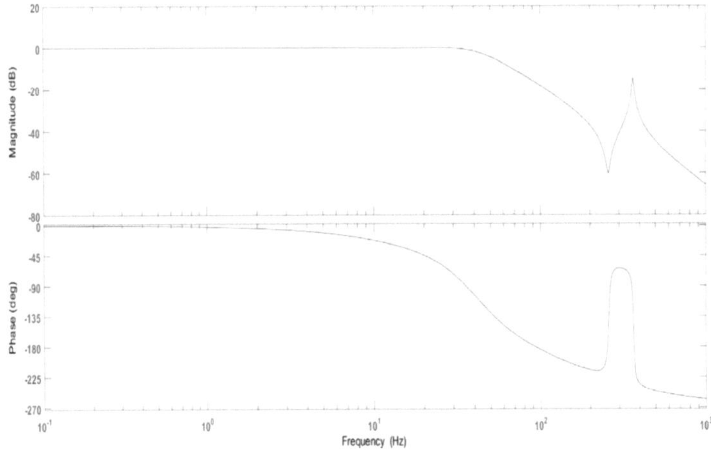

Figure 15. Bode diagram of the system.

http://dx.doi.org/10.5772/63573

Figure 16 shows the time response for a constant input force. The electric SEA shows an 8% overshoot and it settling time is 0.03 s which is very fast. While hydraulic SEA of reference work (Leal Junior et al., 2015) has an overshoot about 12% and a settling time of 0.07 s. Although hydraulic SEA has larger overshoot and settling time than electric SEA of this section, these two parameters are compatible with the reference works results (Paine et al., 2013; Robinson, 2000). Therefore, even in the Hydraulic SEA case the settling time is tolerated which is discussed in Section 4.

2.4. Final remarks

This section presented the digital prototyping of a linear electric SEA for exoskeletons, wearable robots and mechatronic devices. It was used as a digital prototyping environment for design and assembly of actuators parts, generation of section views images in perspective and exploded views, and all fabrication drawings of the actuator parts. The prototype is driven by a Maxon EC 60 Ø60 mm, brushless 400W DC motor, with a Planetary Gearhead GP 62 A (Maxon Motor Catalogue) and an R Series 10mm diameter rolled steel ball screw (NSK Catalogue).

This section also presented the actuator's performance, evaluated under a PID impedance controller. **Figure 15** shows the frequency response of the magnitude and phase of Fl/Fd. **Figure 16** shows the step response of unit force input. The step response can also be tuned to final displacement or to contact force. Some improvements to the actuator itself will be needed for the next steps of the project, aiming at the fabrication, assembly and testing of the actuator.

Future works are the construction of a real aluminum alloy based prototype and testing its performance in a real time environment to evaluate the actuator's performance under force,

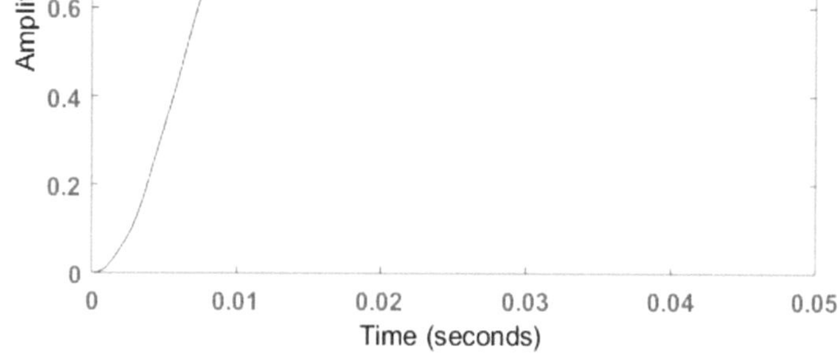

Figure 16. Time response for a 1N step input force.

position, velocity and mixed controllers. It is also important to verify the effect of spring stiffness and gear ratio in load bandwidth. Another further investigation is the development of tuning strategies between the spring stiffness – actuator bandwidth and elastic energy store relationship. Since these effects were neglected in this section, actuator's backslash and viscous friction have to be analyzed in future works.

A broad comparison between the electric serial elastic actuator and the hydraulic SEA has to be made to define which one is the best for a compliant robot application.

3. Hydraulic series elastic actuator

3.1. Hydraulic series elastic actuator design

3.1.1. Methodology

A basic model of a series elastic hydraulic actuator has the serial element, i.e. a rod with the spring's arrangement and a hydraulic circuit which basically comprises of the hydraulic cylinder and the servo valve. The actuator may be understood as a combination of two parts. The first one is the serial elastic element presented. The other one is the hydraulic circuit which has some construction alternatives as well.

Once the hydraulic circuit and serial elastic element of the actuator are defined, the next step is to select the hydraulic components and design the spring based on the application. This is done by defining a condition for this system which, in this case, is the actuator output force to be up to 5000 N. To achieve this output force, the springs are selected based on consolidate knowledge presented by mechanical engineering design bibliography. Moreover, the rod had to be able to support the load without buckling. A special attention must be given to the actuator's springs because they affect system's impedance and bandwidth. After setting these two parameters, iteration could be necessary to evaluate the desirable spring constant.

After the selection of the rod and the spring, it is possible to use the force caused by the spring under allowable extension-compression limits as the force output of the hydraulic cylinder. This force and the maximum speed of 0.5 m/s, as defined by Robinson and Pratt (2000), led to the selection of the correct cylinder by a commercial data sheet; otherwise it would be possible to manufacture a custom cylinder. After the parts design, they are assembled in Siemens Solid Edge (Siemens PLM Software, Germany) digital prototyping environment which allows obtaining inertial properties of the assembly.

3.1.2. Digital prototyping design

The design was done by the methodology mentioned on the previous section. Following these steps, the actuator achieved a good volume-output force ratio for the available components combination. Also all the internal components had been enclosed to avoid accidents, components contamination and make the actuator suitable to human interaction (Bento Filho et al., 2014).

Figure 17 shows the prototype of the actuator. There is a hydraulic cylinder (1) that moves forward and dislocates the movable rod (2) fixed in it with a screwed connection. The model has two low friction sliding cylinders (6) and (8) fixed on the base tube (3) and one guide (7) fixed on the movable rod (2). There are two springs (4) and (5) fixed between the low friction sliding cylinders (6) and (8). A connection guide (10) is used to connect the hydraulic cylinder rod (1) to the base tube (3). The load terminal (9) is also connected directly on the base tube (3). The encapsulation tube (12) is the last part of the assembly. It is connected on the hydraulic cylinder (1) and base tube (3). A guide bolt (11) was applied to make the connection between the base tube (3) and the encapsulation tube (12), but without interfere on the base tube's (3) linear movement.

Figure 17. Actuator's components: (1) hydraulic cylinder; (2) movable rod; (3) base tube; (4) and (5) springs; (6), (7) and (8) low friction sliding cylinders; (9) load terminal; (10) connection guide; (11) guide bolt; (12) encapsulation tube.

There are four low friction sliding cylinders (6), (7), (8), and (10) in this actuator model. The application of the four sliding cylinders is to give to the system the necessary stability. Without these components, the movable rod (2) would oscillate creating additional forces on the springs instead of only an axial force. Moreover, the load terminal is connected with the base tube only.

The springs are responsible to the hydraulic SEA stiffness reduction. However, this component has to be carefully assembled since it will stretch and compress depending on the movement of the actuator and the load attached to it. Choosing the spring is always challenging, because it has to be a balance between a large bandwidth requiring a high stiffness and impedance that needs a low spring constant. Therefore, after choosing all system characteristics and components, a minimum acceptable point must be defined between the largest force bandwidth and the minimum tolerance impedance level. After setting these two bounds, iteration is required to estimate the spring constant.

The movement of the actuator can be analyzed in three different cases:

Case 1: Movement without load

This is the simplest case, because it is a free movement. The hydraulic cylinder moves forward or backward and the movable rod and all the others components attached to it such as base tube, guides and load terminal follow the movement of the hydraulic cylinder. In this case, there is only a little deflection in the springs due to system's inertia.

Case 2: Movement with load

When a load is attached in the load terminal, the movement of the actuator depends on the springs. For a forward movement, the spring (5) deflects and spring (4) stretches due to springs preload. These deflections and stretching depends on the load mass. The actuator moves forward only when the spring's deformation resulting force equals the applied load. This system characteristic can reduce the frequency range and makes a limit of bandwidth of the system, which is the major drawback of a series elastic actuator.

For the backward movement the same characteristics can be observed, but the spring (4) deflects and spring (5) stretches.

Case 3: Contact of the load terminal with the load

When the actuator reaches an obstacle or the load during its free movement, the springs deform and the value of the resulting force makes the actuator continues its forward movement or stop.

The prototype shows outstanding characteristics of output force and compactness. The actuator has 200 mm length when retracted with 100 mm of range. The maximum diameter is 26 mm. Moreover, with 100 bar of working pressure, this SEA produces output force up to 4.9 kN which is a great power to weight ratio since the estimated weight of the actuator is 1.5 kg. In order to ease the manufacturing, model structure material chosen is the carbon steel. Actuator's parameters are measured for modeling and control purposes. Important parameters are the deflection of the springs and the displacement of the movable rod, which are due the movement of the hydraulic cylinder.

A linear potentiometer was chosen to evaluate the movable rod's displacement. Since the potentiometer did not have the resolution to evaluate the deflection of the springs, an optical sensor used on a printer was chosen to evaluate this deflection. This sensor has a scaled tape with black lines, with distance between the lines known, which does not allow light beam passage. The tape movement produces pulses, which are proportional to spring deflection.

The assembly of the sensors was shown in **Figure 18**. The linear potentiometer (3) was connected directly on the hydraulic cylinder rod (2) and the hydraulic cylinder bore (1). The optical sensor has a circuit board (7) with a sensor (9). This circuit board was connected on the sliding cylinder (6) by a bolt connection. The optical tape (8) was connected on the sliding cylinders (4) and (5) by a bolt connection and positioned on the optical sensor (9).

http://dx.doi.org/10.5772/63573

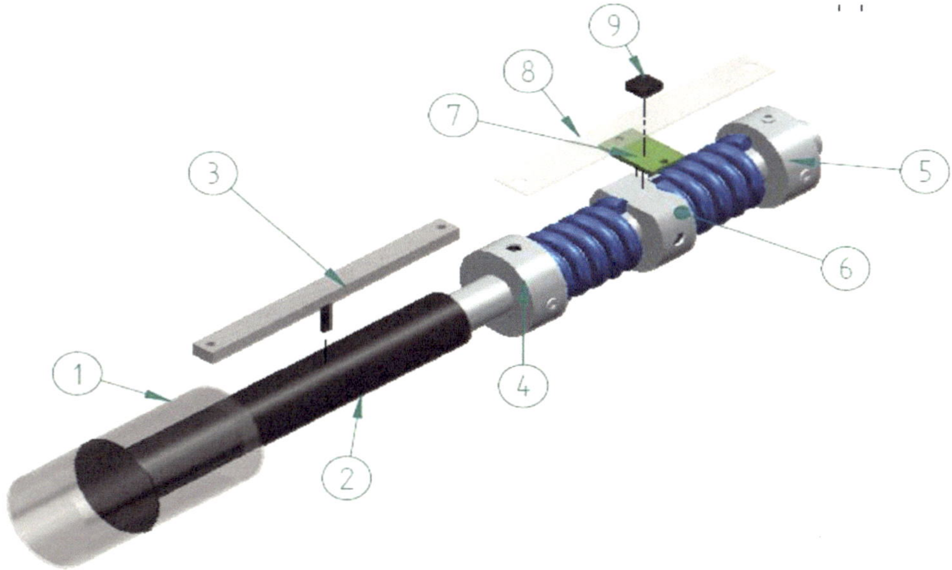

Figure 18. Sensors assembly: (1) hydraulic cylinder bore; (2) hydraulic cylinder rod; (3) linear potentiometer; (4), (5) and (6) guides; (7) circuit board; (8) optical sensor tape; (9) optical sensor.

3.2. Dynamic model

The system consists of a servo-hydraulic system with a serial elastic element. The overall system has two degrees of freedom. One degree is the hydraulic cylinder displacement and the other is the serial elastic element displacement. This displacement and the spring deflection were used to define the output force of the system and to control it. **Figure 19** illustrates the mechanical model of this system discussed above.

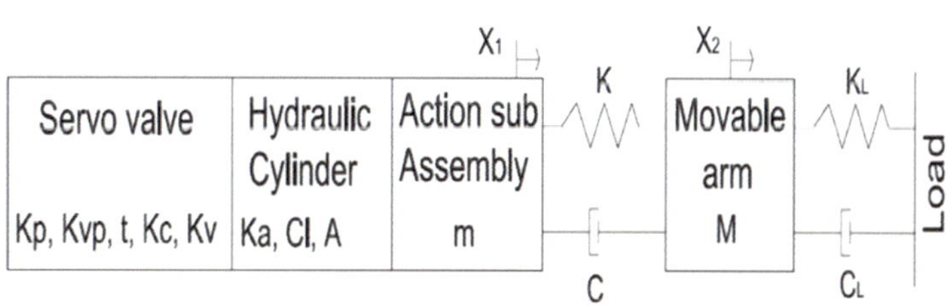

Figure 19. Actuator's mechanical model.

Although most of the servo-hydraulic system is nonlinear, each component of this model has been linearized (Qian et al., 2014). The linearization occurs at an operating point and also can be made at multiple operating points. The controller is tuned based on the linearized model. Although an actuator does not work at a fixed operation point, the controller based on a linearized system has good response even when it is off the operation point.

The servo valve's pilot stage is assumed to be of first order. Since high order dynamics in the valve are more than an order of magnitude above the frequency range of 40 Hz which is the frequency of interest (Robinson and Pratt, 2000), the higher order dynamics can be neglected. Eq. (17) represents the servo valve pilot stage equation as seen in Qian et al. (2014).

$$\frac{Q_s(s)}{v_i(s)} = \frac{K_{vp}}{\tau s + 1} \times K_p \times K_C \times K_v \tag{17}$$

where K_{vp} is the flow gain of the pilot-stage servo valve and τ pilot-stage's equivalent time constant, v_i valve control signal transmitted from the hydraulic system internal proportional controller, K_p represents inner loop proportional gain, K_C is the valve pressure gain which relates flow from the pilot stage with spool position. The coefficient K_v is the flow gain of the servo valve and Q_s is the supply flow. Rearranging the flow terms and applying the well-known cylinder force equation gives Eq. (18) (Qian et al., 2014).

$$\frac{F_A(s)}{Q_s(s)} = \frac{\dfrac{4}{\beta V s + C_l} \times A}{1 + \dfrac{4}{\beta V s + C_l} \times A^2 \times X_1(s) s} \tag{18}$$

This β is the fluid bulk modulus, V is the fluid volume, A is the cross sectional area, $X_1(s)$ is the action sub-assembly displacement.

The serial elastic element has to be modeled by Newton's second law (Eq. (19)). The properties of the movable rod and the springs are considered on the stiffness, damping and mass terms. The load characteristics are also considered in the actuator's dynamic model (Eq. (20)).

$$X_1(s)(ms^2 + Cs + K) = F(s) \tag{19}$$

where $X_1(s)$ is the action sub-assembly displacement, m is the action sub-assembly mass, C is the damping coefficient, K is the stiffness coefficient and F is the output force.

$$X_2(s)(Ms^2 + C_L s + K_L) = F(s) \tag{20}$$

Referring to Eq. (20), M is the movable arm mass, C_L and K_L are the load interface viscous friction coefficient and load interface elastic constant respectively. $X_2(s)$ is the movable arm displacement. The simplified closed loop transfer function block diagram for the dynamic model is presented in **Figure 20**. Each subsystem presented in the figure has a transfer function according the equation discussed on this topic. The actuator was modeled to have a natural velocity feedback (Qian et al., 2014). The leakage of the servo valve stages was neglected. The assumptions made in this model are constant fluid properties, servo valves are not saturated, supply pressure is much greater than the load pressure, friction force can be modeled as viscous damping, main stage spool opening is proportional to pilot stage flow (Qian et al., 2014) **Figure 20** shows the hydraulic SEA block diagram.

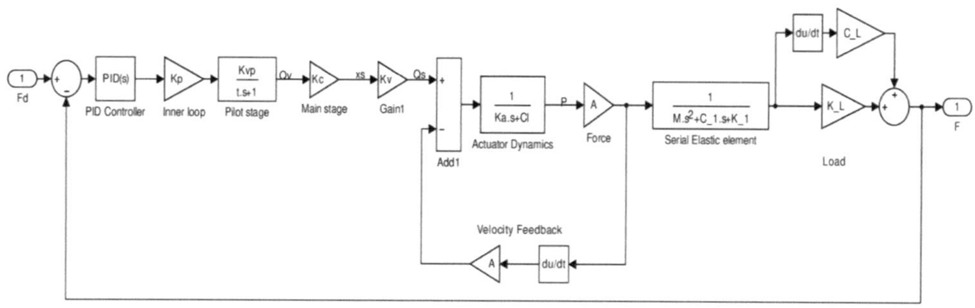

Figure 20. Closed loop transfer function block diagram.

3.3. Results and discussion

Once the actuator dynamic model is described, the parameter values had to be established. The parameters were obtained by the digital prototype and data sheets of some components i.e. servo valve and hydraulic actuator. **Table 3** summarizes the HSEA parameters applied on dynamic model.

Parameter	Value	Units
Area (A)	4.91E–04	(m^2)
Stiffness (K1)	7.5E+07	(N/m)
Damping (C1)	18.4	(Ns/m)
Mass (M)	1.5	(Kg)
Bulk modulus (B)	1.23E+09	(Pa)
Leakage coefficient (Cl)	2E–06	(m^3s/kPa)
Fluid volume (Vt)	2.60E–05	m^3
Load stiffness (KL)	3.30E+09	(N/m)
Load damping (CL)	8.08	(Ns/m)
Inner loop gain (Kp)	8.40E+02	(s^{-1})
Valve pressure gain (Kc)	4.00E+06	(m^3s/m)
Valve flow gain (Kv)	6.24E–05	(m^3s/m)
Servo valve time constant (t)	0.0015	(s)

Table 3. HSEA parameters.

Spring constant is the most important parameter in the system. To define the spring constant, the minimum impedance level and a maximum bandwidth were set with respect to the previously selected components of the hydraulic system. Iteration was made to choose the best value with upper and lower bounds.

After knowing all the actuator's parameters, the PID controller could be designed. The controller was tuned by applying the pole placement and phase margin method presented in Section 2. A PID controller is able to produce good responses. This controller is widely applied on industrial robots and processes due to its robustness and simple implementation.

To achieve a robust control with large bandwidth, an overshoot of about 10% and a settling time of 0.02 s were set. Phase margin was set to 60°. **Table 4** shows the controller parameters for the system's characteristics aforementioned.

Parameter	Value
Proportional gain (Kp)	54.9946
Integral gain (Ki)	4412.1219
Derivative gain (Kd)	0.0575

Table 4. Controller parameters.

The bode diagram of the system (**Figure 21**) shows an excellent frequency band of about 100 Hz which was greater than all references works presented (Bento Filho et al., 2014; Paine et al., 2013; Robinson and Pratt, 2000). The phase angle for this frequency band was around 45°. The resonance frequency of the system was 1100 Hz which is far enough to the system's operation frequency. This implies that the effects due to vibrations are efficiently filtered into the frequency band of the actuator.

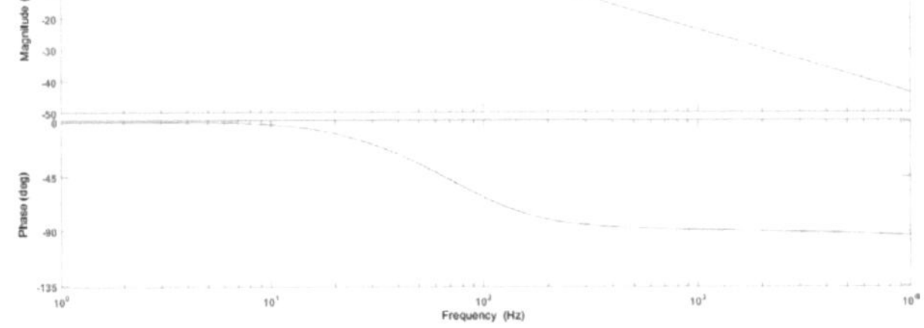

Figure 21. Bode diagram of the system.

To show the system's response, a Unit step input was applied. **Figure 22** shows the system's response where it could be seen that the settling time is around 0.02 s which was a very fast convergence to this system. The maximum overshoot was a little higher than 10% which was an acceptable value.

http://dx.doi.org/10.5772/63573

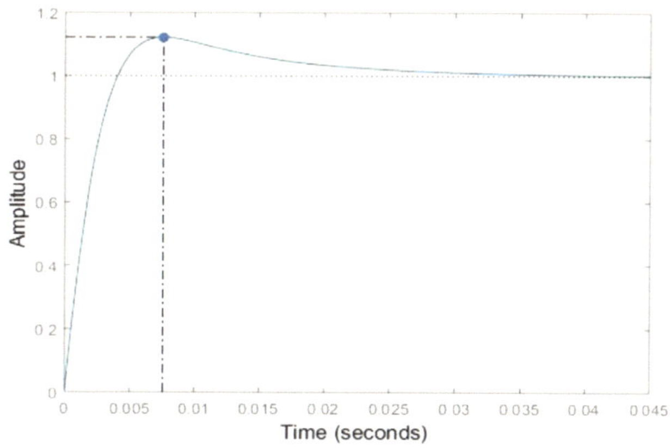

Figure 22. Time response of the system.

3.4. Final remarks

This section presented a digital prototyping of a linear series elastic hydraulic actuator for compliant robots. It was used in digital prototyping environment for design and assembly of actuators parts, generation of images of section views in perspective and exploded views, and all fabrication drawings of the parts of the actuator. The model obtained was very compact with a high output force which is desirable for all applications. A dynamic model with a PID controller was also presented which described the linear characteristics of the actuator. The system showed good time and frequency responses. The results obtained help the development of robots and manipulators for non-structured environments with another level of compactness and force capacity.

Future work is the construction of the actuator and investigation of benefits and limitations of this actuator on a compliant robot with human interaction.

4. Comparison between series elastic actuators

4.1. Comparison parameters

As mentioned in Section 1, the analysis of an actuator comprises the investigation of force bandwidth, mechanical output impedance, dynamic range, force density and the time response for the system without load. Each of these parameters has its own importance for the system and its own method to be determined. Robinson (2000) first introduced the comparison methodology adopted in this chapter. In these analyses, the load stiffness and viscous damping presented for ESEA and HSEA presented in **Tables 1** and **3** respectively is not considered on the following analyses. Therefore, it is expected that the systems responses have deviations between the ones presented in Sections 2 and 3.

4.1.1. Force bandwidth

The force bandwidth of an actuator is how quickly it can generate the desired forces. In order to be able to perform the desired task, the bandwidth must be sufficiently high to transmit the forces through the machine structure. To control a human joint, only a few Hertz of bandwidth is necessary. Moreover, the bandwidth needed for a four-legged dynamic robot for weight carrying is small too. Reference authors estimate that only 20 Hz is more than necessary for the given application. The bandwidth analysis is divided in the closed loop force bandwidth and the large force bandwidth. Closed loop bandwidth is evaluated with a step input of a small force in the actuator closed loop model (**Figure 23**). The large force bandwidth is obtained with an oscillation of the full steady-state output force of the actuator as an input on the closed loop model. HSEA has 4.9 kN of output force and ESEA has only 450 N. Due to this difference between the actuators capabilities, only the closed loop bandwidth is compared in this chapter. Referring to **Figure 23**, X_m is the motor displacement, F_d is the desired force, F_l is the output force, k_s is the spring stiffness, K and V_c are model characteristics.

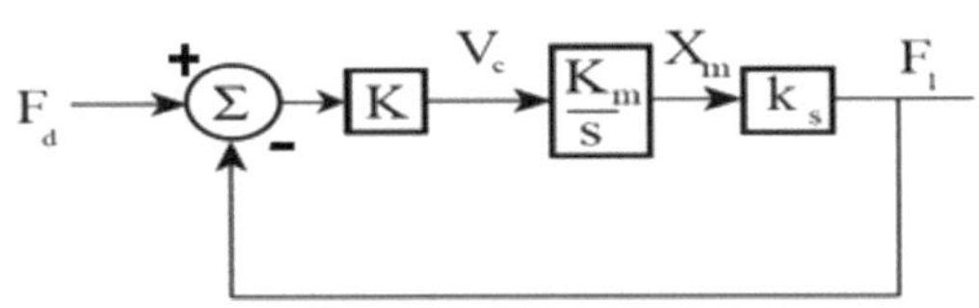

Figure 23. Simplified block diagram for closed loop bandwidth evaluation (Robinson, 2000).

4.1.2. Mechanical output impedance

The mechanical output impedance is defined as the minimum force needed to move a certain load. An actuator with low impedance can be considered a pure force source since its internal dynamics are negligible. Low impedance actuators can also be referred as back-drivable (Robinson, 2000). The analysis of the output impedance is made by applying to each actuator model the case of a load, which is free to move, attached to the actuator's output. The impedance is evaluated around a constant desired output force as an analogy of the concept of impedance on electrical circuits (**Figure 24**). Referring to **Figure 24**, the x_l is the load displacement, x_s is the spring displacement and F_l is the force output.

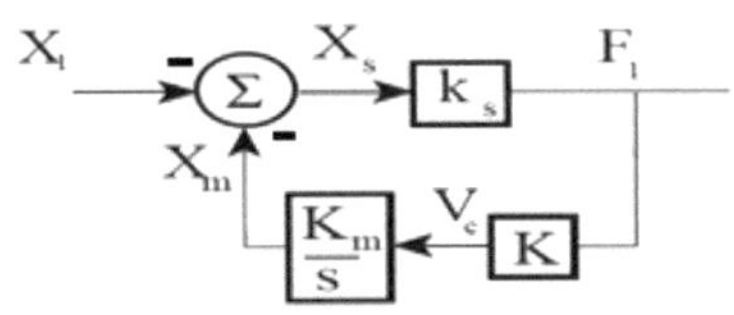

Figure 24. Simplified block diagram for mechanical output impedance evaluation (Robinson, 2000).

http://dx.doi.org/10.5772/63573

4.1.3. Dynamic range

The dynamic range is the ratio between the maximum output force of the actuator and the minimum resolvable force of this device. The dynamic range analysis on an actuator is very important since it gives a measure of how sensitive the actuator is to small force compared to the actuator maximum force. For example, if an actuator has high maximum force and high minimum resolvable force, its dynamic range is low. Therefore, this actuator can only make tasks of high force and it is not able to manipulate fragile objects or have direct contact with humans. Moreover, a large dynamic range is desirable for almost all robotic applications, since it allows the robot to be versatile to perform high sensitive operations and large force demanding operations. Since it can sense or grasp almost all fragile objects and manipulate it in precise tasks and, on the same time, can perform high force demanding tasks such as weight lift, one of the largest dynamic range actuator is the human muscle.

4.1.4. Force density

Force density is defined as the ratio between the output force and the mass or the volume of an actuator. An actuator with large force density allows the application in lighter robots with good force output characteristics since it not overburden the robot structure and allows the quick response of the mechanism.

4.2. Results and discussion

The two SEA presented in previous sections are compared based on the key parameters of the series elastic actuators aforementioned which are bandwidth, output impedance, time response, power density, dynamic range.

The spring stiffness affects all parameters presented above. In order to evaluate the differences between the two systems with the same parameter for the compliant sensor, the spring stiffness is 3.336E+03 kN/m for both actuators.

4.2.1. Force bandwidth results

Figure 25 shows the HSEA and ESEA Bode diagram which allows evaluating the two systems closed loop force bandwidth.

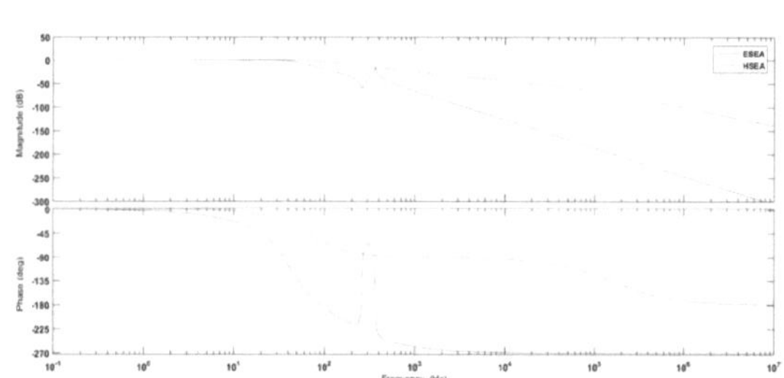

Figure 25. Closed loop bode diagram.

The ESEA frequency band is about 40Hz with a phase angle of 90° which is quite high. On the other hand, the HSEA has 30Hz but the phase angle is about 55° which is more appropriate for practical applications. Since a four-legged robot works on small bandwidths, the smaller bandwidth range of the HSEA is not an issue for this application. In addition, the two actuators present resonance frequency above 100Hz which is far enough to the required frequency.

4.2.2. *Time response results*

The time response for a Unit step input of the system without the load is given in **Figure 26**. In the ESEA and HSEA design made in Sections 2 and 3, respectively, the system dynamic has a load damping of 8 Ns/m and a load stiffness of 3E+06 N/m. ESEA shows an 8% overshoot and settling time of 0.03 s which is very fast. While HSEA has an overshoot about 12% and a settling time of 0.07 s. Although HSEA has larger overshoot and settling time than ESEA, these two parameters are compatible with the reference works results (Paine et al., 2013; Robinson, 2000). Therefore, even in the HSEA case the settling time is tolerated.

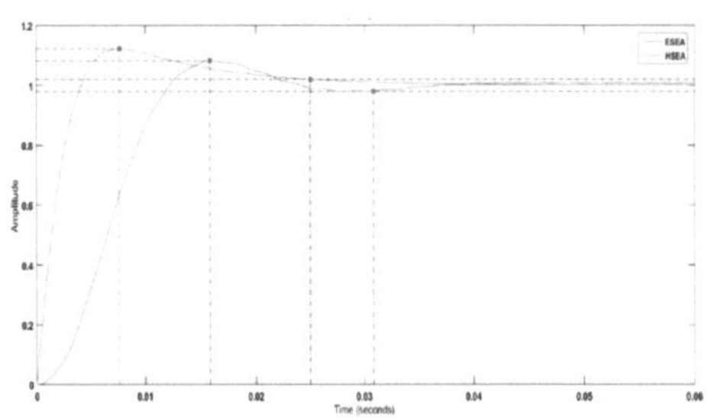

Figure 26. Time response of the actuators.

4.2.3. *Mechanical output impedance results*

Since output impedance is the force at an actuator output given a moving load position, the impedance analysis was made by setting a load position with a constant desired force. An ideal actuator has zero impedance. In other words, the actuator is a pure force source. If this happens, the dynamics of the actuator are completely decoupled from the dynamics of the load motion (Robinson, 2000). Although the compliant element of the series elastic actuator limits the bandwidth, it reduces the output impedance which is desirable for many actuator applications (Pratt et al., 2002).

Nevertheless, these two actuation systems have the same spring stiffness. Therefore, the differences between the systems output impedance are due to the actuator's moving system which are the motor with gear reduction for the ESEA and servo valve with hydraulic piston for the HSEA (Robinson, 2000).

Figure 27 shows the output impedance for both hydraulic and electric sources. The output impedance of the series elastic actuator is small at low frequency and it achieves the spring stiffness magnitude (which is −130dB for the spring previous selected) when the frequency rises. This behavior was proved by simulations and physical SEA experiments (Robinson, 2000).

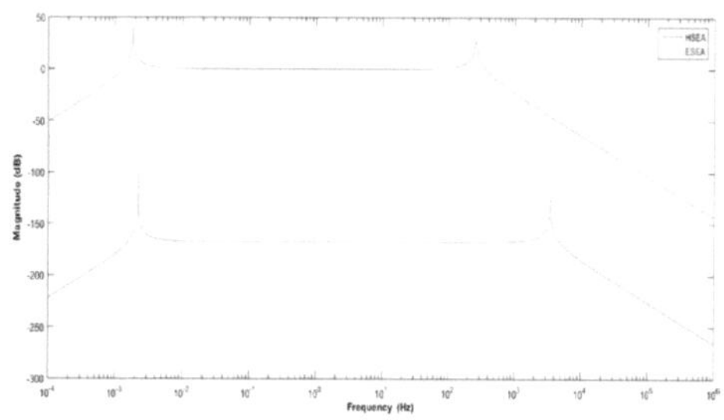

Figure 27. HSEA and ESEA output impedance.

Since the both systems have the same compliant element, they achieve the same value with frequency increase. Therefore, the difference on the impedance is given by the curve slope. Examining the slope of the both curves, the HSEA has the smallest value of the impedance when the frequency increases. It can be seen two peaks of the magnitude, the first one due to controller gains and the second occurs in the systems resonance frequency.

4.2.4. Power density results

Power density is the actuator power per mass unit. High power density actuators can delivery great power with a small package. It prevents over-burdening the actuator structure with excessive mass allowing a responsive robot performance. The lightweight structure with a quick performance is very important for a weight carrying dynamic robot because it provides a compact robot which is suited to a non-structured environment and has a great output force.

As shown in previous section, the hydraulic series elastic actuator has more output force than the electric SEA. Moreover, the ESEA is heavier than the HSEA.

The output force of the HSEA is 4.9kN and it weighs 1.5kg given the velocity of 0.5m/s, the HSEA has a power density about 1600W/kg which is a great performance. On the other hand, the ESEA has 450N of output force and weighs 3.7kg. The ESEA is also slower than HSEA, the electric SEA velocity is 13.4cm/s which gives a power density of 16W/kg. This result shows a poor performance of ESEA for weight carrying dynamic robot.

Even though the HSEA mass applied on the power density calculation did not include the mass of servo valve, reservoir, pump, hoses and pipes which are heavier than the ESEA batteries that also are not included on the electric SEA mass. Even if these accessories mass is considered, the HSEA still has way more power density even with these items included.

4.2.5. Dynamic range results

Dynamic range is the ratio between the maximum output force and the minimum resolvable output force. It gives quantitative information of how sensitive the actuator is with respect of its maximum output force capability. It is a crucial parameter for dynamic robots because it allows the actuator to be used in sensitive activities such as human contact, equipment operation, and non-structured environment exploration. In addition, the actuator with great dynamic range can do hard work activities such carrying, push or pull weights. In summary, high dynamic range is an important characteristic for a four-legged dynamic robot application.

The minimum resolvable output force is evaluated by the resolution of the spring deflection sensor. Resolution gives the minimum spring deflection that can be detected. This deflection times the spring stiffness gives the actuator minimum resolvable force.

Both the SEAs have the same spring stiffness in this comparison. Also, both actuators have the same spring deflection sensor which is a linear encoder. Therefore the minimum output force is the same for both which is 3N.

The HSEA dynamic range is 1500:1 while the ESEA range is 150:1. It shows that the hydraulic SEA has a range 10 times bigger than electric one.

5. Conclusions

This chapter presented a comparison between the two series elastic actuator (electric SEA and hydraulic SEA) for four-legged dynamic application. The actuators were digitally prototyped and the comparison was made considering actuator's bandwidth, output impedance, time response, power density, and dynamic range.

Although the ESEA has a better time response, both actuators have its responses within desired range for a dynamic robot application. The bandwidth also is not a problem for both actuators because all systems examined have large bandwidth compared with the given application. HSEA and ESEA presented good output impedance with a slightly better behavior from hydraulic SEA. Both actuators have two peaks on the output impedance, one peak due to controller gains and another due to the resonance frequency of the system.

Power density and dynamic range are the two key parameters that differs the weight carrying dynamic robot application from others dynamic robots applications. In these two parameters, the HSEA showed a better performance than ESEA. The HSEA's power density is 100 times higher than the ESEA's. Moreover, the dynamic range of hydraulic SEA is 10 times higher than the electric SEA dynamic range. The results showed the HSEA as the better series elastic actuator for this task and it could contribute for the research of SEA applied on robot links.

Further work is required on the simulation of SEA, specially the HSEA, on a dynamic robot. With this simulation, the advantages and drawbacks of the implementation can be analyzed and quantified. Also, a construction of a real aluminum alloy based prototype and testing its performance in a real time environment for evaluating the actuator's performance under force, position, velocity and mixed controllers. Another future work is on the development of tuning strategies between the spring stiffness – actuator bandwidth and elastic energy store relationship.

http://dx.doi.org/10.5772/63573

Author details

Arnaldo Gomes Leal Junior[1], Rafhael Milanezi de Andrade[1,2*] and Antônio Bento Filho[1]

*Address all correspondence to: rafhael.andrade@ufes.br

1 Department of Mechanical Engineering, Universidade Federal do Espirito Santo, Vitória, ES, Brazil

2 Department of Mechanical Engineering, Bioengineering Laboratory, Universidade Federal de Minas Gerais, Belo Horizonte, MG, Brazil

References

[1] Arumugom, S., Muthuraman, S., and Ponselvan, V., "Modeling and application of series elastic actuators for force control multi legged robots," J. Comput., vol. 1, no. 1, pp. 26–33, 2009.

[2] Au, S., Weber, J., Herr, H., "Biomechanical design of a powered ankle-foot prosthesis" in IEEE 10th International Conference on Rehabilitation Robotics, 2007.

[3] Bento Filho, A., Andrade, R., Mattos, M., (2014). "Digital prototyping of a series elastic actuator for exoskeletons" in CONEM 2014.

[4] Brooks, R., Breazeal, C., Marjanovic, M., Scassellati, B., Williamson, M., "The Cog project: Building a humanoid robot" in Springer-Verlag Lecture Notes in Computer Science Volume, 1999.

[5] Curran S., and Orin, D., "Evolution of a jump in an articulated leg with series-elastic actuation," in Proc. IEEE Int. Conf. Robot. Autom., May 2008, pp. 352–358.

[6] Diftler, M., Mehling, M., Abdallah, M., Radford, N., Bridgwater, L., Sanders, A., Askew, R., Linn, D., Yamokoski, J., Permenter, F., Hargrave, B., Piatt, R, Savely, R., and Ambrose, R., "Robonaut 2—The first humanoid robot in space," in Proc. IEEE Int. Conf. Robot. Autom., May 2011, pp. 2178–2183.

[7] Edsinger-Gonzales, A., and Weber, J., "Domo: A force sensing humanoid robot for manipulation research," in Proc. IEEE/RAS 4th Int. Conf. Humanoid Robot., Nov. 2004, vol. 1, pp. 273–291.

[8] Garcia, E., Arevalo, J., Munoz, G., Gonzalez-de-Santos, P. "Combining series elastic actuation and magneto-rheological damping for the control of agile locomotion" in Robotics and Autonomous Systems vol. 59, pp. 827–839, 2011.

[9] Gregorio, P., Ahmadi, M., and Buehler, M., "Design, control, and energetics of an electrically actuated legged robot," IEEE Trans. Syst., Man, Cybern. B, Cybern., vol. 27, no. 4, pp. 626–634, 1997.

[10] Grimmer, M., Seyfarth, A., "Stiffness Adjustment of a Series Elastic Actuator in a Knee Prosthesis for Walking and Running: The Trade-off between Energy and Peak Power Optimization" in 2011 IEEE/RSJ International Conference on Intelligent Robots and Systems, 2011.

[11] Hutter, M., Remy, C., and Siegwart, R., "Design of an articulated robotic leg with nonlinear series elastic actuation," in Proc. 12th Int. Conf. Climb. Walking Robot. Supp. Technol. Mobile Mach., Sep. 2009, pp. 645–652.

[12] Hutter, M., Remy, C., Hoepflinger, M., and Siegwart, R., "ScarlETH: Design and control of a planar running robot," in Proc. IEEE/RSJ Int. Conf. Intell. Robot. Syst., Sep. 2011, pp. 562–567.

[13] Iyer, S., "Modeling and Testing of a Series Elastic Actuator with Controllable Damping" Master's thesis, Worcester Polytechnic Institute, 2012.

[14] Johansson, R., Robertsson, A., Nilsson, K., Brogårdh, T., Cederberg, P., Olsson, M., Olsson, T., Bolmsjö, G., "Sensor integration in task-level programming and industrial robotic task execution control", Industrial Robot: An International Journal, vol. 31, no. 3, pp.284 – 296, 2004

[15] Kong, K., Bae, J., and Tomizuka, M., "A compact rotary series elastic actuator for human assistive systems," IEEE/ASME Trans. Mechatronics, vol. 17, no. 2, pp. 288–297, Apr. 2012.

[16] Kong, K., Bae, J., and Tomizuka, M., "Control of rotary series elastic actuator for ideal force-mode actuation in human-robot interaction applications," IEEE/ASME Trans. Mechatronics, vol. 14, no. 1, pp. 105–118, Feb. 2009.

[17] Kong, K., Bae. J., Tomizuka, M., "A compact rotary series elastic actuator for knee joint assistive system," in Proc. IEEE Int. Conf. Robot. Autom., May 2010, pp. 2940–2945.

[18] Krupp, B., "Design and control of a planar robot to study quadrupedal locomotion". Master's thesis, Massachusetts Institute of Technology, 2000.

[19] Lagoda, C., Schouten, A., Stienen, A., Hekman, E., and van der Kooij, H., "Design of an electric series elastic actuated joint for robotic gait rehabilitation training," in Proc. IEEE 3rd RAS and EMBS Int. Conf. Biomed. Robot. Biomechatron., Sep. 2010, pp. 21–26.

[20] Leal Junior, A., Andrade, R, Bento Filho, A.(2015). "Linear serial elastic hydraulic actuator: digital prototyping and force control", in IFAC-Papers Online, vol. 8, no. 6, pp. 279-285.

http://dx.doi.org/10.5772/63573

[21] Maxon Motor Catalogue, available in >www.http://www.maxonmotor.com/medias/sys_master/root/8816809443358/15-320-EN.pdf<. Accessed in 04/02/2014.

[22] NSK Catalogue, available in >http://www.nsk.com.br/upload/file/B02.pdf<. Accessed in 04/02/2014.

[23] Pestana, J., Bobin, R., Arevalo, J. C., and Garcia., E., "Characterization of emerging actuators for empowering legged robots," presented at the 13th Int. Conf. Climbing and Walking Robots Support Technologies Mobile Machines, Nagoya, Japan, 2010.

[24] Paine, N., Oh, S., and Sentis, L., "Design and control considerations for high-performance series elastic actuators" IEEE/ASME Trans. Mechatronics, vol. PP, no. 99, pp. 1–12, 2013.

[25] Paluska, D. and Herr, H., "Series elasticity and actuator power output," in Proc. IEEE Int. Conf. Robot. Autom., May 2006, pp. 1830–1833.

[26] Parietti, F., Baud-Bovy, G., Gatti, E., Riener, R., Guzzella, L., and Vallery, H.,"Series viscoelastic actuators can match human force perception," IEEE/ASME Trans. Mechatronics, vol. 16, no. 5, pp. 853–860, Oct. 2011.

[27] Pratt J., and Pratt, G., "Intuitive control of a planar bipedal walking robot," in Proc. IEEE Int. Conf. Robot. Autom.,May 1998, vol. 3, pp. 2014–2021.

[28] Pratt, G. and Williamson, M., "Series elastic actuators," in Proc. IEEE/RSJ Int. Conf. Intell. Robot. Syst. Human Robot Interact. Cooper. Robot., Aug. 1995, vol. 1, pp. 399–406.

[29] Pratt, G., Willisson, P., Bolton, C., and Hofman, A., "Late motor processing in low-impedance robots: Impedance control of series-elastic actuators," in Proc. Amer. Control Conf., Jul. 2004, vol. 4, pp. 3245–3251.

[30] Pratt, J., Krupp B., and Morse C., "Series elastic actuators for high fidelity force control", Industrial Robot: An International Journal, vol. 29, no 3, p. 234–241, 2002.

[31] Qian, Y., Ou G., Maghareh, A., Dyke, S.J.."Parametric identification of a servo-hydraulic actuator for real-time hybrid simulation," in Mechanical Systems and Signal Processing, vol. 48, pp 260–273, 2014

[32] Ragonesi, D., Agrawal, S., Sample, W., and Rahman, T., "Series elastic actuator control of a powered exoskeleton," in Proc. IEEE Annu. Int. Conf. Eng. Med. Biol. Soc., Sep. 2011, pp. 3515–3518.

[33] Robinson, D., Pratt, J., Paluska, D., Pratt, G., "Series Elastic Actuator development for a biomimetic walking robot" in IEEE/ASME AIM99, 1999.

[34] Robinson, D.,"Design and analysis of series elasticity in closed-loop actuator force control", Ph.D. Thesis, Massachusetts Institute of Technology, 2000.

[35] Siemens PLM Software – Solid Edge ST7, available in >http://www.plm.automation.siemens.com/en_us/products/solid-edge/index.shtml<. Accessed in 04/02/2014.

[36] Soyguder, S., Alli, H., "Design and prototype of a six-legged walking insect robot", Industrial Robot: An International Journal, vol. 34, no. 5, pp. 412-422, 2007

[37] Tang, W., Huang, J., Wu, J., Wang, Q., "A PID tuning method based on dominant poles and phase margin," in Proceedings of the 29th Chinese Control Conference, 2010.

[38] Taylor, M. D., "A compact series elastic actuator for bipedal robots with human-like dynamic performance," Master's thesis, Robotics Inst., Carnegie Mellon Univ., Pittsburgh, PA, USA, 2011.

[39] Thurfjell, L., McLaughlin, J., Mattsson, J., and Lammertse, P., "Haptic interaction with virtual objects: the technology and some applications", Industrial Robot: An International Journal, vol. 29, no. 3, pp. 210–215, 2002.

[40] Torres-Jara, E., and Banks, J., "A simple and scalable force actuator," presented at the Int. Symp. Robot., Paris, France, Mar. 2004.

[41] Wang, Z., "Interactive virtual prototyping of a mechanical system considering the environment effect. Part 1: Modeling dynamics", Comptes Rendus Mécanique, vol. 339, no. 9, pp. 591–604, 2011.

[42] Williamson, M., "Series elastic actuators" Master's thesis, Massachusetts Institute of Technology, 2000.

Chapter 11

Fish-Like Robot Encapsulated by a Plastic Film

Mizuho Shibata

Additional information is available at the end of the chapter

http://dx.doi.org/10.5772/63506

Abstract

Underwater robots are currently utilized to evaluate water quality and the undersea landscape. Small-sized underwater robots are especially useful in improving the spatial resolution of the measurements, yielding high-quality data. This chapter describes a small-sized fish-like robot, with its surface composed of a flexible thin plastic film. Its internal components, including an actuator, could be encapsulated in the plastic film using a vacuum packaging machine. To simplify the waterproofing and pressure resistance properties of the fish-like robot, its internal components can be filled with insulating fluid. The plastic film on the surface has electromagnetic-wave-transmitting properties, allowing sensors to be arranged within the device, enabling assessment of its autonomous locomotion using infrared sensors. Robot attitude can be altered, based on geography of its internal components, floating blocks, and insulating fluid. This attitude could be especially determined by the differences in densities between the floating block and insulating fluid. Evaluation of attitude control showed that an insulating fluid heavier than water allows a large variation.

Keywords: fish-like robots, underwater robots, flexible mechanism, vacuum packaging, plastic film

1. Introduction

In this chapter, we develop a small-sized lightweight fish-like robot, with its surface composed of a flexible thin plastic film. In robotics, novel designs have been led to advances. For example, designs of tensegrity structures [1], which are composed of a set of disconnected rigid elements connected by continuous tensional members and have been used to develop lightweight robots such as a crawling robot [2], a robotic arm [3], an underwater vehicle [4], and a fin mechanism [5]. Stream-lined designs of gliding wings have allowed underwater robots to energy-efficient wide-area observations [6–8]. Some designs such as a manipulator unit [9] and a mechanical contact mechanism [10], fixed onto a commercial remotely operated vehicle (ROV), have improved inspection efficiency of underwater vehicles for undersea landscape inspection. Novel fabrication methods can also lead to advances in robotics. For example, the microfabrication of soft material has been shown to produce a gecko robot that can climb a wall [11]. Moreover, the use of a composite material resulted in a bee robot that could fly [12].

We have applied a vacuum packaging method to fabricating a lightweight fish-like robot [13, 14]. The internal components of this robot consisted of a motor, a drive circuit, a battery, a microcontroller, and an oscillation plate to generate thrust (**Figure 1**). These components were encapsulated by a plastic film bag using a vacuum packaging machine. Vacuum generators have been studied in various industrial settings, including food packaging [15], object gripping by a mechanical hand [16], material formation [17, 18], and casting materials [19]. In robotics, engineering the utilization of a vacuum has included the construction of robots with suction cups for wall climbing [20–23] and a handling tool for nanorobots [24]. This research should not only contribute to the development of a fish-like robot but represents the application of a novel fabrication method to robotics.

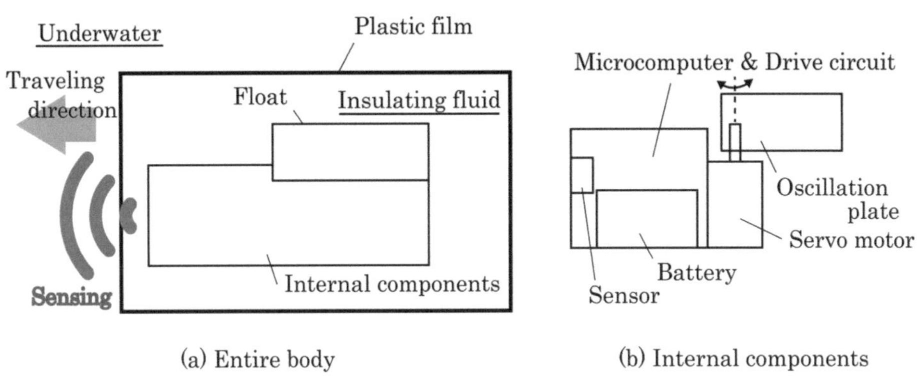

Figure 1. Concept of a fish robot encapsulated by a plastic film.

Recently, small-sized underwater robots are especially required to improve the spatial resolution of these measurements, resulting in high-quality data. Biomimetic designs to improving small-sized underwater robots have included the development of a mechanical pectoral fin [25], fish-like robots [26–29], and snake robots [30]. These robots can swim through water by creating undulations oscillating their bodies. The entire body of the fish-like robot we proposed also generates thrust by the body flexure. The plastic film encapsulating the internal components is inflected by the oscillation plate fixed on the servo motor. To improve the lubricity between the oscillation plate and the plastic film and to simplify the waterproofing and pressure resistance properties of the fish-like robot, its internal components can be filled with insulating fluid.

http://dx.doi.org/10.5772/63506

Most of the underwater robots are encased in a solid, pressure-resistant structure made of metal, such as a stainless-steel and titanium alloy to improve the waterproofing features. The weight of these robots will therefore tend to be greater due to the density of these metal components. To overcome this drawback, we have designed a fish-like robot, the entire outer layer of which is composed of a plastic film, resulting in a lightweight body with low elasticity. In developing the prototype, we selected a low force/torque actuator by utilizing a thin plastic film. This film was flexible, but had lower elasticity for bending than deformable materials such as silicone.

To achieve autonomous control, underwater robots must detect obstacles under water. In traditional underwater robots, sensors such as a camera [31] and a photodetector [32] to detect obstacles are arranged in pressure tight cases. This study was designed to evaluate the electromagnetic-wave-transmitting properties of the thin plastic film. These properties can enable noncontact sensors to be arranged within the encapsulating plastic film (see **Figure 1**). Similar to the other internal components of our robot, these sensors did not require special waterproofing. Additionally, we were able to easily determine the arrangement of these noncontact sensors because the entire surface of the fish-like robot was composed of an electromagnetic-wave-transmitting film, thus enhancing the design flexibility of its internal components.

Underwater robots also require three-dimensional nonholonomical movement to move over wide areas under water. For underwater robots, several vertical depth control techniques must be implemented, including throwing the ballast [33] and changing the volume [34]. Difficulties may be overcome by attitude changing schemes, including use of a movable weight in the body [35], a movable float on the body [10], the reaction force of internal rotors [36], the gyro effect of a flywheel in the body [37], and thruster forces for a neutral buoyant underwater robot [38]. This study involved changing the position of the floating block in the robot body, allowing the selection of a low torque motor.

This chapter is organized as follows: the next section briefly outlines the fabrication of an underwater robot encapsulated by a plastic film. This film was applied by a vacuum packaging machine used in the food industry. We also utilized insulating fluid to simplify the pressure resistance properties of the robot. Section 3 discusses the methods used to control our fish-like robot with a plastic-filmed body. We first investigated the performance of an infrared sensor, taking into account the influence of water and a plastic film. We showed that the signals from the infrared sensors could direct simple autonomous locomotion of our robot. We also developed an attitude control mechanism, based on the geography of the floating block in the body filled with insulating fluid. We showed that changes in the densities of the floating block and the insulating fluid can change attitude. Section 4 summarizes our conclusions.

2. Fabrication

We utilized a vacuum packaging machine to fabricate a fish-like robot, the entire body of which was composed of a flexible plastic film. We called this fabrication robot packaging [13, 14].

Figure 2 shows the process used to fabricate robot packaging. The process can be classified into four steps: (a) encapsulation of the internal components, including a microcontroller, a drive circuit, a battery, a servomotor, and an oscillation plate, in a plastic film bag used to package foods; (b) pouring of insulating fluid, specifically industrial oil [13] or cleaning fluid for semiconductors [14], into the plastic bag. This would reduce the quantity of air in the package after the insulating fluid was defoamed and packaged; (c) defoaming the inside of the robot using a vacuum packaging machine; (d) sealing of the plastic film by a sealer within the chamber of the vacuum packaging machine after defoaming. The drive circuit in the body of the robot is not shortened by the insulating fluid surrounding the circuit. Using this method, we were able to easily fabricate the entire body of a fish-like robot at low cost and in a short time because the body of the robot consisted of only a thin plastic film, which was sealed by a vacuum packaging machine to form the entire body of the robot.

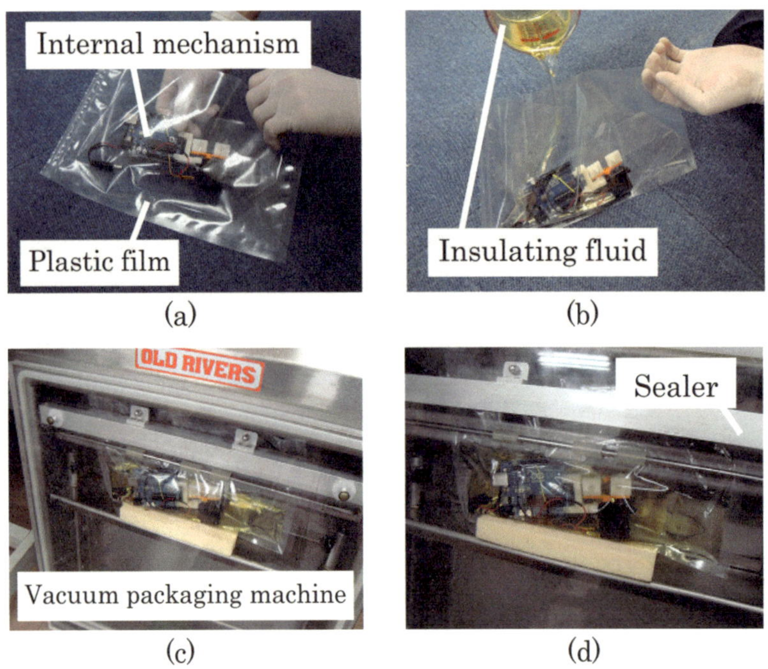

Figure 2. Fabrication process of the robot packaging method.

Ideally, a plastic film fabricated by a robot packaging method does not break in response to water pressure because the pressure inside the robot is equal to the environmental pressure. The plastic film and the insulating fluid are deformed slightly by water pressure; however, the volume of the insulating fluid does not change markedly due to its high incompressibility. To assess the validity of robot packaging, we can test the pressure resisting feature of a servo motor encapsulated by a transparent plastic film. The pressure test is performed using a transparent acrylic cylindrical pressure tight case, to which a pump funneled water. The pressure tests are performed using images captured by a camera due to the transparencies of the plastic film and the tight case [14]. These images are used to investigate the pressure resistance properties of these robots, based on frequency analyses of movement of the servo motor. **Table 1** shows the motion characteristics of a servo motor (RS304MD; Futaba) with a servo horn at 1 MPa pressurization steps. The angle of the servo horn was determined by the positions of the center of rotation and the LED mounted onto the tip of the servo horn. The amplitudes in **Table 1** were the average amplitudes and the frequencies of the servo horn were computed by frequency analysis utilizing fast Fourier transformation (FFT). As shown by the amplitude in **Table 1**, however, the motor could not move in an environment pressurized at 10 MPa. The frequency calculated by FFT at 10 MPa was not used because the power spectrum was much smaller than the other estimated frequencies at up to 9 MPa in **Table 1**.

	Amplitude (rad)	Frequency (Hz)
0 MPa	1.49	0.8
1 MPa	1.54	0.8
2 MPa	1.52	0.8
3 MPa	1.58	0.8
4 MPa	1.50	0.8
5 MPa	1.54	0.8
6 MPa	1.48	0.8
7 MPa	1.54	0.8
8 MPa	1.56	0.8
9 MPa	1.51	0.8
10 MPa	0.01	–

Table 1. Example of motion characteristics of a pressured servo motor.

3. Control

3.1. Autonomous control by sensors embedded in the outer body

This section describes the sensing system used in developing an autonomous fish-like robot with a body constructed of plastic film. As stated in Section 1, the electromagnetic-wave-transmitting properties of the thin plastic film can enable noncontact sensors to be arranged within the encapsulating plastic film (see **Figure 1**). This robot is filled with insulating fluid so that these devices did not require special waterproofing. In this section, we used an off-the-shelf infrared sensor module (GP2Y0A710K, SHARP) as a noncontact sensor to detect obstacles under water.

As the sensor system of our robot was encapsulated by a plastic film, we evaluated the performance of the infrared sensor under three conditions (**Figure 3**), during which infrared rays were reflected by a stainless steel plate, as an example of obstacles, and fixed on a jack. The infrared sensor is fixed on a rigid plate as shown in **Figure 3**. To measure the output voltage of the infrared sensor, the distance *d* between the sensor and the plate was changed by altering the height of the jack. In Condition 1 (**Figure 3(a)**), only air was present between the sensor and the plate. In Condition 2 (**Figure 3(b)**), a plastic film in contact with the sensor was positioned between the sensor and the plate. In this condition, the infrared ray was passed through the film and the air. In Condition 3 (**Figure 3(c)**), a plastic film was placed in contact with the sensor; water was positioned between the film and the plate. Hence, the plastic film was also in contact with water. In this condition, the stainless plate was positioned in the water so that the infrared ray passed through the film and the water.

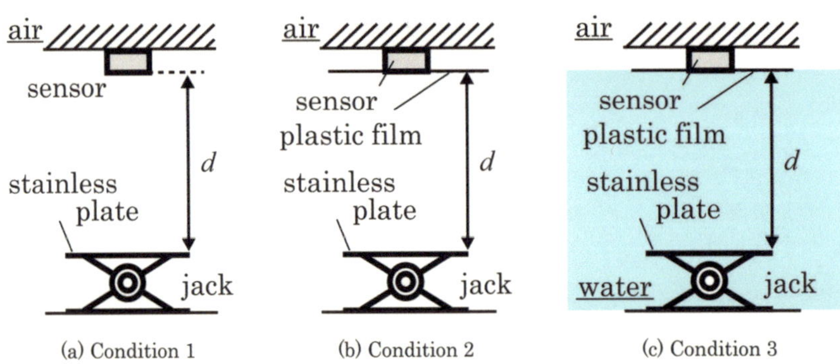

Figure 3. Experimental setups for measuring sensor performance.

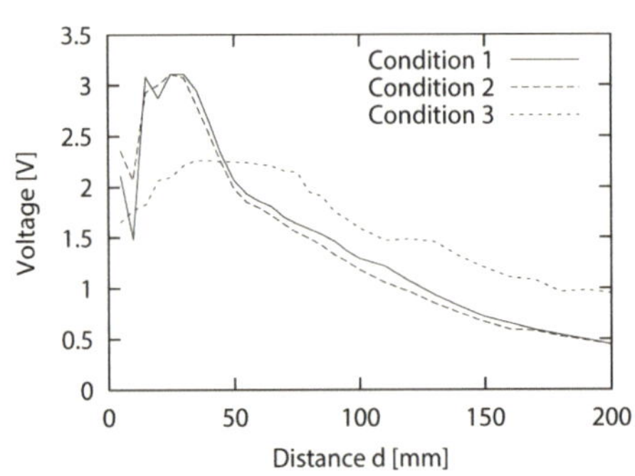

Figure 4. Experimental measurements of sensor performance.

Figure 4 shows experimentally measured sensor performance. The plastic film utilized in Conditions 2 and 3 was the multilayered film described in Section 2. This film is commercially available and used to cover foods such as meats and vegetables. Under each condition, a multimeter was used to measure the output voltage of the infrared sensor for each distance *d*. The measured range was within 200 mm. The distance *d* was changed by 5 up to 100 mm and 10 up to 200 mm under each experiment. Performance was measured in the perpendicular direction. As shown in **Figure 4**, the results observed using Conditions 1 and 2 showed similar tendencies. The tendency of Condition 3 differed, however, as the magnetic permeability of water was dominant. Although the maximum output voltage was lower during Condition 3 than during Conditions 1 and 2, a peak was observed in the graph of Condition 3. This finding indicates that our robot can detect obstacles under water utilizing this sensor.

The performance of this infrared sensor indicated that our prototype was able to move autonomously. **Figure 5** shows the prototype robot with one infrared sensor each mounted onto the two sides of its head section. This head section consists of the copolymer foam as the floating material. A polyethylene plate was utilized as an oscillation plate to generate the thrust force of the prototype in water. A microcontroller (Arduino Pro Mini 328 5 V, 16 MHz), a battery (9 V), a servo motor (RS304MD, Futaba), and insulating fluid (Fluorinert FC-3283, 3M) was encapsulated into a plastic film bag. The plastic film covering the internal components was composed of three layers: 12 μm thick antistatic polyethylene terephthalate (PET), 20 μm thick low-density (LD) polyethylene, and 50 μm thick antistatic linear low-density (LLD) polyethylene. The film was sealed by thermal adhesion of a vacuum packaging machine (TM-HV, Furukawa Mfg. Co., Ltd.). The prototype in **Figure 5** measured 200 mm × 100 mm × 100 mm and weighed approximately 700 g.

Simple autonomous control of our prototype was implemented using the following strategy:

- if $V_1 \geq V_{th}$ turn clockwise,
- otherwise, if $V_2 \geq V_{th}$ turn counterclockwise,
- otherwise going forward.

Let V_1 and V_2 be the output voltages of Sensors 1 and 2, respectively, in **Figure 5(b)**, and V_{th} be the threshold voltage to detect obstacles under water. Based on **Figure 4**, the voltage V_{th} was set at 1.65 V. Using this strategy, three moving modes were implemented in advance. During turning motions, the fin driven by the servo motor was moved 90° in either direction, at a

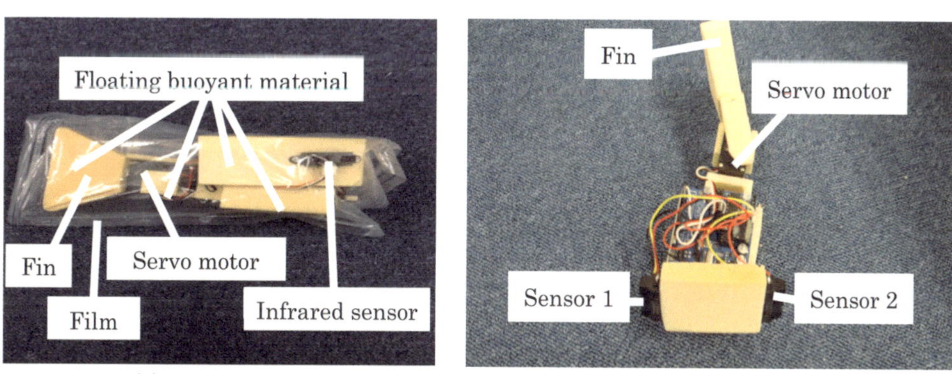

(a) Entire body (b) Internal components

Figure 5. Prototype robot with two infrared sensors.

frequency of approximately 2 Hz. During forward motions, the fin was moved ∓30°, at a frequency of approximately 1 Hz. Based on this strategy, the robot will turn clockwise when the combined voltages of Sensors 1 and 2 are greater than V_{th}.

Figure 6 shows the experimental environment. The experimental pool was 450 mm long and 600 mm wide. An aluminum plate, 300 mm long and 2 mm thick, was placed in the center of the pool as an obstacle. The body of the robot was at neutral buoyancy, allowing the robot to swim at a certain height. **Figure 7** shows a typical experimental result captured by a camera. Cartesian coordinates were assigned to this pool to assess the movements of the prototype. Data were captured by the camera, and the position of the robot measured once per second. As shown in this figure, the robot could swim using combinations of forward and turning motions.

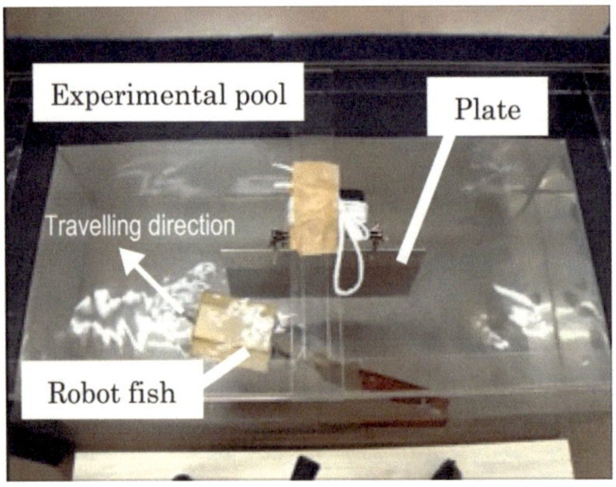

Figure 6. Experiment environment.

This section describes how to achieve autonomous locomotion using the electromagnetic-wave-transmitting properties of a plastic film. To detect an obstacle, two infrared sensors were mounted onto the body of the robot encapsulated by the plastic film. This film had electro-

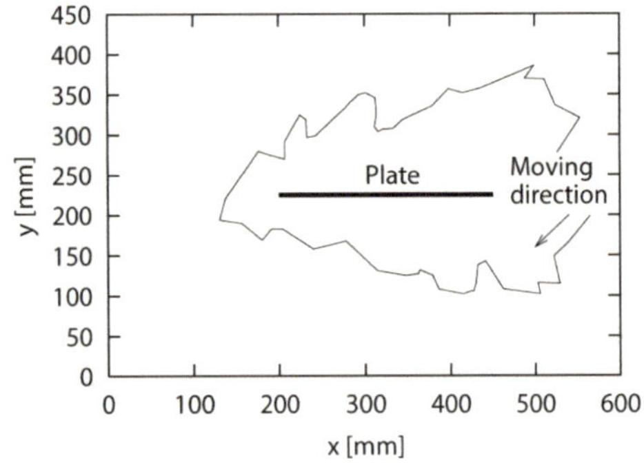

Figure 7. Experiment results of autonomous control.

magnetic-wave-transmitting properties, enabling not only noncontact sensors but other noncontact devices, such as wireless charging modules, and communication devices to be arranged within the device [39]. Using these devices, we will develop a sophisticated fish-like robot that can feed itself under water and cooperative with other, similar robots in performing operations.

3.2. Attitude control mechanism using floating blocks

This section describes the design and control of robot attitude, especially its trim angle. The attitude of traditional underwater robots with bodies made of pressure tight casing can be changed or controlled by a movable weight within the body [35]. In the control scheme called a trim mechanism, attitude is altered by a difference in density between the weight and air. Difficulties were also overcome by dynamic approaches, such as control schemes using the reaction force of internal rotors [36], the gyro effect of flywheels [37], and thruster forces [38] for a neutral buoyant underwater robot. We used a static approach, based on the equilibrium between gravitational and buoyant forces, to control the attitude of our underwater robot. Specifically, we developed an attitude control system using movable floats on a dual-armed underwater robot [10]. The attitude of the robot depends on the position of the floating blocks attached to a bar fixed onto the motor, with low-density floating blocks allowing the development of a lightweight underwater robot with attitude control. The attitude of our fish-like robot could be changed by the floating block and insulating fluid, with the placement of these components inside the robot determining its attitude under water. Because the mass of the insulating fluid differs from that of the floating block, the robot centers of gravity/buoyancy change as the positions of its components changes (**Figure 8**). As the low force/torque actuator can move the low-density floating block, the size of the robot body will tend to be small.

Based on the static properties of the robot, we were able to calculate the angle of attitude. These calculations require knowledge of the static properties of the internal components, floating blocks, and insulating fluid. The static properties of fluid used to fill the robot body are generally ignored in determining traditional trim angle control mechanisms because air is the usual insulating fluid and its density is negligible. To determine the angle of attitude, we considered a body-fixed reference frame attached to the robot in three-dimensional space (**Figure 9**). Within the body-fixed frame, the centers of gravity and of buoyancy of the internal

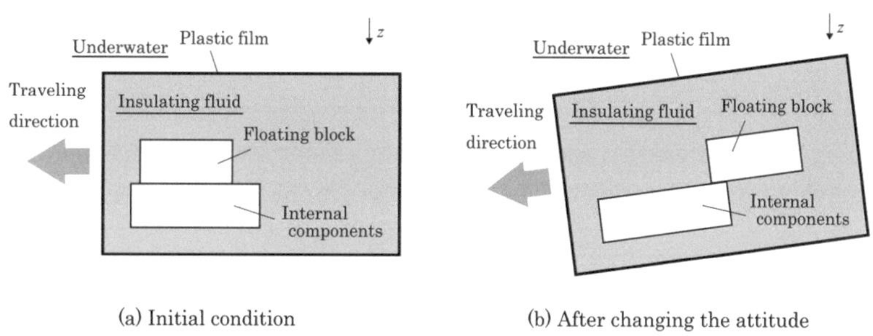

Figure 8. Attitude control using a float arranged in a flexible body.

components of the robot, including the fin, are defined as the vectors $\boldsymbol{r}_{rg}$ and $\boldsymbol{r}_{rb}$, respectively; the centers of gravity and of buoyancy of the floating block are defined as the vectors $\boldsymbol{r}_{bg}$ and $\boldsymbol{r}_{bb}$, respectively; and the centers of gravity and of buoyancy of the insulating fluid within the body are defined as the vectors $\boldsymbol{r}_{ig}$ and $\boldsymbol{r}_{ib}$, respectively. Based on these definitions, the center of gravity $\boldsymbol{r}_g$ and the center of buoyancy $\boldsymbol{r}_b$ of the entire system can be described using the equations:

$$\boldsymbol{r_g} = \frac{m_r\boldsymbol{r_{rg}}+m_b\boldsymbol{r_{bg}}+m_i\boldsymbol{r_{ig}}}{m_r+m_b+m_i}, \tag{1}$$

$$\boldsymbol{r_b} = \frac{V_r\boldsymbol{r_{rb}}+V_b\boldsymbol{r_{bb}}+V_i\boldsymbol{r_{ib}}}{V_r+V_b+V_i}. \tag{2}$$

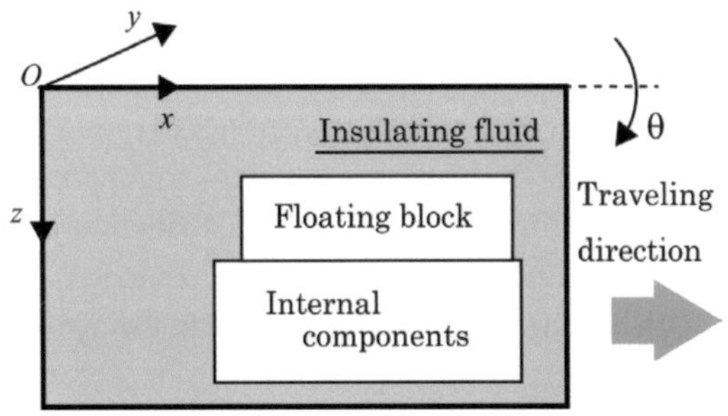

Figure 9. Coordination of a fish-like robot.

In traditional underwater robots with bodies encased in pressure tight cases, the vectors $\boldsymbol{r}_{ig}$ and $\boldsymbol{r}_{ib}$ can be ignored because the fluid filling the body is air. However, we could select a unique $\boldsymbol{r}_{ig}$ because the shape of the insulating fluid was dependent on the arrangement of the internal components and floating block within the plastic-filmed body. The desired attitude angle θ about the x axis (see **Figure 9**) can be analytically or numerically calculated to satisfy the equation:

$$\tan\theta = \frac{r_{gx} - r_{bx}}{r_{gz} - r_{bz}} \tag{3}$$

where r_{gx} and r_{gz} are the x and z vector components, respectively, of vector $\boldsymbol{r}_g$; and r_{bx} and r_{bz} are the x and z vector components, respectively, of vector $\boldsymbol{r}_b$.

Using Eqs. (1)–(3), we can determine the change in attitude changing based on the position of the floating block within the robot body. For simplicity, we assumed the following:

- A shift of floating material in the x direction in **Figure 9** changes the attitude.
- The shape of the plastic-filmed body does not change when the attitude is changed.

http://dx.doi.org/10.5772/63506

- The positions of the internal components do not change when the attitude is changed.

Under these assumptions, r_{gz} in Eq. (3) is a constant because the arrangement of internal components in the z direction does not change during changes in attitude. In addition, r_{bx} and r_{bz} in Eq. (3) are constants because the positions of the centers of buoyancy of the internal components, floating block, and insulating fluid do not change during changes in attitude. The angle θ will therefore depend on r_{gx}. Using Eq. (1), r_{gx} can be calculated as:

$$r_{gx} = \frac{m_r r_{rgx} + m_b r_{bgx} + m_i r_{igx}}{m_r + m_b + m_i} \tag{4}$$

where r_{rgx}, r_{bgx}, and r_{igx} are the x components of vectors $\boldsymbol{r}_{rg}$, $\boldsymbol{r}_{bg}$, and $\boldsymbol{r}_{ig}$, respectively. Based on the above assumptions, the angle θ is dependent on the terms $m_b r_{bgx} + m_i r_{igx}$ because r_{rgx} is a constant. As stated above, we cannot select a unique $\boldsymbol{r}_{ig}$ because the shape of the insulating fluid depends on the arrangement of the internal components and the floating block within the plastic-filmed body. A simple physical model (**Figure 10**) was used to investigate the magnitude of change in $m_b r_{bgx} + m_i r_{igx}$ as a function of the position of a floating block in insulating fluid. This figure shows a floating block within a massless rigid case filled with insulating fluid. Let the volumes of the floating block and insulating fluid be V_i and V_b, respectively; the masses of the floating block and insulating fluid be m_i and m_b, respectively; and the densities of the floating block and insulating fluid be ρ_i and ρ_b, respectively. The origin was set at the centroid of the massless case. The distances between the origin and the centers of gravity of the floating block and insulating fluid were expressed as $|x_1|$ and $|x_2|$, respectively, with both centers of gravity assumed to be on the x axis. If the densities of the insulating fluid and the floating block are the same, the center of gravity of the whole system, including the insulating fluid and floating block, would correspond to the origin. Therefore, to balance the moment around the origin, the following equation should be satisfied:

$$|x_2| \rho_i V_i g = |x_1| |\rho_i - \rho_b| V_b g \tag{5}$$

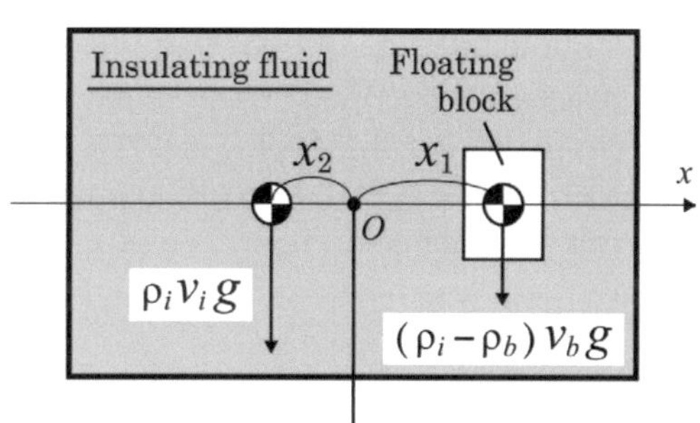

Figure 10. A floating block within a massless case filled with insulating fluid.

Hence,

$$|x_2| = \frac{V_b |x_1| |\rho_i - \rho_b|}{V_i \rho_i}. \tag{6}$$

According to Eq. (6), the magnitude of the shift of the center of gravity of the entire system, or attitude, depends on the difference in density between the insulating fluid and the floating block. Therefore, if the density of the floating blocks is equal to that of the insulating fluid, the attitude could not be altered, even by changing the position of the floating blocks. Using Eq. (6), we also found that a greater difference in density would result in a greater change in attitude of our prototype despite its high weight. Thus, the proper fluid must be selected. For example, most industrial oils are less dense than water, whereas most fluids used to clean semiconductors are denser than water.

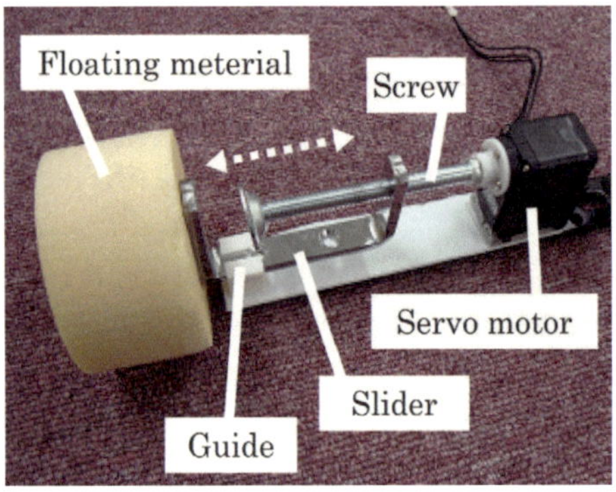

Figure 11. Attitude control mechanism using a screw.

To validate our approach, we developed an attitude control mechanism in which a floating block was moved with a screw mechanism (**Figure 11**). This screw mechanism used friction to generate a high external holding force, resulting in a small-sized attitude control mechanism. The cylindrical floating block, measuring 78 mm in diameter and 50 mm in height, was made of copolymer foam (NiGK Corporation) with a specific gravity of approximate 0.2. A servomotor (AX-12A, Dynamixel) was attached to the actuator of the attitude control system. This system used an Arduino UNO (ver. R3) as a microcontroller. The floating block was mounted onto a metal slider. A guide mechanism made of a resin material was used to regulate the direction of movement of the slider, allowing the floating block to move in a straight line along the guide mechanism. Two 9 V dry-cell batteries were used to drive the servomotor and the microcontroller, respectively.

The performance of the prototype was assessed by performing several experiments in air and under water. In air, the maximum displacement of the floating block was 40 mm, and its speed was 0.73 mm/s. For tests under water, we utilized a TM-HV (Furukawa Mfg. Co.,

Ltd.) to seal the plastic film of the robot body and Fluorinert FC-3283 (3M), with a specific gravity of approximately 1.8, as the insulating fluid (**Figure 12**). A cylindrical pipe made of resin was not in direct contact with the control mechanism or the plastic film, allowing smooth motion during the experiment. The prototype, including the resin pipe, was 260 mm in length and 80 mm in diameter, and weighed 1785 g. **Figure 13** shows the time transition of the angle of attitude. The initial angle was set at 0 rad, and a camera was used to map its position every 3 s. The convergent value and average speed of the change in attitude were approximately 0.22 rad and 3.6×10^{-3}rad/s, respectively.

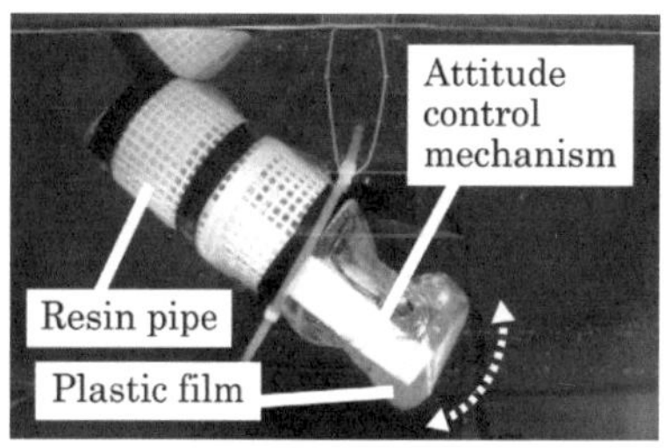

Figure 12. Motion test (under water).

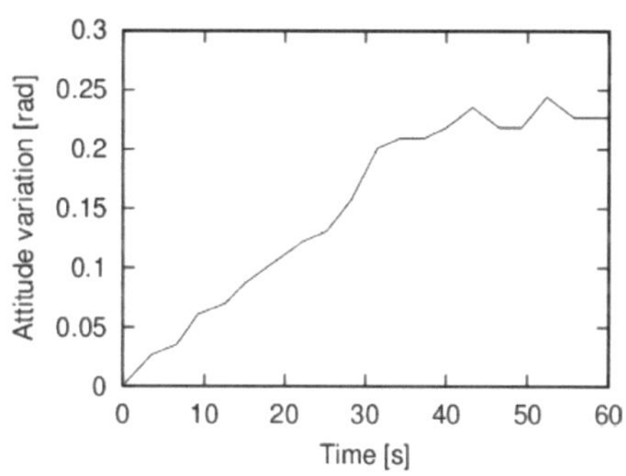

Figure 13. Change in angle of attitude over time.

This section describes the use of a movable floating block within the outer surface of the fish-like robot to achieve attitude control. The arrangement of the floating block determines the attitude depending on the differences in density between the insulating fluid and the floating block. If the density of the floating block is low, a low force/torque actuator can be used to move the floating block, allowing the development of a small-sized body. The arrangement of the internal components, insulating fluid, and floating block could be altered not only by the movement of the float within the body but also by the inflection of the outer surface itself [40]. One motor may generate sufficient propulsion force to maintain the attitude of the robot body. We intend to develop a sophisticated fish-like robot that can search autonomously for underwater structures, with the type of robot depending on the situations in which it will be utilized.

4. Conclusion

This chapter has described a fish-like underwater robot with an outer surface composed of a flexible thin plastic film. The internal components of the robot, including a servo motor, microcontroller, battery, and floating blocks, were encapsulated in a plastic-filmed bag using a vacuum packaging machine. This vacuum packaging machine enabled rapid, low-cost fabrication of fish-like robots with a plastic-filmed outer surface. To simplify its waterproofing and pressure resistance properties, the internal components of the fish-like robot was filled with insulating fluid. Autonomous and attitude control schemes were proposed based on the characteristics of these fish-like robots. As the encapsulating thin plastic film has electromagnetic-wave-transmitting property, a noncontact sensor could be arranged within the robot. We also designed an attitude control mechanism employing floating blocks, with the attitude of our prototype determined by the arrangement of the floating block and dependent on the differential densities of the floating block and insulating fluid. Attitude control was found to vary more when the insulating fluid was denser than water. In literature [14], we have pointed that insulating fluids, less dense than water, could be used as floating materials to achieve neutral buoyancy. Therefore, the volume of insulating fluid within a prototype should be carefully determined according to the moving performance.

Acknowledgements

This work was partially supported by Furukawa Mfg. Co., Ltd., and partially by the Center of Innovation Program from Japan Science and Technology (JST) Agency. This work was also partially supported by JSPS KAKENHI Grant Number 15K18011. I am grateful to Prof. Sakagami for helpful discussions.

Author details

Mizuho Shibata

Address all correspondence to: shibata@hiro.kindai.ac.jp

Department of Robotics, Kindai University, Higashi-Hiroshima, Hiroshima, Japan

References

[1] MotroR. Tensegrity. Kogan Page; London, UK 2003.

[2] Paul C, Valero-Cuevas F J, Lipson H. Design and control of tensegrity robots for locomotion. IEEE Transactions on Robotics. 2006;22(5):944–957.

[3] Aldrich J B, Skelton R E, Kreutz-Delgado K. Control synthesis for a class of light and agile robotic tensegrity structures. In: Proc. of the American Control Conf.; 4–6 June; Denver, USA. 2003. pp. 5245–5251.

[4] Shibata M, Miyamura T, Sakagami N, Miyata S. Use of a deformable tensegrity structure as an underwater robot body. Journal of Robotics and Mechatronics. 2013;25(5):804–811.

[5] Bliss T, Iwasaki T, Bart-Smith H. Central pattern generator control of a tensegrity swimmer. IEEE/ASME Transactions on Mechatronics. 2013;18(2):586–597.

[6] Eriksen C C, Osse T J, Light R D, Wen T, Lehman T W, Sabin P L, Ballard J W, Chiodi A M. Seaglider: a long-range autonomous underwater vehicle for oceanographic research. IEEE Journal of Oceanic Engineering. 2001;26(4):424–436.

[7] Sherman J, Davis R E, Owens W B, Valdes J. The autonomous underwater glider "Spray". IEEE Journal of Oceanic Engineering. 2001;26(4):437–446.

[8] Webb D C, Simonetti P J, Jones C P. SLOCUM: an underwater glider propelled by environmental energy. IEEE Journal of Oceanic Engineering. 2001;26(4):447–452.

[9] Sakagami N, Ibata D, Ikeda T, Shibata M, Ueda T, Ishimaru K, Onishi H, Murakami S, Kawamura S. Development of a removable multi-DOF manipulator system for man-portable underwater robots. In: Proc. 21th Int. Conf. on Offshore and Polar Eng.; 19–24 Jun.; Hawaii, USA. 2011. pp. 279–284.

[10] Sakagami N, Ishimaru K, Kawamura S, Shibata M, Onishi H, Murakami S. Development of an underwater robotic inspection system using mechanical contact. Journal of Field Robotics. 2013;30(4):624–640.

[11] Kim S, Spenko M, Trujillo S, Heyneman B, Santos D, Cutkosky M R. Smooth vertical surface climbing with directional adhesion. IEEE Transactions on Robotics. 2008;24(1): 65–74.

[12] Wood R J, Finio B, Karpelson M, MaK, Perez-Arancibia N O, Sreetharan P S, Tanaka H, Whitney J P. Progress on "pico" air vehicles. The International Journal of Robotics Research September. 2012;31(11):1292–1302.

[13] Shibata M, Sakagami N. A fish-like underwater robot with flexible plastic film body. In: IEEE International Conference on Robotics and Biomimetics; 12–14 December; Shenzhen, China. 2013. pp. 68–73.

[14] Shibata M, Sakagami N. Fabrication of a fish-like underwater robot with flexible plastic film body. Advanced Robotics. 2015;29(1):103–113.

[15] Ahvenainen R. Novel Food Packaging Techniques. Woodhead Publishing; Cambridge, UK 2003.

[16] Tella R, Birk J R, Kelley R B. General purpose hands for bin-picking robots. IEEE Transactions on Systems, Man and Cybernetics. 1982;12(6):828–837.

[17] Greiner J H, Kircher C J, Klepner S P, Lahiri S K, Warnecke A J, Basavaiah S, Yen E T, Baker J M, Brosious P R, Huang H C W, Murakami M, Ames I. Fabrication process for Josephson integrated circuits. IBM Journal of Research and Development. 1980;24(2): 195–205.

[18] Lee C C, Wang C Y, Matijasevic G S. A new bonding technology using gold and tin multilayer composite structures. IEEE Transactions on Components, Hybrids, and Manufacturing Technology. 1991;14(2):407–412.

[19] Wakimoto S, Ogura K, Suzumori K, Nishioka Y. Miniature soft hand with curling rubber pneumatic actuators. In: Proc. of IEEE Int. Conf. on Robotics and Automation; 12–17 May; Kobe, Japan. 2009. pp. 556–561.

[20] Longo D, Muscato G. The Alicia(3) climbing robot: a three-module robot for automatic wall inspection. IEEE Robotics & Automation Magazine. 2006;13(1):42–50.

[21] Hayakawa T, Nakamura T, Suzuki H. Development of a wave propagation type wall-climbing robot using a fan and slider cranks. In: Proc. of Int. Conf. on Climbing and Walking Robots; 9–11 Sep.; Istanbul, Turkey. 2009. pp. 439–446.

[22] Yoshida Y, Ma S. Design of a wall-climbing robot with passive suction cups. In: IEEE International Conference on Robotics and Biomimetics; 14–18 Dec.; Tianjin, China. 2010. pp. 1513–1518.

[23] Manabe R, Suzumori K, Wakimoto S. A functional adhesive robot skin with integrated micro rubber suction cups. In: IEEE International Conference on Robotics and Automation; 14–18 May; St. Paul, USA. 2012. pp. 904–909.

[24] Zesch W, Brunner M, Weber A. Vacuum tool for handling microobjects with a Nano-Robot. In: Proc. of IEEE Int. Conf. on Robotics and Automation; 20–25 April; Albuquerque, USA. 1997. pp. 1761–1766.

[25] Kato N, LIUH. Optimization of motion of a mechanical pectoral fin. JSME International Journal Series C Mechanical Systems, Machine Elements and Manufacturing. 2003;46(4):1356–1362.

[26] Yamamoto I, Terada Y. Robotic fish and its technology. In: SICE Annual Conf.; 4–6 August; Fukui, Japan. 2003. pp. 342–345.

[27] Liu J, Hu H, Gu D. A layered control architecture for autonomous robotic fish. In: Proc. of IEEE/RSJ Int. Conf. on Intelligent Robots and Systems; 9–15 Oct.; Beijing, China. 2006. pp. 9–15.

[28] Conte J, Modarres-Sadeghi Y, Watts M, Hover F S, Triantafyllou M S. A faststarting mechanical fish that accelerates at 40ms(-2). Bioinspiration and Biomimetics. 2010;5(3): 035004.

[29] Yu J, Tan M, Wang L. Cooperative control of multiple biomimetic robotic fish. In: Lazinica A, editor. Recent Advances in Multi Robot Systems. Intech; Rijeka, Croatia 2008. pp. 263–290.

[30] Ohashi T, Yamada H, Hirose S. Loop forming snake-like robot ACMR7 and its serpenoid oval control. In: IEEE/RSJ Int. Conf. on Intelligent Robots and Systems; 18–22 Oct.; Taipei, Taiwan. 2010. pp. 413–418.

[31] Eustice R M, Pizarro O, Singh H. Visually augmented navigation for autonomous underwater vehicles. IEEE Journal of Oceanic Engineering. 2008;33(2):103–122.

[32] Sumoto H, Yamaguchi S. Development of a motion control system using photoaxis for a fish type robot. In: Proc. of the Int. Offshore and Polar Engineering Conf.; 20–25 Jun.; Beijing, China. 2010. pp. 307–310.

[33] Takagawa S, Takahashi K, Sano T, Kyo M, Mori Y, Nakanishi T. 6,500m Deep Manned Research Submersible "Shinkai 6500" System. In: Proc. of OCEANS; 18–21 Sep.; Seattle, USA. 1989. pp. 741–746.

[34] Shibuya K, Kishimoto Y, Yoshii S. Depth control of underwater robot with metal bellows mechanism for buoyancy control device utilizing phase transition. Journal of Robotics and Mechatronics. 2013;25(5):795–803.

[35] Woolsey C A, Leonard N E. Moving mass control for underwater vehicles. In: Proc. of the American Control Conference; 8–10 May; Alaska, USA. 2002. pp. 2824–2829.

[36] Woolsey C A, Leonard N E. Stabilizing underwater vehicle motion using internal rotors. Automatica. 2002;38(12):2053–2062.

[37] Thornton B, Ura T, Nose Y, Turnock S. Zero-G class underwater robots: unrestricted attitude control using control moment gyros. Journal of Oceanic Eng. 2007;32(3):565–583.

[38] Doniec M, Vasilescu I, Detweiler C, Rus D. Complete SE(3) underwater robot control with arbitrary thruster configurations. In: Proc. Int. Conf. on Robotics and Automation; 3–8 May; Alaska, USA. 2010. pp. 5295–5301.

[39] Shibata M, Sakagami N. A robot fish encapsulated by an electromagnetic wave-transmitting plastic film. In: Proc. of Conference of the IEEE Industrial Electronics Society; 9–12 Nov.; Yokohama, Japan. 2015. pp. 2729–2734.

[40] Shibata M, Sakagami N. Attitude control mechanism for underwater robot with flexible plastic film body. In: Proc. of Int. Ocean and Polar Engineering Conf.; 21–26 Jun.; Hawaii, USA. 2015. pp. 558–563.

Chapter 12

CODA Algorithm: An Immune Algorithm for Reinforcement Learning Tasks

Daniel R. Ramirez Rebollo, Pedro Ponce Cruz and Arturo Molina

Additional information is available at the end of the chapter

http://dx.doi.org/10.5772/63570

Abstract

This document presents the design of an algorithm that takes on its basis: reinforcement learning, learning from demonstration and most importantly Artificial Immune Systems. The main advantage of this algorithm named CODA (Cognition from Data). Is; it can learn from limited data samples- that is given a single example and the algorithm will create its own knowledge. The algorithm imitates from the Natural Immune System the clonal procedure for obtaining a repertoire of antibodies from a single antigen. It also uses the self-organised memory in order to reduce searching time in the whole action-state space by searching in specific clusters. CODA algorithm is presented and explained in detail in order to understand how these three principles are used. The algorithm is explained with pseudocode, flowcharts and block diagrams. The clonal/mutation results are presented with a simple example. It can be seen graphically how new data that has a completely new probability distribution. Finally, the first application where CODA is used, a humanoid hand is presented. In this application the algorithm created affordable grasping postures from limited examples, creates its own knowledge and stores data in memory data in memory in order to recognise whether it has been on a similar situation.

Keywords: artificial immune system, artificial intelligence, neuroscience, biologically inspired algorithm, cognition

1. Introduction

Learning in humans involves approximately 100 billion of neurons (brain cells). Neurons in humans are useless individually but extremely useful and powerful when working together.

Each neuron is formed by a body or soma, and axon that sends information and thousands of dendrites for receiving data. The more dendrite connections there are, the more learning there would be. It can be seen than at its basic functional level, learning is without a doubt a complex activity. It involves not only the brain cells and their connections but also certain factors such as attention, memory, motivation and stress. There are even different ways in which learning can occur, such as empiricism, innatism and constructivism [1].

In humans, learning is not only directly engaged with the amount of connection in the neurons but is also influenced by external factors relative to the subject state. There is a necessity to learn in an entity free of unnecessary passions and no underpinnings that could hinder or limit its learning ability. Machine learning is the response to this paradigm, several algorithms have arisen having as main objective; making computers learn as can be seen in [2], even they may have different particular objectives according to their taks. Machine learning algorithms have already demonstrated their competence at learning in different engineering and science problems [3–5]. However, there is still no algorithm that is more versatile and generalised, i.e., a "do-it-all" algorithm that can be used in several situations no matter the nature of the task itself.

In machine learning, there is no single algorithm that could solve all the problems [6]. To address this, a machine learning algorithm from three principles is created: Learning from Demonstration, Reinforcement Learning and Artificial Immune System. This aims to obtain an algorithm with several advantages from the mixture of those techniques. The resulting algorithm would keep all the advantages from the techniques used plus some of the Artificial Immune System characteristics from **Table 2**. **Table 1** shows advantages of CODA.

CODA'S advantages
Low training samples
Short training time
Produces new knowledge
Unnecessary/unused data is deleted
Reward function assures search for maximum reward
Reward function is task specific
Self-organised memory mechanism reduce searching time in repertoire

Table 1. Advantages that characterise CODA algorithm.

This document is organised as follows. Section 1 contains a discussion on the theory on how learning from demonstration, reinforcement learning and Artificial Immune Systems has been used to develop CODA algorithm in order to help the reader understand how useful these methods are for the algorithm presented in Section 3.

http://dx.doi.org/10.5772/63570

Section 5 explains the CODA algorithm by presenting the pseudocode and simulations. Section 6, on the other hand, explains the application where CODA will be used. Sections 7 and 8 give further discussion and conclusion, respectively, on this chapter.

2. Learning from demonstration

Defining Learning from Demonstration (LfD) should not be difficult. In [7], a simple yet complete sentence clarifies the concept: Learning from Demonstration is learning from watching a demonstration of the task to be performed. Embedded in this sentence is the main goal, which is to learn a skill from demonstrative examples. What is needed is for the computer to learn from demonstrations and this is what it is all about. Learning from demonstration is also known as "programming by demonstration", "imitation learning" or "teaching by showing". As the second name exposes, another goal is to replace certain programming that would be time-consuming and that would require a specialised person in order to modify the programs within the robot. Programming by demonstration promises an automatic self-programming process by showing a robot the task to perform in itself.

Learning from demonstration is not a new topic. It is a well-known discipline in robotics and has been studied in [8–12], where their approaches needed direct teaching to the entity in order for the entity to imitate certain human actions. Recent studies focus on developing the learning from demonstration theory in order to produce better systems that would lead to better demonstration and feedback, resulting in better teaching and learning [13]. Other studies propose a system using a Bayesian nonparametric reward in order to assign rewards to subgoals (more than one reward function) instead of a unique task [14]. And there are still other studies focusing on enhancing the quality of the demonstrations in order to decrease the number of demonstrations and learn more effectively [15].

Learning from demonstration is commonly applied in robotics, the robot assumes the student role and the human is the expert teacher. Thus the goal is to demonstrate the task to the robot, so it should learn from watching the demonstration, so the robot could use the skill when necessary. Once a robot has acquired certain skill it could also teach other robots.

2.1. Learning

Learning supposes the generalisation of a task. Let's suppose you have to learn how a bird looks like. The first step would probably be to list down the characteristics of a bird. You should have a list describing the wings, eyes, feathers, tail and beak. Without focusing on specifics like the species, geography or colours, we can say that animals that fit the description can be called birds. The list is a guideline for you to be able to distinguish a bird from any other animal or object.

Once we have learned the characteristics that are uniquely assigned to certain things, it is now easy to distinguish objects and animals around us. They can be clustered in simple term such as, in this particular case, a bird. Our brain learns to generalise tasks in order to obtain the desired results every time the action is performed regardless of the variability of the environment at each try.

Once a robot is able to reproduce certain skills with the least error possible or even better, non-error, it can be said that the robot has learnt a skill correctly. Therefore, it can be said that it has certain intelligence embedded that let it learn.

Robots can be controlled by different methods and techniques that have been used for several years and produce excellent results in certain conditions and applications [16, 17]. But it is important to note that robots are also being used outside of factories in applications and environments that require high adaptability, reliability and constant learning. These may demand robots that are capable of handling uncertainty and variability in fast and dynamic environments.

It can be assured now that learning is a valuable and almost necessary skill for robots, almost since Alan Turing wrote *Computing Machinery and Intelligence,* concluding that "*We can only see a short distance ahead, but we can see plenty there that needs to be done*" [18].

2.2. Hard programming vs learning

A simple but effective definition of learning from demonstration has been presented previously. In this section, a more formal and structured definition is presented. Learning from demonstration is a mapping built by examples between environment states and actions to perform. This is called policy, and it gives the robot the skill of selecting which action to select if found in a certain environment configuration. The policy is built by all the demonstrations the robot obtains.

There are two paths in learning from demonstration. The first one is very simple. It basically let the robot mimic the motion of the instructor; in other words, the computer must learn how the instructor acts, reacts and handle errors in different situations, which is a process called learning the policy.

The second path, and is more important in this work, is where the robot may not be doing exactly the same task as the demonstrator and there is a small number of task demonstrations available. This means the robot should learn under uncertainty and with low loads of data. In these cases, algorithms should extract the most out of the information in order to be able to process all the information and then create correct cognitive responses topic to be studied in the CODA section.

2.2.1. Pendant programming and kinetic teaching

Pendant programming and kinetic teaching are techniques that could be considered the simplest form of learning from demonstration. Both techniques have their advantages and disadvantages, but they are outside the aim of this book. It is of most importance to mention how they work.

Figure 1 shows an example of pendant programming where the human uses a programming pendant in order to move the robot thru desired points, this technique can produce two paths,

http://dx.doi.org/10.5772/63570

a polynomic or linear path, but learning procedure never takes place during this action, the robot will always move on the chosen path, this limits the generalization capacity of the robot. This is why this technique belongs to a hard programming techniques to control a robot even it uses demonstration to hel the user program the robot.

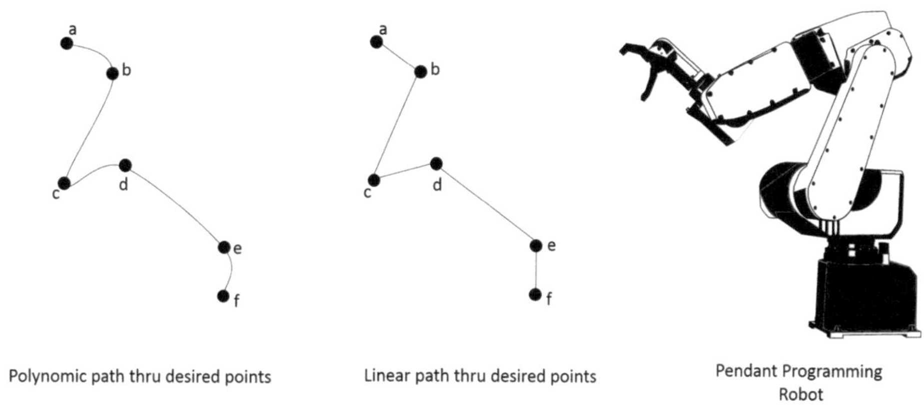

Figure 1. Pendant programming example, same trajectory with two interpolation options.

2.2.2. Learning from data

The previous section exemplified how a robot can be programmed by a demonstration and still lacks intelligence. Intelligence is formed by several different bits such as reasoning and logical deduction but for our purpose, the most basic animal intelligence characteristics that matter are remembering, adapting and generalising.

Remembering, adapting and generalising are part of learning, and these characteristics are the most important to embed in machines. Machines should be able to learn with examples, data and experience. First the machine should obtain an example which contains data, and at each try, the machine should acquire experience. Using the concepts presented earlier, the machine should be able to recognise when was the last time it was in this situation (encountered the same data), tried a particular action (produce this output) and it worked (the output action(s) was/were correct), so it can be tried again. But if it does not work properly, then something different should be tried.

Learning is a skill that gives us the ability to be flexible in our day to day activities. We adapt and adjust our self to new circumstances, recognise similarity between different situations in order to use knowledge in one place or another, and therefore use acquired knowledge in different places and situations.

Learning by demonstration is a powerful tool that has already shown its competence in the robotics field. However, it may be impossible to complete any desired task if it will be the only tool to be used in the entire system. To address this, a hybrid system was built where it will use a non-model based algorithm such as Q-learning or SARSA (explained in the next section) where the policy will be represented by a Q matrix and adjusted directly according to the reward obtained during each try. These consequences of the adjustment in the policy are measured by the reward function to guide future adjustments in the policy.

The reward function does not actually tell the algorithm if the output is correct or not; instead it tells how much correct it is, and the stored values of the Q matrix should serve as a repertoire of knowledge in order to recognise certain "*encountered data–output action*" pairs for future reference.

3. Reinforcement learning

Taking this extract from [19]:

> *"Of several responses made to the same situation, those which are accompanied or closely followed by satisfaction to the animal will, other things being equal, be more firmly connected with the situation, so that, when it recurs, they will be more likely to recur; those which are accompanied or closely followed by discomfort to the animal will, other things being equal, have their connections with that situation weakened, so that when it recurs, they will be less likely to occur. The greater the satisfaction or discomfort, the greater the strengthening or weakening of the bond".*

As it can be read from the extract, desired behaviours will prevail over the undesired ones. This is due to the reward or satisfaction feeling, and this is precisely part of what CODA is willing to imitate in a computational environment. And since no machine can feel any kind of emotion, the reward will be numerically represented with a function.

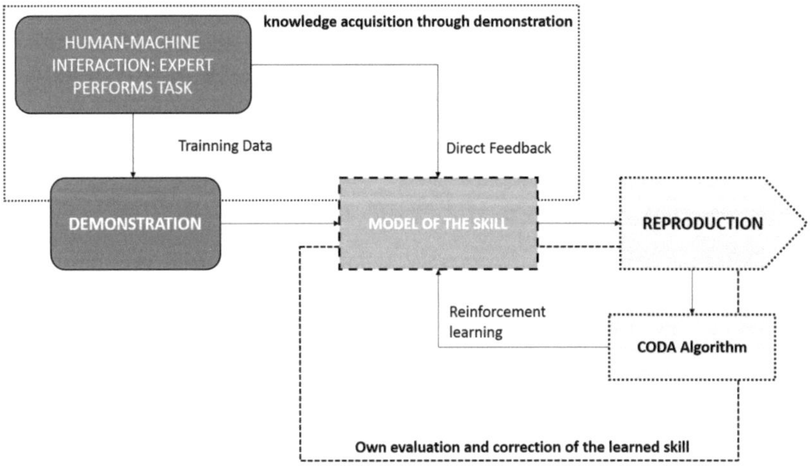

Figure 2. Block diagram showing the cognition model proposed by the authors with the CODA algorithm.

http://dx.doi.org/10.5772/63570

Figure 2 shows a block diagram of the learning by demonstration iterative process taken and modified from [11]. The diagram shows the process of learning a task with the proposed approach and the CODA algorithm.

The interaction between the entities is initiated by the human–machine interaction where the human expert performs the task or gives an example of the task in order for the system to record the information as a training example. In this manner, it gives direct feedback on how the skill is being modelled. The CODA algorithm will have its own evaluation and correction process in order to acquire the knowledge and reproduce the task.

During this process, a reinforcement learning algorithm is the most important tool. SARSA, which is a modification of Q-learning, was chosen. The main difference between the two algorithms is that Q-learning always attempts to search for the optimal path, which will always coincide in that it is the shortest, but this does not necessarily mean it is the best one on each task. In contrast the SARSA algorithm will not always take the shortest path but the safest, because it will try to avoid large negative rewards taking in consideration those "dangerous" situations.

Mathematically, Q-learning and SARSA differ in their expressions as the following two equations state:

$$Q(s_t,a_t) \leftarrow Q(s_t,a_t) + \mu[r_{t+1} + \gamma_{a_{t+1}} max\, Q(s_{t+1},a_{t+1}) - Q(s_t,a_t)], \tag{1}$$

$$Q(s_t,a_t) \leftarrow Q(s_t,a_t) + \mu[r_{t+1} + \gamma\, Q(s_{t+1},a_{t+1}) - Q(s_t,a_t)], \tag{2}$$

Eq. (1) shows the Q-learning algorithm equation for the actualisation of the Q matrix and Eq. (2) shows that of the SARSA algorithm. It can be inferred that mathematically the main difference between them is that Q-learning is off-policy since it takes the maximum value of $Q(s',a')$ instead of computing a'. SARSA on the other hand makes this possible by the simple modification made where it takes care not only of the actual state and action but also the next possible state, action and reward, making SARSA an on-policy algorithm, something explained previously.

Since the inputs and outputs of the reinforcement learning algorithms are discretised if the size of the action space and state space can be reduced, it is always a good idea since the reinforcement learner algorithm is basically a search method. Therefore it is good to search for certain techniques that would let us handle these spaces in a proper manner, and it is upon the application and the expertise of the reader how this could be approached.

Figure 3 describes how the algorithms search for the goal state given a state space, action space and a reward function. It also explains how the learning agent, in our case the robotic hand, performs action in the state s_t and receives a reward $r_{(t+1)}$ from the environment, after this the agent will move to state $s_{(t+1)}$ because of the reward given. Basically the sensors on the hand define the state, because they are the representation of the environment around the robot, telling us if the hand is touching any surface or not. The actions are how the robot moves its actuators. The reward could be the similarity between the learnt (desired) data and the produced data on the present state, and could be bigger if similarity is higher and reward should be lower if similarity is not likely.

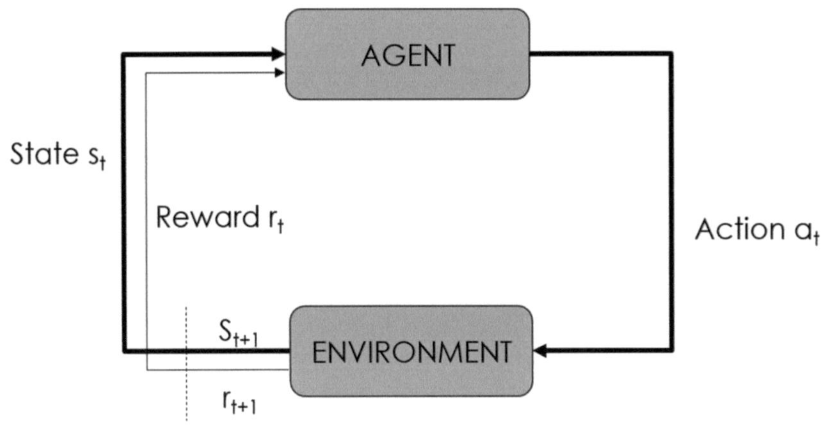

Figure 3. Reinforcement learning cycle, agent performs action, receives reward and ends up on a new state.

4. Artificial immune systems

In the human body, the immune system is the collection of cells, tissues and molecules that defend the body against infections [20]. Its basic function is to eradicate and prevent infections.

Naturally in our body there are two types of immunity: innate immunity and adaptive immunity, both share the same objective, but have different tasks and time reactions. Innate immunity acts in the first hours whereas adaptive immunity operates through days.

Adaptive immunity can further be classified as humeral immunity and cell-mediated immunity, containing different responding, effectors cells and functions.

A detailed biologically inclined overview of the Artificial Immune System can be found in [20]; while a computationally inclined overview can be found at [21].

Simulating active immunity is of great interest for the CODA algorithm. The most common example is vaccination of an individual, in this case the "naive individual" is exposed to antigens in order to mount an active response and be able to eradicate the infection. After this process the individual will be immune to that microbe and that is because it has already built resistance for a later infection. In contrast, passive immunity is shown in new-borns that do not have an immune system mature enough to fight against pathogens, but are protected by their mother's antibodies through the placenta and milk.

Artificial Immune Systems where developed with active immunity in mind since at the very first they were implemented for intrusion detection in computer servers. The main goal of the system was to detect insiders and external intruders that were committing abuse or misuse of the computer systems.

There are several crucial characteristics that are important for the CODA algorithm to inherit the robustness, adaptability and specificity among other characteristics that are desirable for a real problem-solving algorithm. The characteristics that are considered important and should be applied in the computational version of the immune system are shown in **Table 2**.

Feature	Functional significance
Specificity	Ensures that distinct antigens elicit specific responses
Diversity	Enables immune system to respond to a large variety of antigens
Memory	Leads to enhanced responses to repeated exposures to the same antigens
Clonal expansion	Increases number of antigen-specific lymphocytes to keep pace with microbes
Specialisation	Generates responses that are optimal for defence against different types of microbes
Contraction and homeostasis	Allows immune system to respond to newly encountered antigens
No reactivity to self	Prevents injury to the host during responses to foreign antigens

Table 2. Natural Immune System characteristics that make it such a potent tool in the human body. This features are imitated by the computer counterparts of the NIS the AIS.

Specificity and diversity: This is an important feature that lets the immune system distinguish from similar antigens, helping us specify the response for a certain antigen and no response to any other even if they are quite similar. Therefore lymphocyte repertoire could be over millions of different units, all of them cloned, built specifically for different antigens, leading to a much extended antigen distinguishing.

Memory: When the system is exposed to an antigen for the first time, the response is called primary immune response and is mediated by lymphocytes, called naive lymphocytes. The second encounter is called secondary immune response, and it is supposed to be rapid, larger and more effective.

In order to implement and use an Artificial Immune System, it is quite important to understand deeply how the Artificial Immune systems works for the several processes that occur within the system and manage several tasks. In [22], a very explicit description of the AIS based upon the natural model from the human body of the Natural Immune System (NIS) can be found. According to the explanation of the model, the immune system is an example of a mechanism that is capable of learning and remembering. This memory is capable not only of string previous interactions but also forgetting information with little use.

The NIS is an example of a yet adaptive, decentralised and effective system. The B and T cells are just examples of how the NIS has a working structure that delegates all tasks. All the previous characteristics are desirable for a problem-solving algorithm that can offer novel methods and thus testing is required to compare this paradigm across all the machine learning methods.

In [22] the authors test the algorithm with a simple pattern recognition problem as a first test. In order to have a better overview of the performance of the AIS, it was applied to a real-world problem such as the recognition of the promoters in DNA sequences. According to the results presented, all were consistent with other approaches such as ANN and Quinlan ID3 obtaining similar results. The performance was better than Nearest Neighbour Algorithm.

The diversification of the AIS is an important characteristic since it does not focus on a global optima, instead the antibodies evolve in order to handle all the variety that the antigens can represent. **Figure 6** shows how the antigen is taken and mutated to produce all the antibodies represented by the blue circles. It is also quickly adaptable to changing situations, naturally the system can handle event-response situations. This is one of the most important features of the NIS as well, and the response must be active in a matter of minutes or hours in order to protect the human body.

According to [22] a remarkable characteristic of the AIS is the genetic mechanism that can mutate and reproduce the antibodies, memory and its self-organising properties. In CODA algorithm these concepts are implemented [22], for example, the self-organising characteristic implemented with a clustering step running over the antibodies in order to organise all data.

It is important to notice that the computational model presented in this document or in any of the articles that were revised does not aim to model the human NIS perfectly, nor it is an attempt to provide explanations of how the system works within the body. Rather, the stated features are emulated in order to solve problems that need such characteristics presented in **Table 1** and **Table 2**, such as specificity, diversity and memory, among others.

Farmer et al. presented Eq. (3) as a mathematical model of the immune system, in which N represents the antibodies, n represents the antigens, c is a rate constant that depends on the number of comparisons how antibodies are being stimulated, a represents the current B cell, xe_j represents the j_{th} B cell's epitope, xp_j represents the j_{th} B cell object paratope and y represents the current antigen. Finally the first term in the sum is the affinity between antibodies and neighbours; the second term is the enmity between the previous named objects; the third represents how well the antibodies are capable of binding with the antigen; and the last term models the tendency of cells to die if no interactions are present.

$$Stimulation = c\left[\sum_{j=1}^{N} m(a, xe_j) - k_1 \sum_{j=1}^{N} m(a, xp_j) + k_2 \sum_{j=1}^{N} m(a, y)\right] - k_3, \tag{3}$$

The model used by Farmer et al. presented three types of mutation presented as follows.

Multi-point mutation: Each element in the antibody is processed in turn. If a randomly generated number is above the mutation threshold, then the element is mutated.

http://dx.doi.org/10.5772/63570

Substring regeneration: Two points are selected at random in the antibody's paratope. Then all the elements between these two points are replaced by randomly generated elements, resulting in a partial regeneration of the antibody.

Simple substitution: Operator uses the roulette wheel [23] algorithm to select another B cell object from which elements will be substituted into the current B cell object.

The mutation procedure is an important procedure since it has been designed to introduce diversity into the system, as antibodies are created, the previous examples are just a few techniques that can be used during the mutation process, but there are several ways in which the mutation can be implemented.

The AIS systems is compared with other several machine learning approaches such as ANN, LCS, CBR and others showing that AIS offers certain advantages over some of them specifically the self-organising features and the unsupervised nature of the algorithm [22]. It also let us know that it can be noise tolerant and that the algorithms inherently generalises such a problem with ANN that can over fit. Also the self-organising feature makes it easier to handle rather than ANN that can be tedious and time consuming in order to tune them for a certain application checking the bias and variance of a certain configuration of the ANN.

The AIS is a paradigm that yet seems to have certain characteristics that attract engineers and scientist as well. It is a powerful example of a learning system that not only adapts rapidly but it also is a non-linear network, has a content addressable memory and it is self-organised. With this in mind, it is important to stop and analyse about the contributions that the paradigm has brought to different areas where it has been applied. CODA wants to use these features and apply them to the grasping problem in robotics, this is the main reason why the algorithm uses previous released machine learning tools plus it emulates the AIS in order to solve real problems and use AIS in important applications.

Precisely the study [24] explores how the AIS has been applied in “out of the lab” applications, where real world problems state new challenges, this question of how have the AIS has perform in the real world applications or industry leads us to the importance of focusing the efforts on developing theory that will serve as hard evidence of this paradigm.

Since according to the no-free lunch theorem [6], there is no one “do-it-all algorithm” that can outperform all the machine learning methods available so, it is important for any paradigm to contain features that may not be present in any of the previous techniques. In this manner the new algorithm can be described as a truly novel method.

Authors of [22] disagree with the definition proposed in [25] in which effectiveness is measured in terms of how an algorithm performs better or not, compared to other in benchmark test. For example they measured the time it took to complete the task. The same work exposes certain problems where AIS has done a magnificent job, these applications are presented in **Table 3**.

Major	Minor
Clustering/classification	Bio-informatics
Anomaly detection	Image processing
Computer security	Control
Numeric function optimisation	Robotics
Combinatoric optimisation	Virus detection
Learning	Web mining

Table 3. Major and minor areas where AIS have been used.

Table 3 reflects how the articles can be grouped according to the main areas where AIS has been applied. The studies in [3, 26–28] talk about Anomaly Detection, [29–31] expose works on optimisation, image processing [5], robot control [32–35] and web mining [36]. It is important to note that this table does not represent all publications related to AIS. It is a general picture based on information contained in ICARIS 2004 [37], ICARIS 2005 [38] and in the bibliography produced by de Castro [39].

Something important that must be noticed about most of the applications in several papers is that the AIS is tried in benchmark problems and robotic applications tend to be simulated rather than in real environments and almost all of them are small and simplified.

The article [22] concludes that there is a necessity to construct a robust framework that will allow the AIS to have a more biologically grounded theory, but it also remarks the need of multidisciplinary work between computer scientist, engineers, biologists and mathematicians.

Another important aspect discussed is that AIS algorithms are not so generic and most of them are quite specific to the task on duty. More generalised algorithms would be develop if the theoretical aspects of the AIS are improved in such a manner that generalisation would be easier. The authors close their work [22] with a phrase that is important for the development of the AIS algorithms:

> *"However, all this futuristic discussion is interesting, but what is needed is well-grounded immune inspired techniques that are applied in a logical and coherent matter".*

CODA algorithm pretends to handle data that could contain noise or that may be a small amount. Therefore the AIS should be an algorithm that can handle data in those circumstances, from the several bibliography revised it was found that in [40], the use of the AIS helps us in visualising how the AIS can be an unsupervised machine learning method but also how it can handle an analysis on the data and help visualise information.

The article describes the procedures that are internal to the AIS such as the primary and secondary response, how the antibody/antigen binding occurs, the B cell stimulation, the immune network among others that can serve as a really good introductory text for the reader. But it is not until the description of the B cell cloning were interest in such concept was really awaken.

http://dx.doi.org/10.5772/63570

The text is very explicit on how the AIS clones and mutates the B cells to build up the memory mechanism where all the information will be stored and will help on the recognition tasks where similar patterns must be identified. With all this data it is necessary for the memory to have an organising mechanism to form clusters that later will identify patterns.

As it is stated before in this document, there is a need for the AIS to be implemented in a simple yet logical manner, and the authors of [40] build up a really simple method to measure the affinity between the B cells with the Euclidean distance shown in Eq. (4).

$$Affinity(a,b) = \sqrt{\sum_{n=1}^{ND}(a(n) - b(n))^2}, \tag{4}$$

This implementation was tested with the "Fisher iris data set", which consist of 150 instances from three different classes, specifically plants. The attributes taken in consideration for the three classes are: sepal length, sepal width, petal length and petal width. The dataset cannot be separated linearly in two groups (Iris Virginica and Iris Versicolor) but it is possible to segregate the third group linearly corresponding to the Iris Setosa class. First principal component analysis was used to represent the data in a lower dimensional space in order to make possible a two dimensional graph. The AIS was able to recognise three classes independently from the distribution of the data set that may facilitate to classes and have more problems in the recognition of the third class.

The created network was capable of receiving unseen data and still able to recognise between classes, concluding that the networks were effective as a simple classification, one of the main objectives for using this AIS. The second valuable characteristic is that they seem to generalise in a wider region of the input space making it quite interesting since this feature could be from great importance in environments where little data can be collected, where the data could be corrupted with noise or demonstrations of the task are limited.

It has been seen all over the bibliography that a common point is the development of a more reliable theoretical point of view for the AIS, this was one of the first statements when approaching the AIS. The main reason the ABBAS book of Basic Immunology was revised was most of the documents found are a general overview of the NIS and difficult to understand if you have not had any previous experience. One interesting work is [40] and "Where" should be "where". Where the clonal selection mechanism was treated in a really interesting and straight forward manner. The clonal selection theory (CST) [41] is basically used to explain how the NIS response to any antigen stimulus. The main idea of the CST is that those cloned cells are in charge of recognising antigens.

In [20], a recognition region is presented but it is defined as spherical, a characteristic that could limit how the recognition ball could work. In [40] the author does not limit this region with a spherical geometry, leading to advancement in the theoretical framework for the AIS.

In [40] the B cell Algorithm (BCA) is defined as an iterative process that improves candidate solutions in a specific problem, with the use of tools such as cloning, mutation and selection.

Again the Euclidean distance is used as a measure for affinity, which suggests it could be a standard for this process. But how the BCA can be differentiated from one of the most popular evolutionary algorithm such as genetic algorithms? The reader should notice that no cross over is employed in the cloning process, there is no necessity for this method to be applied in the BCA in order to increase diversity.

Specifically talking about AIS algorithms and more precisely about clonal selection ones, the authors consider it really important to make the population variation according to probabilistic rules and follow the nature of the model so this probabilities for the transitions to a new state depend only on the current state so the Markov chain could be satisfied. This also means that the algorithm should converge and find at least one global optimum solution with probability equal to one as $t \to \infty$. One of the very first papers that introduced this was [42] and should be revised for a detailed lecture on the theme.

5. CODA

Cognition from Data is what names the proposed algorithm, as its name can tell, the objective of the algorithm is to aim for cognition in a computational manner – analysing data, extracting more information if possible and acting in an environment to achieve a task with the most accurate action even with few data as training examples. This is done by the NIS when a "naive individual" for the first time recognises an antigen. In order to comply to these characteristics and objectives, an algorithm should be flexible yet specific, should learn from previous experiences and store the knowledge in an organised manner. In doing so, it would respond faster the next time the situation occurs, just as the NIS in the secondary immune response responds to a previous known antigen.

In a search for a system that could have these characteristics, the NIS is one of the most impressive natural systems that not only adapts itself to new and unknown situations. It also learns from every experience and creates its own knowledge by an awesome mechanism where antibodies are created and stored in order to create a fastest response if a second interaction with the antigen occurs. **Table 1** in the previous section talked about certain desirable characteristics that CODA emulates from its biological counterpart – the NIS. This mechanism lets the NIS keep safe our body and it has been doing a great job for millions of years. Therefore CODA should provide a great solution to the grasping problem shown in Section 6, using AIS exceptional features.

The main reason why the NIS was studied in [20] is to have a more detailed panorama from this system that was taken as spinal vertebrae to design an algorithm. The non-free lunch theorem presented in [6] was also taken into consideration since it mentions that there is not one tool that could do all the possible tasks available in the problem space. It is important to create machine learning pipelines that could handle a complex problem and solve it with different tools not just one that would try to be the "do it all".

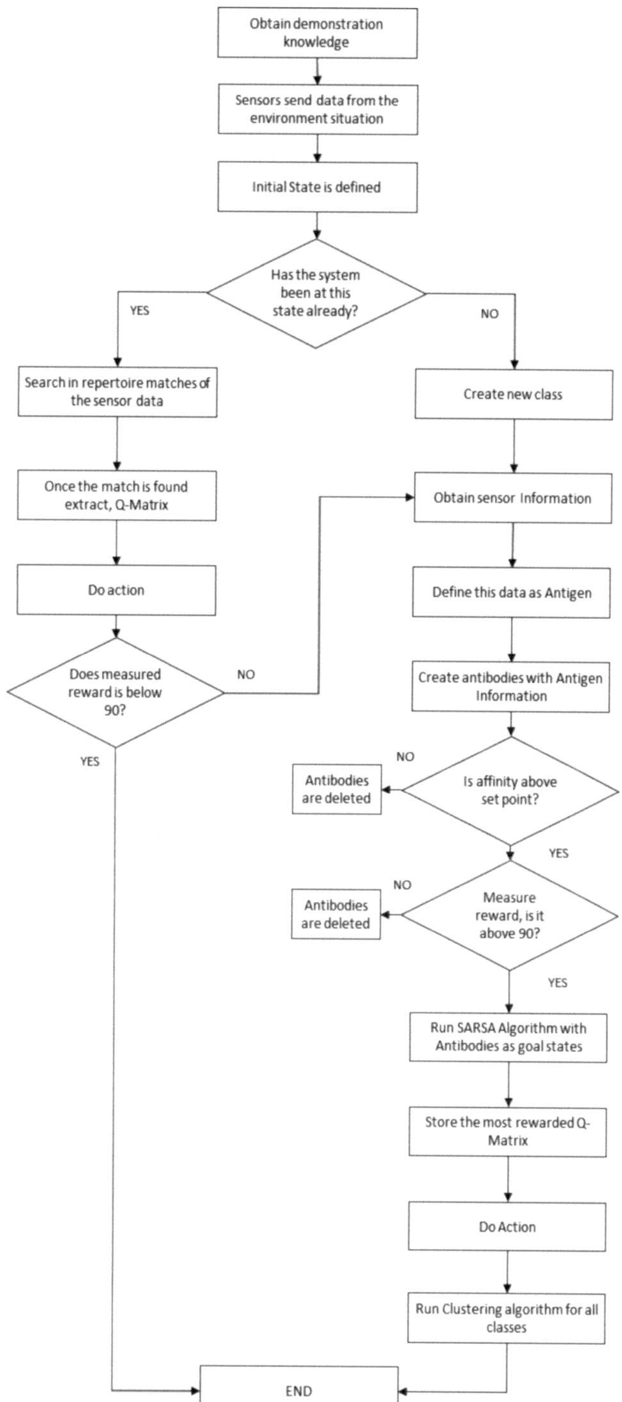

Figure 4. Block diagram showing the cognition model proposed by the authors with the CODA algorithm.

Algorithm 1 in **Figure 9** (Appendix) shows the pseudocode for the CODA algorithm, where it can be seen it has three elements that were presented previously on this document, the Reinforcement learner and the AIS. The learning from demonstration is an external part of the algorithm that acquires the training information and then presents this data to the CODA algorithm. **Figure 4** shows a flow chart that presents in a simpler way the pseudo code steps that may take place and adds the part of the algorithm that takes place outside of the learning system which is the demonstration of the skill.

The sequence of steps shown in **Figure 2** are of great importance since any change in the order will modify how the entity will learn and measure its performance according to the desired state. Also it is important to notice that at the beginning, there is a human–machine interaction that would recover the data from the example, and that it is necessary to be careful on which platform is chosen for this interaction in order to be positive and recover the most of the data possible without causing any discomfort in humans or any stress. The most positive the interaction the easier will be for the humans to share data and for the algorithm to have more and better examples of the task to be done.

In our case a data glove is the hardware chosen in order to acquire data (antigen). The glove is a well-known tool and almost any human has wear one, in this manner the acceptance will be faster, easier and will lead to a better interaction. This is really important because as mentioned before, there will be scenarios where data could be hard to obtain and may be corrupted with noise without mentioning that it may also be limited.

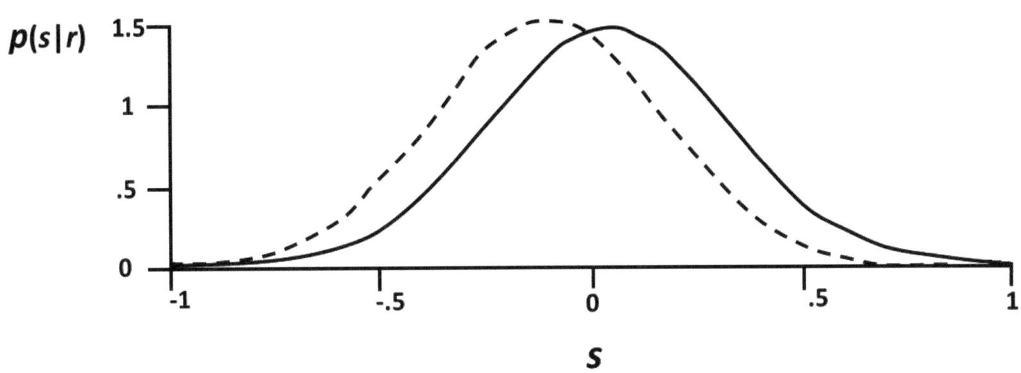

Figure 5. Probability densities from 11 neurons showing their preferred firing rates to certain stimulus, notice the true value stimulus is at $s = 0$.

The flow chart is specific for the presented application but it lets the reader follow the algorithm forward on its run. On the first stage, the knowledge should be given in certain manner. It could be demonstrated (in our case with the data glove) or it could be set with some previous recorded data. CODA will obtain this antigen from the data glove and produce clones from this single example. A simulation result can be found in **Figure 7**, where red squares are the training example and the blue circles are the cloned antibodies produced by CODA's clonal/mutation procedure. Then the entity should measure its state in order to define this information as the initial state, of course a task to be performed by the sensors embedded in the hand as explained in **Figure 8** in Section 6. After the cloning procedure has created the antibodies, the affinity should be measured as explained in Eq. (4). If the affinity criteria are not met the antibodies will be discarded and deleted.

The clonal procedure was inspired from the response of certain neurons given a set of stimulus. The outputs and the probability densities are shown in **Figure 5**. The graph represents the conditional probability presented by certain neurons producing certain firing rates *r* given a specific stimulus *s* [43]. This information is presented in [44] where several experiments are presented using the cercal system of the cricket in order to study neural decoding.

Using a clonal/mutation procedure the algorithm was able to obtain completely different probability distributions maintaining a normal density, this can be shown in **Figure 6** where the original values are compared with the mutated values after the clonal/mutation procedure.

It can be seen that we obtained different values that give much more data where the algorithm can explore and evaluate with the next stages of CODA Algorithm in order to use them if they are suitable for solving the problem.

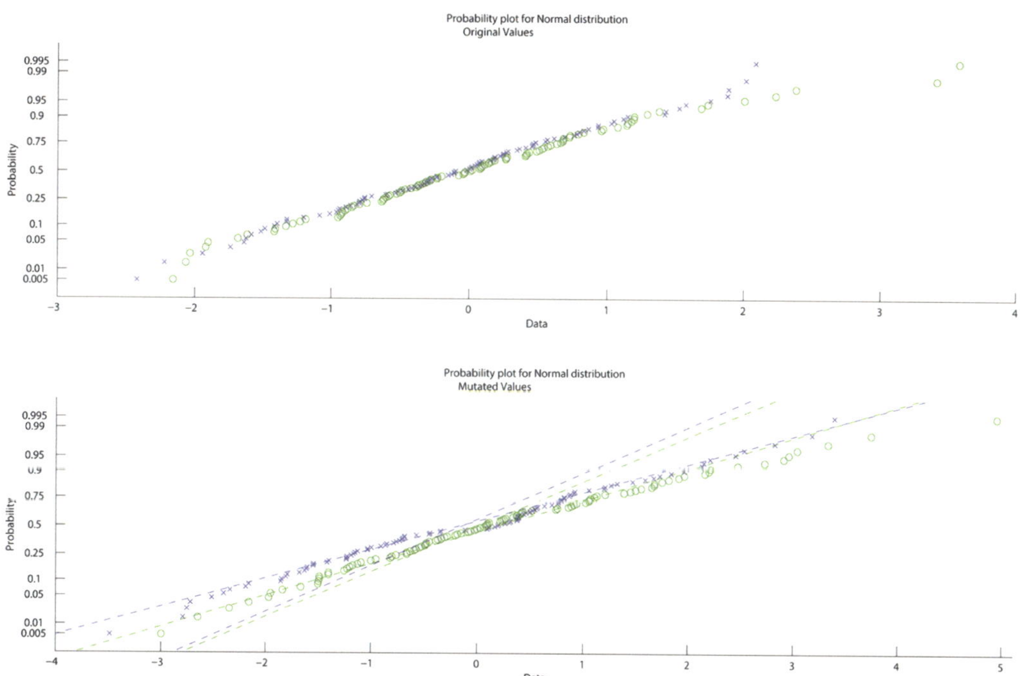

Figure 6. Probability densities comparison from original data and cloned/mutated data using CODA clonal/mutation procedure.

In fact, the clonal/mutation procedure will be given limited data, and it must produce new data samples even if it is given a simple example. **Figure 7** shows precisely that the clonal/ mutation procedure is capable of producing completely new data from a simple example and

Figure 6 shows how this new data has a completely different probability density, providing new opportunities for unseen data. The examples were obtained from the data glove as a demonstration example from the first stage of the block diagram shown in **Figure 8**.

The algorithm was given a five-line two-column array that contains force and angle values for each of the five fingers. With this example the algorithm was able to produce completely new data that before its use should be evaluated (Eq. (4)) in order to be used.

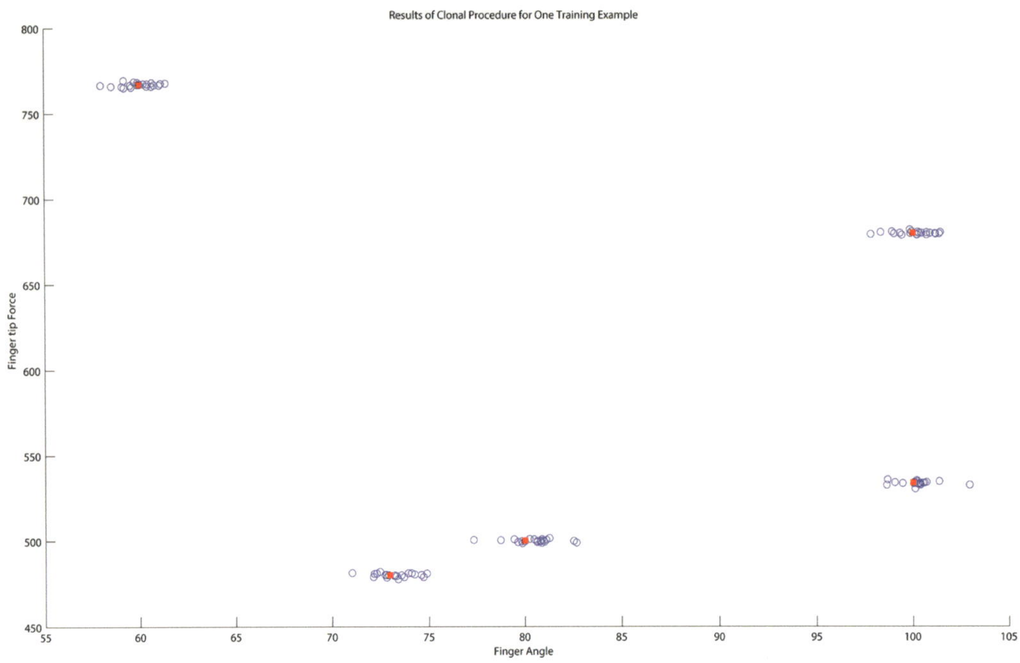

Figure 7. Red squares are antigens that were cloned and mutated producing the blue circles (antibodies) with the CO-DA clonal/mutation procedure. Mutated under a normal distribution set by the user.

Once clonal and mutation procedure is completed the algorithm will take the antibodies information and evaluate their affinity to the original value and with the reward function, this will produce an expected reward before actually using this configuration in the hardware. If values of affinity and reward are both below a threshold, the antibodies are considered to be maintained in the memory.

After the previous procedure is completed, the reinforcement learning algorithm uses the antibodies and defines the most rewarded as the goal state in order to complete the task. Finally from the *M* taken antibodies the algorithm will lead the system to the goal state. It is important to store the antibody-reward pair that corresponds to the one that obtain the highest value by the reward function doing the action, since this was the one that complete the task as desired. Finally a matrix should be constructed in order to store the Q matrix and the antibody-reward pair, for future reference in the memory. The last step will run the clustering in order to segregate the different classes that can be found and traduced as different tasks for the system.

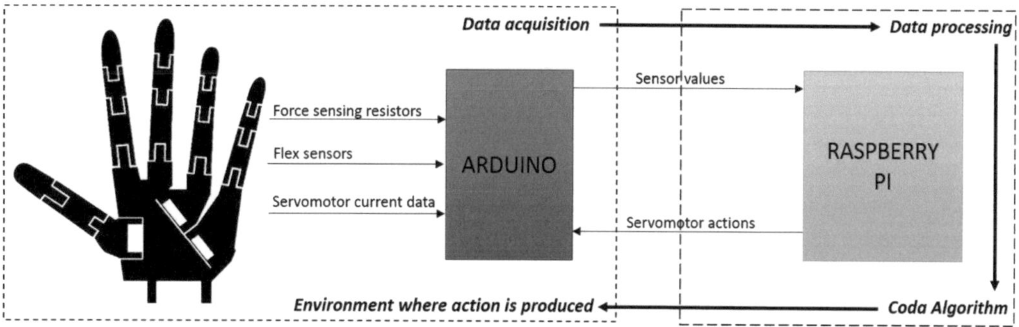

Figure 8. Hand system with its elements, where arrows show how data are managed between hardware components.

6. Application

The CODA algorithm in our laboratory will be implemented in a humanoid hand which is part of the Inmoov open source project. The application is one of the most difficult tasks in robotics – grasping objects. The system components are as follows and it is shown in the picture below.

The humanoid hand, all 3D printed in the laboratory consist of six standard servomotors, flex sensors for position information and force-sensing resistors for obtaining information about grasping forces.

The main brain for the entire system is a Raspberry pi where the algorithm will be saved and implemented. The Raspberry pi is connected with an Arduino which acquires the data from the sensors and the servomotors in order to send it to the Raspberry pi for processing.

Figure 3 will help the reader understand how the algorithm will be used in the application. The main idea is to produce an affordable grasping for an object. To exploit the advantages of CODA, the training examples will be limited, and CODA should produce more data as the one showed in **Figure 6**.

The first stage will demonstrate how to grasp the object (training example or antigen), where the human will use the data glove in order to grasp the object and record this training data. Once the training data is obtained, CODA will then define it as an antigen, since it is emulating the NIS, the information from the antigen will be taken for the cloning procedure.

The clonal procedure, affinity measuring, reward measuring and reinforcement learning will all take place in the Raspberry pi, which is the dedicated hardware for processing all data and store the knowledge or repertoire of antibodies.

The previous mentioned steps shown in **Figure 3** and in Algorithm 1 will then produce an affordable grasp that can be used by the hand in order to grasp the same object grasped by the human hand previously.

In this manner CODA will let the hand grasp objects with minimum data examples and with no knowledge of the physical characteristics of the human or robot hand. Instead it will use a reward function specifically designed for the application that will give numerical incentives in order to reach certain finger angles and force values.

Angle	Force	Mutated angle	Mutated force	Affinity
80	500	82.6499	498.8566	2.886058483
100	534	100.1784	534.696	0.7185500216
73	480	74.0844	481.2094	1.624368099
60	767	60.427	767.4406	0.613561211
100	680	100.8786	680.0866	0.882857588

Table 4. Single example given to the CODA algorithm, showing a simple mutated pair of angle and force values for each finger.

Table 4 shows a simple example of the original pair obtained from the data glove including angle and force. It can be seen how the clonal/mutation procedure in CODA modifies the values producing new data. Within the same table in columns three and four contain a simple set of cloned values that are also shown in **Figure 7**. And finally, the last column exposes the measured affinity computed with Eq. (4). Now the reward should be measured before the antibodies can be stored and used by the SARSA algorithm as goal states.

The Discussion section talked about the reward function that needs to be designed taking in consideration several factors in the system in order to make it possible to measure how suitable are the final antibodies for reproducing the task.

From this point on, the algorithm is straightforward. SARSA is used in order to reproduce the grasping posture which was set as a goal by the antibodies that meet the criteria. And finally, the Q-Matrix will be stored with all the related values such as antigen, antibodies, reward and affinity.

This application was selected because it happens to have low amounts of examples available about the task and a really difficult task that most of the time needs the use of kinematics and dynamics in order to produce grasping postures. Thus grasping is a great application area, since it is desired to create new knowledge about the grasping postures and remember those executed grasping in order to have faster responses for further interactions with the same object just as the AIS does with the secondary response with an antigen that has already been in the body.

The more objects the hand grasp, the more knowledge it will acquire. To simulate the self-organised memory, a clustering algorithm should run through the knowledge database and cluster similar postures, suggesting which of those postures correspond to similar shaped objects but with different dimensions.

Figure 9. Grasping a cube with data-glove, simple example of grasping posture obtained is on the graph below.

To demonstrate the functionality of the algorithm, a simple test was done. From a set of geometric figures, one was chosen for the test. The test protocol is simple: first the object is grasped by any user wearing the the data-glove system, once the sensor reading remain stable during grasping the data is saved to MATLAB workspace in order to be handle by CODA as antigen. As explained during previous chapters the algorithm will generate diversity from this given simple example (limited data) and produce certain amount of antigens for later evaluation by the emulated clonal selection and negative selection procedures embedded in the algorithm using the reward and affinity functions. For this specific test the cube figure was selected, and a simple example was recorded. **Figure 9** shows the cube grasped by the human teacher wearing the data-glove and the data obtained.

Finally the obtained information shown in **Figure 9** is feed into CODA. From this, a set of 500 antibodies was created that were evaluated by the reinforcement learning reward function and filtered in order to leave the most rewarded antibodies in the repertoire for the hand to grasp the object. **Figure 10** (left) shows the total antibodies created after the learning from demonstration procedure and on the right, the remaining repertoire data result of the reward and affinity evaluation procedures. This repertoire is considered the most capable data to produce affordable grasping postures, therefore, all of them were delivered to the hand system to confirm the functionality. Resulting to 100% of positive results means all the repertoire pairs were able to grasp the object.

With this simple test, the hand, data-glove and CODA algorithm have proven to function correctly and to produce useful data from limited samples (one simple example). They were able to produce new information that after being filtered and evaluated could reproduce the task without any parametric information regarding the hand size or link dimensions. The algorithm prove to be able to grasp the cube 100% of the time but when tested with an hexagonal prism it worked with 83% of the cases only, but still is a good result. **Figure 11** shows the hand grasping the cube and the hexagonal prism during the tests under the same protocol.

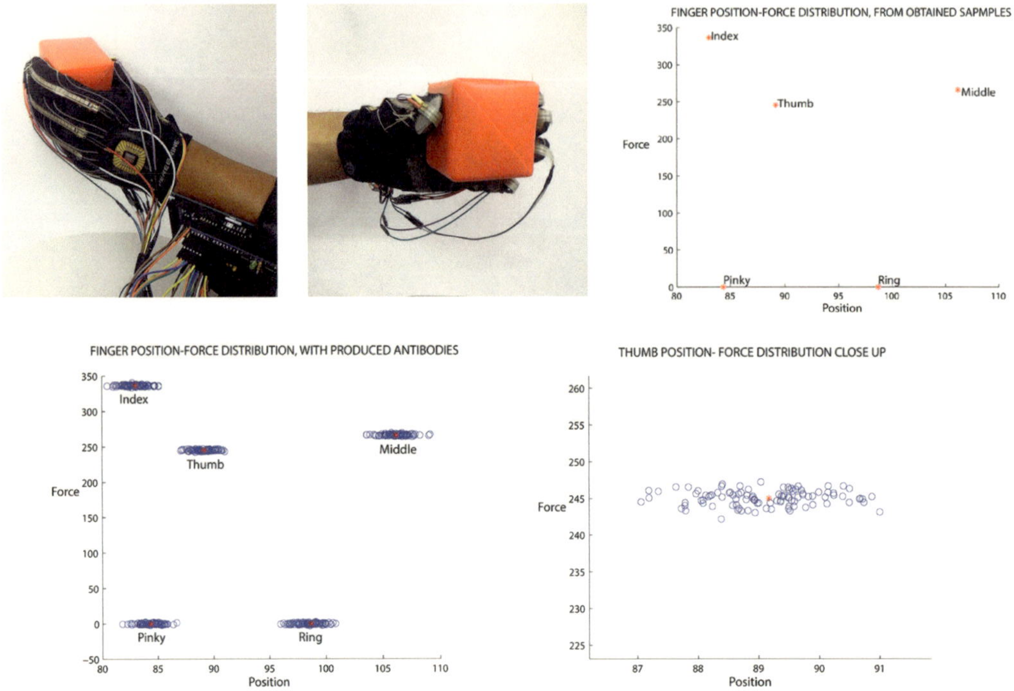

Figure 10. CODA algorithm cube test data obtained. Left, green antibodies created on red the unique example provided. Right, red the unique example, in green the repertoire of the most rewarded antibodies thus the most capable of producing grasping postures.

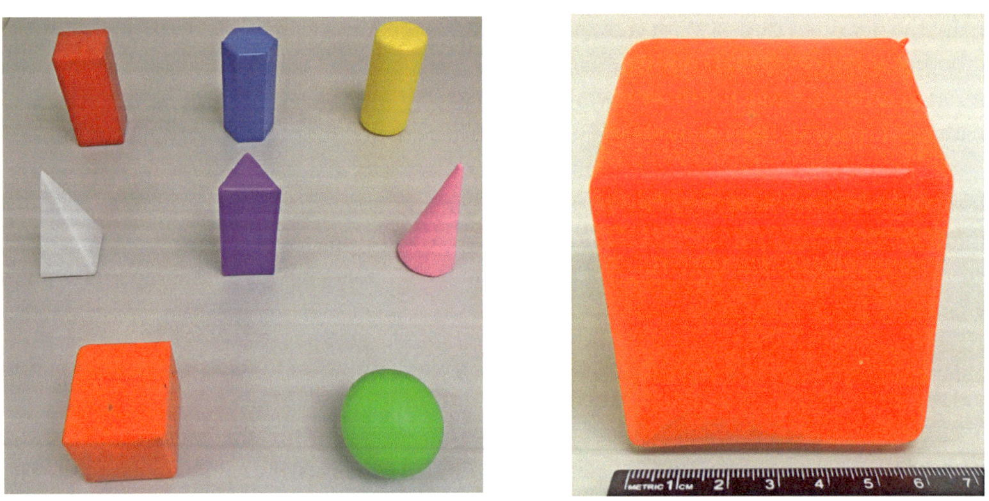

Figure 11. Hand correctly grasping objects from the data obtained from the CODA algorithm.

7. Discussion

CODA algorithm has been described and all the theories that involve this new design have been presented. Also the pseudocode and the flowchart have been exposed in order to explain how this algorithm will first learn from a demonstration, clone and mutate this data in order to create new knowledge (data never ever introduced to the algorithm before, created by the clonal/mutation process over the input data), evaluate its possible success within the system and finally use this new information as a goal.

The algorithm uses NIS as a basis in order to emulate several tools. The NIS is one of the most important and reliable systems within the human body. The developed algorithm uses a machine learning pipeline in order to create an algorithm, since certainly there is no one "do-it-all" method in machine learning, but several tools that need to be used together in order to exploit the different characteristics of each method.

CODA algorithm needs to be evaluated first in a simple environment with a simple task such as a 2R robot, and then with the main applicationexplained in the previous section. The hand grasping test would let usoptimise the algorithm and consider which methods fit better for measuring, affinity, stimulation and if the cloning and mutation processes are suitable for the tasks.

Affinity and stimulation techniques presented would be used in order to evaluate if for the presented algorithm they are suitable or not. If not, they should be modified or even design new ones that would let the algorithm work with much more personalised functions, increasing the algorithm's reliability and positive results.

The reward function is under experimentation. It involves two aspects: the forces and the angles. The algorithm should be given a reward as soon as the hand starts closing that means when angles are bigger and the force starts to increase in the finger tips. But if the hand just closes, there is the possibility that the grasp will not be affordable since the object could be damaged. Therefore the function should give high reward when the force starts increasing but reduce the reward if the force is beyond a certain threshold. Something similar happens with the angles since they may not be bigger and completely close the hand. This procedure needs to be done experimentally in order to obtain values from the sensors and set the thresholds for the angles and forces.

The three methods that conform CODA algorithm were chosen to be part of CODA in order to increase their efficacy at solving problems rather than using them individually, in order to exploit their advantages. They complement each other and are used for a specific objective for the proposed learning process, each one specialises on the one task it can perform better.

The very first contact with information is provided by demonstrations, which is a crucial aspect for the algorithm. Providing these examples simplifies the learning of the task, if no examples were given, an extremely well-designed algorithm was to be designed and programmed in the agent in order for it to reproduce the desired task, and the algorithm would need to be really descriptive providing every possible detail in order to obtain correct results. This procedure would require specially trained personnel and it would be time consuming. Compared to the demonstration process explained in this document that requires the sensors on the robot or in the person and almost anyone would be able to teach numerous skills just by reproducing them, without specialised studies or any coding knowledge.

In the previous paragraph it was said that if programming with a complete, well-designed and correct algorithm, an entity (in this case a robot) could reproduce a task. This is completely true if conditions and the environment remain constant. The truth is that for dynamic environments and uncertainty situations this would not be the case. Today's robots are being more and more embedded in our society. In these dynamic environments, sensor noise is more frequent. For these reasons a robot programmed as mentioned before would not be able to commit to its objective easily and would probably fail. This is due to a lack of learning capabilities, which means, robot would not be able to learn from past mistakes and acquire new abilities or perform the same task at different scenarios. This is where reinforcement learning used by CODA becomes an important method.

As explained in previous sections, reinforcement learning would produce a change in an entity's behaviour in order to obtain the highest reward, therefore since the reward function is designed upon the task's most crucial and important aspects, the algorithm will always search for the states that will produce the highest reward, if not, actions would be needed in order to achieve the task. These actions have the advantage that they can be low-level (i.e., voltages applied to motors) or high-level decisions as complex concepts or complex decisions (i.e. don't move).

In the CODA Algorithm the reinforcement learning helps the Artificial Immune System (AIS) to produce the best antibodies possible given a task-specific reward function, meaning the new antibodies would produce higher rewards. If bad antibodies are being produced, actions would alter the clonal/mutation procedure in order to modify the diversity generation.

Finally, the third method is the AIS which produces new data from single example – from the pathogen's antigen reducing the necessity of big data set, that may result expensive, difficult to obtain and construct thus time consuming. It also handles all the information and evaluates the produced data in order to find and produce the best possible antibodies. Finally it constructs a self-repertoire of state-action arrays in order to produce a faster response the next time the agent is encountered in the same state.

As an example of the functionality of CODA the system was tested, the test methodology is as follows: First wearing the data-glove a person grasped one of the geometric shapes in **Figure 12**. For this specific example the grasp was done with the cube. Once grasped, the sensor values were stored.

With these stored values, CODA produces 100 new data samples (antibodies). They are evaluated in order to select the most rewarded. Once filtered, the data set was reduced to maintain the most capable antibodies to reproduce the grasping task, then these antibodies were used to reproduce the task.

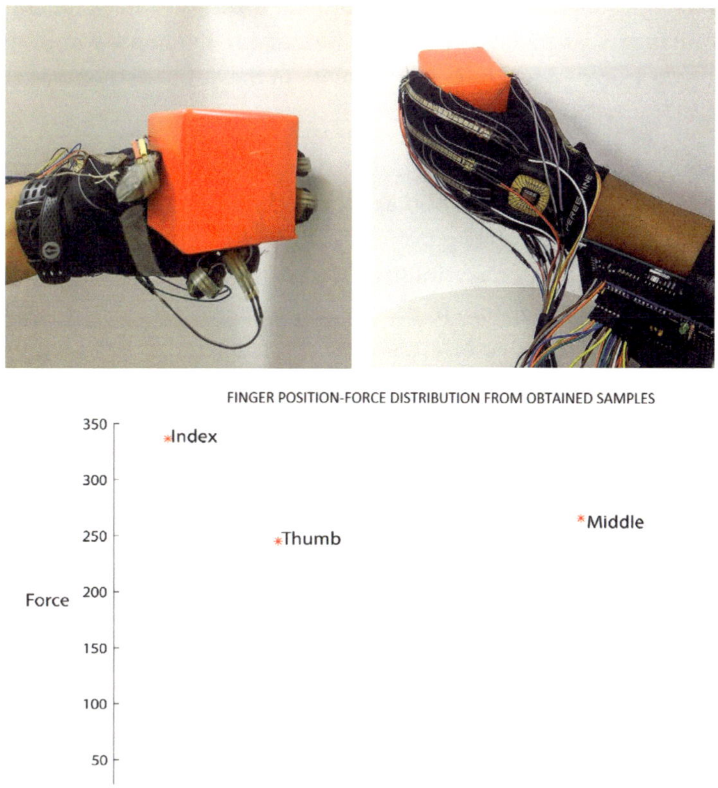

Figure 12. Test result from grasping cube, top left figures show the data-glove system used to acquire grasping data. Top right graph shows the single example acquired and provided to CODA. Bottom left show the example and the antibodies produced by CODA, finally bottom right show a close up of the thumb antibodies and sample distribution.

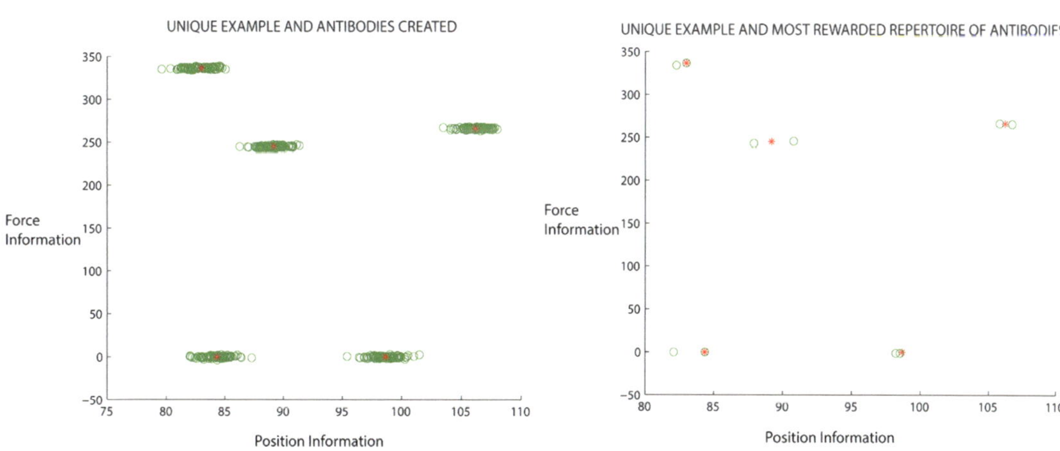

Figure 13. Geometric figures for testing the system, and the cube used for this very first test.

It is important to notice that the only example provided was the one obtained with the data-glove. No information was given to the algorithm about the hand size or any parametric information. It only searched for the best possibility in the antibody population based on the reward function (**Figure 13**).

The CODA algorithm has proved to be able to use one simple example and produce new and useful data from this unique information and from there produce a grasping posture. The requirements are less than some other grasping algorithms that can correctly produce grasping postures as well but need much more information such as 3D objects [52], orientation of the hand and tracking with tags [45], tons of pre-processing hand pose (examples) needed for the algorithm to learn with expensive computational procedures for reducing dimensions [46, 47] or partial object geometry information [48]. These approaches are way more expensive not only computationally but also in the hardware required and rely on the use of big databases and more requirements than CODA does.

8. Conclusion

CODA Algorithm is a Machine learning approach that can solve the grasping problem in robotic systems. In fact, the main algorithm is proposed as general solution when the problem can be expressed in Markov Chains. The algorithm is designed with the Computational model of the Natural Immune System, which is an Artificial Immune System.

The Artificial Immune System is a really broad system that has several task and subsystems as it is decentralised; therefore, numerous algorithms have been modelled in order to solve problems such as classification, fault detection among others. In particular, the AIS memory, the self-organised model, the genetic procedure in which antibodies are created and selected to complete the tasks, are all features that will lead to a robust yet versatile algorithm.

The algorithm presented in the previous sections was designed in order to produce new knowledge from limited data samples in order to reduce the need of big databases of examples of an specific task, that are difficult to build and most of the time have to be paid. In the bibliography it can be often see how there is a trend to reduce the training data sets [53], or studies to select correct data sets that would be representative enough [49], reduce the use of examples in order to learn [50] or reduce the population in order to optimise the algorithm [51]. From the previous articles it can be seen that the advantage of CODA over some mentioned algorithms is that CODA can use a simple example provided by demonstration and use this information to produce more knowledge. For example in [53] training data sizes range from 47 to 17861 samples. This would mean to ask for 17861 samples of grasping postures to people which is not impossible of course but it would be expensive, time consuming and tedious. In contrast CODA will need one demonstration of the task and it would do the hard job of producing new knowledge evaluate it and then use the most capable antibodies in order to reproduce the task.

http://dx.doi.org/10.5772/63570

According to **Table 1** CODA has several advantages. The first one being the quantity of examples provided in order to learn. Some algorithms need a big number of examples in order to learn a task, just as presented in the previous paragraph. CODA lets the algorithm produce new example possibilities distributed in a Gaussian space from a single example, user configurable. This is achieved thanks to the clonal/mutation procedure of the algorithm obtained from the Natural Immune systems. This can be seen in **Figure 7** were CODA was given a single example and proved to produce new data in seconds. Additionally **Figure 13**; shows how the algorithm produced 100 new data samples from a single demonstration of grasping a sphere.

The clonal/mutation procedure produced *N* different data points that can be used in order to solve the task. They have to be evaluated in order to corroborate the affinity to the original example. It is important to notice the importance of the evaluation in the algorithm, since the clonal/mutation can produce as much data as the user configures the process. Therefore, the evaluation process filters the data that may be harmful for the system or non-useful, reducing risks. The algorithm is capable of reducing learning time compared to algorithms that have to adjust weights or parameters in order to learn from a database. It is well known that those algorithms require certain minimum amount of data to effectively model the task, and it is also known that the more data samples, the more learning of the task. With CODA and the way it was designed, the task can be modelled from the very beginning since the learning is acquired by a demonstration of an expert. The clonal/mutation procedure reduces the amount of learning examples needed to be provided to the algorithm. The affinity function evaluates if the data obtained is coherent with the example provided in order to eliminate data that was created by the clonal/mutation procedure that can be harmful or that has no sense.

The next procedures complete the task with a reinforcement learning algorithm that assume as a goal state the data produced by the previous steps. This advantage of CODA creates a new generation of algorithms that should be capable of learning but reduce the necessity of data and training time, in order to model a task. Another advantage of these algorithms should be to produce own knowledge and the capability of evaluating their own, in order to produce reliable data.

CODA is an algorithm that pretends to complete the gap between data and cognition developing trust worthy responses, creating its own knowledge, measure its reliability, memorise and organise the data in order to re-use it when needed based on the immune system. These are procedures that CODA was able to emulate numerically.

Appendices

Appendix

See **Figure 14**.

Algorithm 1 The CODA algorithm

```
INITIALISATION
Set μ ← a value between 0 and 1
Set γ ← a value between 0 and 1
Set ϵ ← a value between 0 and 1
for all classes do
  if antigen belongs to any stored repertoire then
    Obtain repertoire class
    Obtain information for repertoire
    Extract Q-Matrix
    Do action
  else
    Create new class
    Store antigen
  end if
  Clone N antibodies from the antigen information
  if antibody affinity is greater than parameter then
    Keep antibody
  else
    Delete Antibody
  end if
  for all kept antibodies do
    Measure reward
  end for
  Obtain Maximmum Reward-Antibody pair
  Set Q(s,a) ← small random values
  while training scenarios < m do
    s ← random state of all possible states
    a ← according to ϵ-greedy policy
    while s =! antibody do
      Obtain r ← according to s and a
      Replace s' ← a
      Obtain a' ← according to ϵ-greedy using s'
      Update Q(s,a) ← Q(s,a)+ μ (r+ γQ(s',a')-Q(s,a))
      s ← s'
      a ← a'
    end while
    m ← m+1
  end while
  Print Q(s,a)
  Print s
  Store Q matrix, maximum Reward-Antibody pair and reward in repertoire
  Do action
  Run Clustering Algorithm
end for
```

1

Figure 14. CODA algorithm pseudocode.

http://dx.doi.org/10.5772/63570

Author details

Daniel R. Ramirez Rebollo*, Pedro Ponce Cruz and Arturo Molina

*Address all correspondence to: dan.ramirez@itesm.mx

Tecnologico de Monterrey, Mexico, DF, Mexico

References

[1] Lawson, Anton E. The Neurological Basis of Learning, Development and Discovery: Implications for Science and Mathematics Instruction. 18th ed. Springer Science & Business Media; 2006.

[2] Marsland, Stephen. Machine Learning: An Algorithmic Perspective. CRC Press; Boca Raton, FL. 2014.

[3] Pinto, José Carlos L and Von Zuben, Fernando J. Fault detection algorithm for telephone systems based on the danger theory. In: Artificial Immune Systems. Springer; Berlin, Heidelberg 2005. pp. 418–431.

[4] Luh, Guan-Chun and Liu, Wei-Wen. An Immunological Approach to Mobile Robot Navigation. INTECH Open Access Publisher; Taiwan, Republic of China 2008.

[5] Su, Mu-Chun, Yang, Yuan-Shao, Lee, Jonathan and Chen, Gwo-Dong. A neuro-fuzzy approach for compensating color backlight images. Neural Processing Letters. 2006;23(3):273–287.

[6] Wolpert, David H and Macready, William G. No free lunch theorems for optimization. IEEE Transactions on Evolutionary Computation. 1997;1(1):67–82.

[7] Atkeson, Christopher G and Schaal, Stefan. Robot learning from demonstration. In: ICML; 1997. pp. 12–20.

[8] Bakker, Paul and Kuniyoshi, Yasuo. Robot see, robot do: An overview of robot imitation. In: AISB96 Workshop on Learning in Robots and Animals; 1996. pp. 3–11.

[9] Dillmann, Rüdiger, Kaiser, M and Ude, Ales. Acquisition of elementary robot skills from human demonstration. In: International Symposium on Intelligent Robotics Systems; 1995. pp. 185–192.

[10] Ikeuchi, Katsushi and Suehiro, Takashi, Tanguy, Peter and Wheeler, Mark. Assembly plan from observation. Symbolic Visual Learning. 1990;193–224.

[11] Billard, Aude, Calinon, Sylvain, Dillmann, Rdiger and Schaal, Stefan. Robot programming by demonstration. In: Springer Handbook of Robotics. Springer; Berlin, Heidelberg 2008. pp. 1371–1394.

[12] Schaal, Stefan and others. Learning from demonstration. Advances in Neural Information Processing Systems. 1997;1040–1046.

[13] aus der Wieschen, Maria Vanessa, Fischer, Kerstin and Krüger, Norbert. Error feedback for robust learning from demonstration. In: Proceedings of the Tenth Annual ACM/IEEE International Conference on Human-Robot Interaction Extended Abstracts; Portland, Oregon, USA: 2015. pp. 225–226. DOI: 978-1-4503-3318-4

[14] Michini, Bernard, Walsh, Thomas J, Agha-Mohammadi, Ali-Akbar and How, Jonathan P. Bayesian nonparametric reward learning from demonstration. IEEE Transactions on Robotics. 2015;31(2):369–386.

[15] Brys, Tim, Harutyunyan, Anna, Suay, Halit Bener, Chernova, Sonia, Taylor, Matthew E, and Nowé, Ann. Reinforcement learning from demonstration through shaping. In: Proceedings of the International Joint Conference on Artificial Intelligence (IJCAI); 2015.

[16] Spong, Mark W, Hutchinson, Seth and Vidyasagar, Mathukumalli. Robot Modeling and Control. 3rd ed. Wiley, New York; 2006.

[17] de Wit, Carlos Canudas, Siciliano, Bruno and Bastin, Georges. Theory of Robot Control. Springer Science & Business Media; London, UK 2012.

[18] Turing, Alan M. Computing machinery and intelligence. Mind. 1950: 433–460.

[19] Thorndike, Edward Lee. Animal Intelligence: Experimental Studies. Macmillan; New York, USA 1911.

[20] Abbas, Abul K, Lichtman, Andrew HH, and Pillai, Shiv. Basic Immunology: Functions and Disorders of the Immune System. Elsevier Health Sciences; Philadelphia, PA 2012.

[21] Dipankar Dasgupta. Artificial Immune Systems and their Applications. Springer; Berlin, Heidelberg 1998.

[22] Hunt, John E and Cooke, Denise E. Learning using an artificial immune system. Journal of Network and Computer Applications. 1996;19(2):189–212.

[23] Golberg, David E. Genetic Algorithms in Search, Optimization, and Machine Learning. Addion Wesley. Massachusetts, USA 1989;

[24] Hart, Emma and Timmis, Jon. Application areas of AIS: The past, the present and the future. Applied Soft Computing. 2008;8(1):191–201.

[25] Garrett, Simon M. How do we evaluate artificial immune systems? Evolutionary Computation. 2005;13(2):145–177.

[26] Ayara, Modupe, Timmis, Jon, De Lemos, Rogério and Forrest, Simon. Immunising automated teller machines. In: Artificial Immune Systems. Springer; Berlin, Heidelberg 2005. pp. 404–417.

[27] Dasgupta, Dipankar, KrishnaKumar, Kalmanje, Wong, D and Berry, Misty. Negative selection algorithm for aircraft fault detection. In: Artificial Immune Systems. Springer; Sicily, Italy 2004. pp. 1–13.

[28] Taylor, Dan W, and Corne, David W. An investigation of the negative selection algorithm for fault detection in refrigeration systems. In: Artificial Immune Systems. Springer; 2003. pp. 34–45.

[29] Campelo, Felipe, Guimaraes, Frederico G, Igarashi, Hajime, Ramírez, Jaime, Noguchi, So and others. A modified immune network algorithm for multimodal electromagnetic problems. IEEE Transactions on Magnetics. 2006;42(4):1111–1114.

[30] Kalinli, Adem and Karaboga, Nurhan. Artificial immune algorithm for IIR filter design. Engineering Applications of Artificial Intelligence. 2005;18(8):919–929.

[31] Lee, Zne-Jung, Lee, Chou-Yuan and Su, Shun-Feng. An immunity-based ant colony optimization algorithm for solving weapon–target assignment problem. Applied Soft Computing. 2002;2(1):39–47.

[32] Hart, Emma, Ross, Peter, Webb, Andrew and Lawson, Alistair. A role for immunology in "next generation" robot controllers. In: Artificial Immune Systems. Springer; 2003. pp. 46–56.

[33] Lau, Henry YK, Wong, Vicky WK, and Lee, Ivan SK. Immunity-based autonomous guided vehicles control. Applied Soft Computing. 2007;7(1):41–57.

[34] Philip, Ninan Sajeeth. Optimal selection of training data for the difference boosting neural networks. In: Proc. of iAstro., Guan-Chun and Liu, Wei-Wen (editors). An Immunological Approach to Mobile Robot Navigation. INTECH Open Access Publisher; 2008.

[35] Zhong, Yanfei and Zhang, Liangpei, Huang, Bo, and Li, Pingxiang. An unsupervised artificial immune classifier for multi/hyperspectral remote sensing imagery. IEEE Transactions on Geoscience and Remote Sensing. 2006;44(2):420–431.

[36] Nasraoui, Olfa, Rojas, Carlos and Cardona, Cesar. A framework for mining evolving trends in web data streams using dynamic learning and retrospective validation. Computer Networks. 2006;50(10):1488–1512.

[37] Artificial immune systems. In: Proceedings of ICARIS; Springer; 2004. DOI: LNCS3239

[38] Jacob, Christian, Pilat, Marcin, Bentley, Peter and Timmis, Jonathan. Artificial Immune Systems: 4th International Conference Proceedings. In: ICARIS; August 14-17; Alberta Canada. Springer; 2005.

[39] de Castro, L.. Available from: http://www.dca.fee.unicamp.br/lnunes/AIS.html

[40] Timmis, Jon, Neal, Mark and Hunt, John. An artificial immune system for data analysis. Biosystems. 2000;55(1):143–150.

[41] Timmis, Jonathan, Hone, Andrew, Stibor, Thomas and Clark, Edward. Theoretical advances in artificial immune systems. Theoretical Computer Science. 2008;403(1):11–32.

[42] Burnet, Sir Frank Macfarlane and others. The Clonal Selection Theory of Acquired Immunity. University Press Cambridge; 1959.

[43] Villalobos-Arias, Mario, Coello, Carlos A Coello, and Hernández-Lerma, Onésimo. Convergence analysis of a multiobjective artificial immune system algorithm. In: Artificial Immune Systems. Springer; 2004. pp. 226–235.

[44] Peter Dayan and Laurence F. Abbott, Theoretical Neuroscience; Computational and Mathematical Modeling of Neural Systems. MIT Press; 2001.

[45] Cordella, Francesca. Grasping algorithms for anthropomorphic robotic hands inspired to human behavior [thesis]. Faculty of Engineering, University of Naples Federico II. 2011.

[46] Romero, Javier, Kjellstrom, Hedvig, and Kragic, Danica. Modeling and evaluation of human-to-robot mapping of grasps. In: IEEE International Conference on Advanced Robotics ICAR; 2009. pp. 1–6.

[47] Romero, Javier. From human to robot grasping [thesis]. KTH Royal Institute of Technology: 2011.

[48] Dang, Hao and Allen, Peter K. Semantic grasping: planning task-specific stable robotic grasps. Autonomous Robots. 2014;37(3):301–316. DOI: Springer

[49] Borovicka, Tomas, Jirina Jr, Marcel, Jirina, Marcel and Kordik, Pavel. Selecting Representative Data Sets. INTECH Open Access Publisher; 2012.

[50] Tommasi, Tatiana, Orabona, Francesco and Caputo, Barbara. Safety in numbers: Learning categories from few examples with multi model knowledge transfer. In: IEEE Conference on Computer Vision and Pattern Recognition; 2010. pp. 3081–3088.

[51] Brest, Janez and Maucec, Mirjam Sepesy. Population size reduction for the differential evolution algorithm. Applied Intelligence. 2008;29:228–247. DOI: Springer

[52] Goldfeder, Corey and Allen, Peter K. Data-driven grasping. Autonomous Robots. 2011;1:1–20. DOI: Springer

[53] Philip, Ninan Sajeeth. Optimal selection of training data for the difference boosting neural networks. Proc. of iAstro, Nice, France, October; 16–17.